WITHDRAWN
D0276288

Chemical Genomics

METHODS IN MOLECULAR BIOLOGY™

John M. Walker, *Series Editor*

327. ***Epidermal Growth Factor:*** *Methods and Protocols,* edited by *Tarun B. Patel and Paul J. Bertics, 2006*
326. ***In Situ* Hybridization Protocols,** *ThirdEdition,* edited by *Ian A. Darby and Tim D. Hewitson, 2006*
325. **Nuclear Reprogramming:** *Methods and Protocols,* edited by *Steve Pells, 2006*
324. **Hormone Assays in Biological Fluids,** edited by *Michael J. Wheeler and J. S. Morley Hutchinson, 2006*
323. **Arabidopsis Protocols,** *Second Edition,* edited by *Julio Salinas and Jose J. Sanchez-Serrano, 2006*
322. ***Xenopus*** **Protocols:** *Cell Biology and Signal Transduction,* edited by *X. Johné Liu, 2006*
321. **Microfluidic Techniques:** *Reviews and Protocols,* edited by *Shelley D. Minteer, 2006*
320. **Cytochrome P450 Protocols**, *Second Edition,* edited by *Ian R. Phillips and Elizabeth A. Shephard, 2006*
319. **Cell Imaging Techniques**, *Methods and Protocols,* edited by *Douglas J. Taatjes and Brooke T. Mossman, 2006*
318. **Plant Cell Culture Protocols**, *Second Edition,* edited by *Victor M. Loyola-Vargas and Felipe Vázquez-Flota, 2005*
317. **Differential Display Methods and Protocols**, *Second Edition,* edited by *Peng Liang, Jonathan Meade, and Arthur B. Pardee, 2005*
316. **Bioinformatics and Drug Discovery**, edited by *Richard S. Larson, 2005*
315. **Mast Cells:** *Methods and Protocols,* edited by *Guha Krishnaswamy and David S. Chi, 2005*
314. **DNA Repair Protocols:** *Mammalian Systems, Second Edition,* edited by *Daryl S. Henderson, 2005*
313. **Yeast Protocols:** *Second Edition,* edited by *Wei Xiao, 2005*
312. **Calcium Signaling Protocols: Second Edition, edited by David G. Lambert, 2005**
311. **Pharmacogenomics:** *Methods and Applications,* edited by *Federico Innocenti, 2005*
310. **Chemical Genomics:** *Reviews and Protocols,* edited by *Edward D. Zanders, 2005*
309. **RNA Silencing:** *Methods and Protocols,* edited by *Gordon Carmichael, 2005*
308. **Therapeutic Proteins:** *Methods and Protocols,* edited by *C. Mark Smales and David C. James, 2005*
307. **Phosphodiesterase Methods and Protocols,** edited by *Claire Lugnier, 2005*
306. **Receptor Binding Techniques:** *Second Edition,* edited by *Anthony P. Davenport, 2005*
305. **Protein–Ligand Interactions:** *Methods and Applications,* edited by *G. Ulrich Nienhaus, 2005*
304. **Human Retrovirus Protocols:** *Virology and Molecular Biology,* edited by *Tuofu Zhu, 2005*
303. **NanoBiotechnology Protocols,** edited by *Sandra J. Rosenthal and David W. Wright, 2005*
302. **Handbook of ELISPOT: Methods and Protocols,** edited by *Alexander E. Kalyuzhny, 2005*
301. **Ubiquitin–Proteasome Protocols,** edited by *Cam Patterson and Douglas M. Cyr, 2005*
300. **Protein Nanotechnology:** *Protocols, Instrumentation, and Applications,* edited by *Tuan Vo-Dinh, 2005*
299. **Amyloid Proteins:** *Methods and Protocols,* edited by *Einar M. Sigurdsson, 2005*
298. **Peptide Synthesis and Application,** edited by *John Howl, 2005*
297. **Forensic DNA Typing Protocols,** edited by *Angel Carracedo, 2005*
296. **Cell Cycle Control:** *Mechanisms and Protocols,* edited by *Tim Humphrey and Gavin Brooks, 2005*
295. **Immunochemical Protocols,** *Third Edition,* edited by *Robert Burns, 2005*
294. **Cell Migration:** *Developmental Methods and Protocols,* edited by *Jun-Lin Guan, 2005*
293. **Laser Capture Microdissection:** *Methods and Protocols,* edited by *Graeme I. Murray and Stephanie Curran, 2005*
292. **DNA Viruses:** *Methods and Protocols,* edited by *Paul M. Lieberman, 2005*
291. **Molecular Toxicology Protocols**, edited by *Phouthone Keohavong and Stephen G. Grant, 2005*
290. **Basic Cell Culture Protocols**, *Third Edition,* edited by *Cheryl D. Helgason and Cindy L. Miller, 2005*
289. **Epidermal Cells**, *Methods and Applications,* edited by *Kursad Turksen, 2005*
288. **Oligonucleotide Synthesis**, *Methods and Applications,* edited by *Piet Herdewijn, 2005*

METHODS IN MOLECULAR BIOLOGY™

Chemical Genomics

Reviews and Protocols

Edited by

Edward D. Zanders, PhD

CamBP Ltd., Cambridge, UK

HUMANA PRESS ✱ TOTOWA, NEW JERSEY

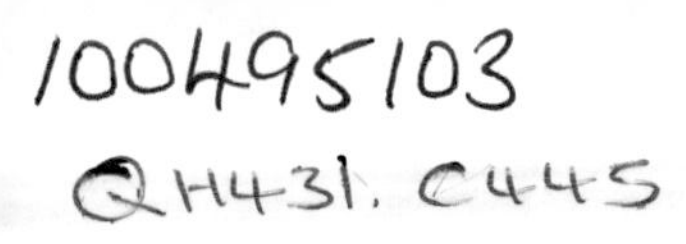

© 2005 Humana Press Inc.
999 Riverview Drive, Suite 208
Totowa, New Jersey 07512

www.humanapress.com

This publication is printed on acid-free paper. ∞
ANSI Z39.48-1984 (American Standards Institute)

Permanence of Paper for Printed Library Materials.
Cover illustration: Background: Imantinib co-crystalized with c-Abl kinase domain. Used with the permission of Dr. P. M. Dean, CSO, DeNovo Pharmaceuticals, Histon, Cambridge, UK. Foreground: Image of protein array. Figure 4 from Chapter 14, "Fabrication of Protein Function Microarrays for Systems-Oriented Proteomic Analysis," by Jonathan M. Blackburn and Darren J. Hart. Used with the permission of the authors.

Production Editor: Tracy Catanese
Cover design by Patricia F. Cleary

For additional copies, pricing for bulk purchases, and/or information about other Humana titles, contact Humana at the above address or at any of the following numbers: Tel.: 973-256-1699; Fax: 973-256-8341; E-mail: orders@humanapr.com; or visit our Website: www.humanapress.com

Printed in the United States of America. 10 9 8 7 6 5 4 3 2 1

e-ISBN: 1-59259-148-6

Library of Congress Cataloging in Publication Data

Chemical genomics : reviews and protocols / edited by Edward D. Zanders.
p. ; cm. -- (Methods in molecular biology ; v. 310)
Includes bibliographical references and index.
ISBN 1-58829-399-8 (alk. paper)
1. Genomics. 2. Proteomics. 3. Pharmacogenetics. 4. Biochemical genetics. 5. DNA microarrays. 6. Combinatorial chemistry.
[DNLM: 1. Molecular Probes--chemical synthesis--Laboratory Manuals. 2. Protein Array Analysis--methods--Laboratory Manuals. 3. Combinatorial Chemistry Techniques--Laboratory Manuals. 4. Genomics--methods--Laboratory Manuals. QU 25 C517 2005] I. Zanders, Edward D. II. Series.
QH431.C45197 2005
572.8'6--dc22

2005013118

Preface

Chemical genomics is an exciting new field that aims to transform biological chemistry into a high-throughput industrialized process, much in the same way that molecular biology has been transformed by genomics. The interaction of small organic molecules with biological systems (mostly proteins) underpins drug discovery in the pharmaceutical and biotechnology industries, and therefore a volume of laboratory protocols that covers the key aspects of chemical genomics would be of use to biologists and chemists in these organizations. Academic scientists have been exploring the functions of proteins using small molecules as probes for many years and therefore would also benefit from sharing ideas and laboratory procedures. Whatever the organizational backgrounds of the scientists involved, the challenges of extracting the maximum human benefit from genome sequencing projects remains considerable, and one where it is increasingly recognized that chemical genomics will play an important part.

Chemical Genomics: Reviews and Protocols is divided into two sections, the first being a series of reviews to describe what chemical genomics is about and to set the scene for the protocol chapters. The subject is introduced by Paul Caron, who explains the various "flavors" of chemical genomics. This is followed by Lutz Weber and Philip Dean who cover the interaction between organic molecules and protein targets from the different perspectives of laboratory experimentation and *in silico* design. The protocols begin with the methods developed in Christopher Lowes' laboratory (Roque et al.) for what could be described as a classical example of chemical genomics, namely the design of small molecules as affinity ligands for specific protein families. The theme is continued with detailed protocols for *in silico* docking by Jongejan et al. that highlights the importance of computational approaches to protein–small molecule interactions. The remaining protocols are directed towards the aim of producing highly diverse collections of proteins, carbohydrates, and small molecules for use in arrays containing large numbers of molecules. This high-throughput approach to screening for interaction between small and large biological molecules is the essence of chemical genomics. The chapters by Ryu, Doyle, Murphy, Sawasaki, Endo, Kohno, and Hoyt cover methods for the production of proteins and carbohydrates using different expression systems. Webster and Oxley give a protocol for analyzing the proteins using mass spectrometry. The techniques for arraying these proteins and carbohydrates on solid supports are detailed in the chapters by Blackburn, Marik, and Wang. Finally an in vivo method for identifying small molecule–protein interactions is described by Khazak et al. using the yeast two-hybrid system.

Although we recognize that no single book on chemical genomics can be totally comprehensive in its coverage, we hope that the protocols here, in covering the key elements of the subject, will be of genuine use to the wide variety of scientists in this rapidly expanding field.

Edward D. Zanders

Contents

Contributors

Jonathan M. Blackburn • *Department of Biotechnology, University of the Western Cape, Cape Town, South Africa; Procognia Ltd, Maidenhead, UK*
Paul R. Caron • *Head of Informatics, Vertex Pharmaceuticals, Cambridge MA*
Min Chen • *The State Key of Microbial Technology, School of Life Science, Shandong University, Jinan, Shandong, People's Republic of China*
Philip M. Dean • *Chief Scientific Officer, De Novo Pharmaceuticals Ltd, Cambridge, UK*
Sharon A. Doyle • *Proteomics Group, DOE Joint Genome Institute, Walnut Creek, CA*
Yaeta Endo • *Cell-Free Science and Technology Research Center, Ehime University, Matsuyama, Japan*
Iwan J. P. de Esch• *Division of Medicinal Chemistry, Leiden/Amsterdam Center for Drug Research (LACDR), Faculty of Sciences, Vrije Universiteit, Amsterdam, The Netherlands*
Erica A. Golemis • *Division of Basic Sciences, Fox Chase Cancer Center, Philadelphia, PA*
Mudeppa D. Gouda • *Cell-Free Science and Technology Research Center, Ehime University, Matsuyama, Japan*
Chris de Graaf • *Division of Molecular Toxicology, Leiden/Amsterdam Center for Drug Research (LACDR), Faculty of Sciences, Vrije Universiteit, Amsterdam, The Netherlands*
Geeta Gupta • *Institute of Biotechnology, University of Cambridge, Cambridge, UK*
Darren J. Hart • *High Throughput Group, Grenoble Outstation, European Molecular Biology Laboratory, Grenoble, France*
Jon Hoyt • *Department of Cell Biology, Institute of Chemistry and Cell Biology, Boston, MA*
Aldo Jongejan • *Division of Medicinal Chemistry, Leiden/Amsterdam Center for Drug Research (LACDR), Faculty of Sciences, Vrije Universiteit, Amsterdam, The Netherlands*
Takayasu Kawasaki • *Cell-Free Science and Technology Research Center, Ehime University, Matsuyama, Japan*
Vladimir Khazak • *Director of Biology, NexusPharma, Langhorne, PA*
Randall W. King • *Department of Cell Biology, Institute of Chemistry and Cell Biology, Boston, MA*

Toshiyuki Kohno • *Laboratory of Structural Biology, Mitsubishi Kagaku Institute of Life Sciences (MITILS), Tokyo, Japan*
Kit S. Lam • *Division of Hematology & Oncology, Department of Internal Medicine, UC Davis Cancer Center, University of California, Davis, CA*
Rob Leurs • *Division of Medicinal Chemistry, Leiden/Amsterdam Center for Drug Research (LACDR), Faculty of Sciences, Vrije Universiteit, Amsterdam, The Netherlands*
Hanfen Li • *Department of Biochemistry, The Ohio State University, Columbus, OH*
Steven Lin • *Department of Biochemistry, The Ohio State University, Columbus, OH*
Shaoyi Liu • *Columbia Genome Center, Columbia University College of Physicians & Surgeons, New York, NY*
Christopher R. Lowe • *Institute of Biotechnology, University of Cambridge, Cambridge, UK*
Jan Marik • *Division of Hematology & Oncology, Department of Internal Medicine, UC Davis Cancer Center, University of California, Davis, CA*
Michael B. Murphy • *Proteomics Group, DOE Joint Genome Institute, Walnut Creek, CA*
David Oxley • *Proteomics Research Group, Babraham Institute, Cambridge, UK*
Ana Cecília A. Roque • *Institute of Biotechnology, University of Cambridge, Cambridge, UK*
Kang Ryu • *Department of Biochemistry, The Ohio State University, Columbus, OH*
Tatsuya Sawasaki • *Cell-Free Science and Technology Research Center, Ehime University, Matsuyama, Japan*
Dhaval Shah • *Columbia Genome Center, Columbia University College of Physicians & Surgeons, New York, NY*
Jun Shao • *Department of Biochemistry, The Ohio State University, Columbus, OH*
Jing Song • *The State Key of Microbial Technology, School of Life Science, Shandong University, Jinan, Shandong, People's Republic of China*
Kazuyuki Takai • *Cell-Free Science and Technology Research Center, Ehime University, Matsuyama, Japan*
Yuzuru Tozawa • *Cell-Free Science and Technology Research Center, Ehime University, Matsuyama, Japan*
Takafumi Tsuboi • *Cell-Free Science and Technology Research Center, Ehime University, Matsuyama, Japan*

Nico P. E. Vermeulen • *Division of Molecular Toxicology, Leiden/ Amsterdam Center for Drug Research (LACDR), Faculty of Sciences, Vrije Universiteit, Amsterdam, The Netherlands*

Denong Wang • *Carbohydrate Microarray Laboratory, Departments of Genetics, Neurology and Neurological Sciences, Stanford University School of Medicine, Stanford, CA*

Peng George Wang • *Department of Biochemistry, The Ohio State University, Columbus, OH*

Ruobing Wang • *Carbohydrate Microarray Laboratory, Departments of Genetics, Neurology and Neurological Sciences, Stanford University School of Medicine, Stanford, CA*

Wei Wang • *The State Key of Microbial Technology, School of Life Science, Shandong University, Jinan, Shandong, People's Republic of China*

Lutz Weber • *President, NexusPharma, Langhorne, PA*

Judith Webster • *Proteomics Research Group, Babraham Institute, Cambridge, UK*

Wen Yi • *Department of Biochemistry, The Ohio State University, Columbus, OH*

Edward D. Zanders • *CamBP Ltd, Cambridge, UK*

I

Reviews

1

Introduction to Chemical Genomics

Paul R. Caron

1. Introduction

Small-molecule drugs are a cost-effective way to treat and prevent disease. A study by the Slone Institute published in 2002 estimated that over 50% of the adult population in the United States used at least one pharmaceutical drug during the preceding week. The positive impact of small-molecule drugs on health care has been well documented *(1,2)*.

The discovery of novel drugs has traditionally been a combination of clever science, brute force, and good fortune. With the advent of high-throughput screening technology, combinatorial chemistry, and the completion of the human genome sequence in the late 1990s, the hope was that technology could address the brute-force aspect and the genome sequence would provide insights into the underlying science, and good fortune would continue. Although there are some exceptions, productivity in the industry overall has gone down. Some of this is owing to higher regulatory standards and more difficult therapeutic areas, but a significant portion is the result of the lack of well-validated targets to apply the technology to. The industry portfolio of pharmaceutical targets of approx 500 in 1996 *(3)* has not been significantly expanded.

The availability of the human genome sequence and novel biological tools, such as siRNA, antisense, knockouts, and transgenics, suggests that over time, the physiological function of many of the genes in the genome may be deciphered. However, the time frame for this may be much greater than most people anticipated. For comparison, the first bacterial genome sequence was completed in 1995 *(4)*, and although we may be able to now classify the majority of the genes by biochemical function, we don't know most of their physiological roles. One

From: *Methods in Molecular Biology, vol. 310: Chemical Genomics: Reviews and Protocols*
Edited by: E. D. Zanders © Humana Press Inc., Totowa, NJ

conceivable way to speed up the molecular dissection of the biology underlying various disease states is to use small-molecule compounds that specifically inhibit individual targets.

There are several key factors required to be successful when using small molecules to explore biology.

- The relative selectivity of the chemical probes that will be used must be known.
- The correlations between the cellular readout(s) used and the pathway or phenotype that is being assessed must be independently validated.
- All data must be fully integrated, allowing the user to navigate through biological pathways and supporting literature, assay results, and detailed information on compounds.

2. Different Flavors of Chemical Genomics

As with any emerging field, there are often differences of opinion on terminology among researchers, sometimes subtle, sometimes not. I will attempt to capture and describe the major variants and apologize if I inadvertently leave out any major themes, or end up misrepresenting some differences in trying to summarize the field.

2.1. Chemical Genetics

Chemical genetics, as described by Schreiber et al. in 1999 *(5)*, refers to the use of small molecules to induce alterations in gene products in mammalian systems, in a manner similar to using mutations. This approach became feasible through the combination of high-throughput cellular assays and diverse libraries of compounds. The ability to perform genetic screens in cellular assays vastly increases the throughput—traditionally a key limitation when studying higher organisms. It also allows the separation of effects in somatic cells from those in development.

It is critical to this chemical genetic approach to have a library of compounds that have a high probability of being relatively selective; otherwise, the ability to interpret the results becomes at least as complex as deciphering highly polygenetic phenotypes. To address this, diversity-oriented synthesis has been proposed to provide arrays of complex small molecules that are easily synthesized. The natural-product basis for many of the molecules and their complexity are believed to contribute to their cellular potency and selectivity *(6)*. This chemical genetic approach has been applied to identify novel inhibitors of alpha-tubulin and histone deactylation *(7)*.

As in classical genetics, a chemical genetic approach involves screening with probes that potentially could interact with any target in the genome, while trying to identify specific phenotypes. An alternate approach, termed *reverse chemical genetics*, is analogous to introducing specific gene disruptions. Here, compounds that are known to specifically interact with a given target are used in broad phenotype screens to help identify the physiological role of that target.

2.2. Reverse Chemical Genetics

The key to a reverse chemical genetic approach is to have a one-to-one link between the small-molecule compound and the target of interest. This can be achieved by optimization of chemical reagents by thorough profiling against other potential targets, or alternatively by altering the target itself, to introduce changes that can be exploited for specificity. This approach has been most broadly applied to members of the protein kinase family, where specific changes can be made to the residues surrounding the active site that don't significantly alter the affinities or kinetics for natural substrates, but now allow the binding of specific inhibitor analogs *(8)*. Replacement of the wild-type copies of a given gene with these engineered mutants allows these compounds to be used to inhibit the function of the gene in cellular assays as well as in adult animals. The in vivo assay, complete with the complications of pharmacokinetics and pharmacodynamics, closely mimics the effect that would be expected from dosing an animal with a selective inhibitor against the wild-type target.

Additionally, introducing mutations into the adenosine triphosphate (ATP)-binding sites of targets allows the binding of labeled ATP analogs; these can then be used to trace biochemical pathways at the molecular level by looking directly at phosphorylated substrates, thus furthering the link between the target and the observed phenotype.

A comparison between standard genetics and chemical genetics is shown in **Fig. 1** (*see also* **Table 1**).

2.3. Screening

Screening of large sets of compounds, often assembled to be quite diverse, is often one of the first steps in a drug discovery project. The assay used for the initial screen will both help define the likelihood of getting potent hits, as well as form the foundation for the follow-up path. Assays that are more physiological require that active molecules pass through additional filters depending on the assay, such as transversing the cell wall, serum binding, bioavailability, and metabolic stability. These factors tend to decrease the hit rate, but result in molecules with better overall properties. The downside to this approach is that further compound optimization may be hindered by a lack of knowledge about the molecular target(s) of the initial hits. Biochemical-based target screening is likely to yield hits that have a clearer path to optimization, but are at higher overall risk because the link between the target and the desired physiological

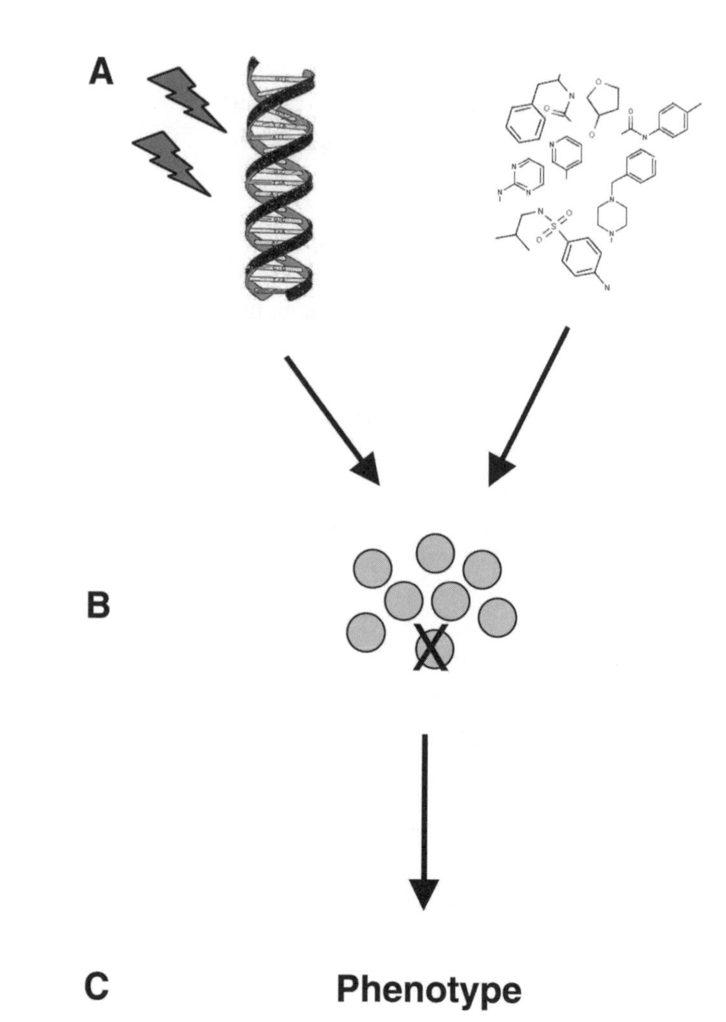

Fig. 1. Comparison of genetic and chemical genetics. In the traditional genetics approach genes are mutated resulting in missing or altered protein products. In a chemical genetic approach, compounds bind to specific proteins modulating their normal physiological functions. The throughput at steps A, B, C is used in Table 1 to help define the different chemical genetic approaches.

changes may not be strong and these compounds may be far from having the desired physical properties.

Smaller compound sets and focused compound libraries can be used to screen more broadly for physiological phenotypes. In the typical high-content screening experiment, the effects of compounds on cellular assays are captured

Table 1
Approaches to Chemical Genetics

		Approach	Goal	Step 1	Step 2	Step 3
Genetics	AcB	Randomly perturbate the system looking for changes in a specific phenotype	Target/pathway identification	Assemble large diverse chemical library	Assay for specific phenotype	Identify targets of compounds that induce the phenotype
Reverse Genetics	baC	Broad search for phenotype associated with perturbations of specific genes	New role for target	Select target	Find specific inhibitor	Broadly assay for phenotypes
Screening	bAc	Search for compounds which modulate a target with a known phenotype	Compounds/drugs for validated targets	Select target with presumed physiological role	Assemble large diverse chemical library	Screen for modulators of target
Genomics	ABC	Screen all targets for specific modulators, then try to identify phenotypes for each target	Target/pathway identification	Assemble large diverse chemical library	Screen for modulators of any targets	Broadly assay for phenotypes
Profiling	AC	Broadly screen for phenotypes associated with compounds, regardless of the target	Link between compounds and phenotype—efficacy or toxicity	Assemble large diverse chemical library	Broadly assay for phenotypes	
Chemo-genomics	BaC	Use information on targets to identify specific modulators, then look for phenotypes	Compounds/drugs for novel targets	Select targets	Identify inhibitor	Broadly assay for phenotypes

The steps in the second column refer to **Fig. 1A–C**. An uppercase letter indicates that the number of compounds, targets, or assays in the approach represents a large set. A lowercase letter is use to denote a small, focused set.

by multiple parameters. These range from cellular and subcellular morphology, intracellular translocation events, changes in state, such as phosphorylation, and proteomic and genomic profiling. The advances in recent years of high-throughput screening, miniaturization, and imaging have made it technically practical to assemble vast databases detailing the effects of specific compounds on a genome-wide scale. The current challenge in the field is to learn how to interpret this information to expand our knowledge of the underlying biological pathways. Whereas the primary goal of the high-content screening in this chemical genomic approach may be elucidation of the link between targets and the underlying physiology, the data collected on specific compounds can be used to drive drug programs.

2.4. Chemogenomics

Chemogenomics refers to the generation of specific sets of compounds with drug-like properties, which are specific for a given set of targets. The distinction between this and chemical genomics is that the knowledge extracted from sequences, 3D structures, assay results, and known chemical properties are used in the compound-design phase so that the compounds have a high likelihood of being selective and the biological consequences of inhibiting the target are known. In a typical chemogenomic approach, there is a very focused set of assays used to drive potency and selectivity, and a broader set of assays used to optimize chemical properties.

The different chemical genetic approaches described in the above subheadings are summarized in **Table 1**.

An example of a chemogenomic approach is shown in **Fig. 2**, where the protein sequences of members of a gene family have been overlayed on the 3D structure of an inhibitor-binding site from a representative of one of the targets in this family. The colored surface suggests areas of the active site where specificity is likely to be obtained, and the superimposed ligand suggests locations where changes to the core scaffold should be directed. The overall similarity in sequence and protein structure allows information obtained from one member of the gene family to be transferred to other members. Using this common frame of reference allows knowledge gleaned from the optimization of one chemical scaffold to be extrapolated to guide the tuning of the potency and specificity of other scaffolds that bind to the same site.

3. Goal

The goal of all of these approaches is to use chemical compounds to interrogate biological systems and ultimately provide insights that lead to improved health. A key limitation in the field is the ability to extrapolate from a molecular

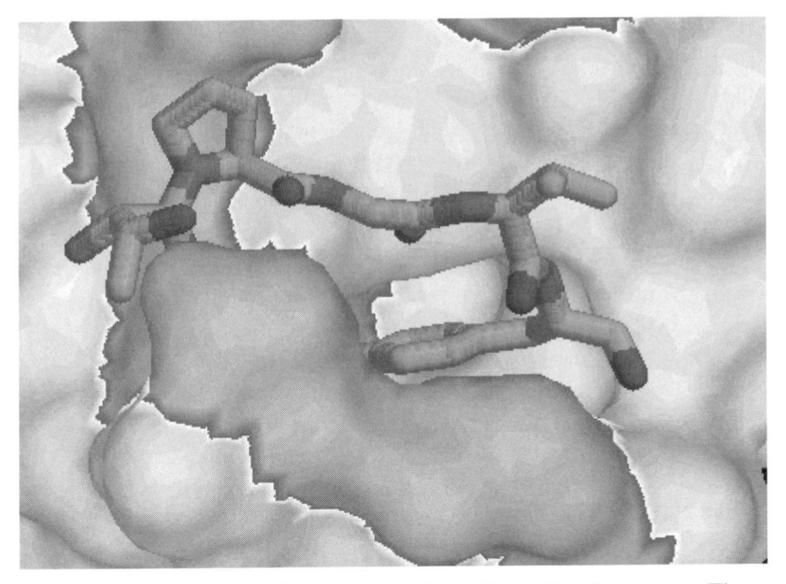

Fig. 2. Mapping of conserved features onto three-dimensional structure. The sequence differences between chymotrypsin and its closest homologs are mapped onto a surface representation of the structure of chymotrypsin (PDB: 1ab9). A peptide ligand (TPGVY) is show in stick format.

phenotype, to cellular response, to animal pharmacology, to efficacy in the clinic. In a traditional target or cell-based screen, there are a limited number of readouts that are used to determine the efficacy of a compound on a system—certainly not enough data to safely or accurately predict any effects in the clinic. Broad screening of specific compounds using technologies such as gene profiling or proteomics can provide vast amounts of data at the cellular detail level, but it is still difficult to translate this into known biological phenotypes. Whereas a bottom-up approach, where events at the molecular level can be used to rationally explain physiology at a macroscopic, is an objective of some in systems biology, insights can be made by more empirical approaches. Correlations discovered between molecular readouts and more phenotypically relevant measurements, such as toxicity and efficacy, are proving to be quite powerful. These relationships will become more robust as more data are gathered and integrated into systems allowing cross-discipline statistical analyses.

Although there is still no substitute for well-controlled clinical trials, the use of these technologies should improve the quality of the molecules entering the clinic and hopefully reduce failure rates, allowing treatments for more difficult diseases to be considered.

References

1. Lichtenberg, F. R. (2003) The economic and human impact of new drugs. *J. Clin. Psychiatry* **64(Suppl 17),** 15–18.
2. Hansen, R. W. (1986) New pharmaceuticals reduce cost of illness. *Can. Pharm. J.* **119(6),** 318–325.
3. Goodman, L. S., Hardman, J., and Limbird, L. (eds.). *Goodman and Gilman's The Pharmacological Basis of Therapeutics,* 9th ed. McGraw-Hill, New York.
4. Fleischmann, R. D., Adams, M. D., White, O., et al. (1995) Whole-genome random sequencing and assembly of Haemophilus influenzae Rd. *Science* **269(5223),** 496–512.
5. Stockwell, B. R., Haggarty, S. J., and Schreiber, S. L. (1999) High-throughput screening of small molecules in miniaturized mammalian cell–based assays involving post-translational modifications. *Chem. Biol.* **6(2),** 71–83.
6. Spring, D. R., Krishnan, S., Blackwell, H. E., and Schreiber, S. L. (2002) Diversity-oriented synthesis of biaryl-containing medium rings using a one bead/one stock solution platform. *J. Am. Chem. Soc.* **124(7),** 1354–1363.
7. Haggarty, S. J., Koeller, K. M., Wong, J. C., Butcher, R. A., and Schreiber, S. L. (2003) Multidimensional chemical genetic analysis of diversity-oriented synthesis-derived deacetylase inhibitors using cell-based assays. *Chem. Biol.* **10(5),** 383–396.
8. Shogren-Knaak, M. A., Alaimo, P. J., and Shokat, K. M. (2001) Recent advances in chemical approaches to the study of biological systems. *Annu. Rev. Cell Dev. Biol.* **17,** 405–433.

2

Chemistry for Chemical Genomics

Lutz Weber

Summary

New methods and strategies have been developed to design and use small molecules that allow the functional dissection of molecular pathways, cells, and organisms by selective small-molecule ligands or modulators. In this overview, we are focusing on diversity aspects, design methods, and chemical synthesis strategies for the application of small molecules as tools for chemical genomics. Examples for different successful chemical-genomics strategies include the selection of diverse drug-like molecules, target family–focused compound libraries, natural-product chemistry, and diversity-oriented synthesis.

Key Words: Chemical diversity; compound design; diversity-oriented synthesis; drug-like compounds; molecular properties; natural products; rule of five; structure–activity relationship; target-focused compound libraries.

1. Introduction

Organic chemistry is the science of the synthesis and properties of molecules that are constructed from only a few atom types, such as carbon, hydrogen, nitrogen, oxygen, and sulfur, with carbon atoms constituting the majority of the core of these chemicals. As these atoms are also the building blocks of naturally occurring peptides or oligonucleotides, this chemistry was termed *organic*, as opposed to other disciplines of chemistry. Such chemicals, commonly referred to as *small molecules*, are valuable as medicines to treat diseases ranging from headache to cancer.

Small molecules have recently proven to be extremely useful tools to explore the functions of the cell at the genome level, giving rise to the new paradigm of chemical genomics. The functional dissection of molecular pathways, cells, and organisms by having a small-molecule ligand or modulator for every gene product, was the vision of Stuart L. Schreiber, one of the pioneers of the chemical genomics field ***(1)***.

From: *Methods in Molecular Biology, vol. 310: Chemical Genomics: Reviews and Protocols*
Edited by: E. D. Zanders © Humana Press Inc., Totowa, NJ

Several recent reviews and books deal with the implication of chemical genomics towards drug discovery *(2–5)*. The chemical genomics paradigm is seen as a logical follow-up of the Human Genome Project, and several public initiatives on chemical genomics have been started on national levels. Thus, in the United States, the National Human Genome Research Institute has set up a chemical libraries plan that moves the National Institutes of Health into high-throughput screening and small-molecule development to "determine function and therapeutic potential of genes and to define molecular networks." In Germany, the Nationale Genomforschungsnetz initiated a chemical genomics platform by assembling synthetic compound libraries as probes for protein function. These compounds will help to validate new targets for novel therapies more rapidly, and will enable researchers in the public and private sectors to take these targets and move them through the drug-development pipeline.

In this overview we would like to focus on some chemical aspects, problems, and solutions to the application of small molecules as tools for chemical genomics.

2. Diversity of Small Molecules

What are small molecules? For the sake of simplicity, let us consider only those organic chemicals as small molecules that have a molecular weight of less than 1000 Dalton. Such a definition is by its nature arbitrary, but allows separating out other classes of organic molecules such as molecules that are oligomers of smaller building blocks such as, for example, proteins, oligonucleotides, or oligosaccharides. The number of all possible and different small molecules with a molecular weight below 1000 is assumed to exceed 10^{60}. To be a specific modulator of a target protein, a small molecule has to act as a ligand, binding to its target. As opposed to to large molecules, small molecules have an average interaction area of up to 400 $Å^2$ with their target—which is, for example, the size of a typical enzyme substrate site. This relatively small interaction area poses a serious problem to the whole concept of chemical genomics—is it at all possible to find a specific small molecule for each gene product? Indeed, it appears more likely that small molecules are promiscuous and may interact rather with a range of targets that have similar binding sites. Many of these "unwanted" interactions might not result in undesired effects, giving way to selective drugs in vivo. In odd cases, such secondary interactions may cause toxicities or other side effects; in lucky cases, these secondary interactions might be the real reason why a particular small molecule is an effective drug. Therefore, the rational design of such dual- or triple-action compounds has emerged as a new paradigm *(6)*.

The functional, biological diversity of small molecules appears to be inherently more limited than that of large molecules. Whereas the diversity of oligomers can be described by metrics like sequence and secondary, tertiary, and quaternary structure, the diversity of small molecules is harder to capture. The development of qualitative and quantitative measures for the chemical diversity of small molecules has only started to evolve with the advent of combinatorial chemistry, which enabled the synthesis of a large number of small molecules in one experiment.

As opposed to large molecules, the required high functional diversity of small molecules has to be packed into a rather small volume made up of only 20–30 non-hydrogen atoms. To bind efficiently at a protein-ligand site, most of the small-molecule ligand's binding energy must come from sites not exploited by the natural protein ligand. The limited molecular volume has to interact with the target protein with a maximum binding energy per atom to achieve both the required affinity and specificity. This binding energy has been calculated for maximal free-energy contributions per non-hydrogen atom with approx 1.5 kcal/mol across a wide variety of macromolecule–small-molecule interactions. The empirical data also revealed a significant trend to smaller contributions per atom as the relative molecular mass of the ligand increases ***(6)***. Thus, small-molecule ligands are binding to their targets with approximately three times more binding energy per unit area than protein ligands ***(8)***. As subnanomolar binding can be achieved with ligands containing as few as 10–20 atoms, the remaining 10–20 atoms could potentially be used to obtain selectivity. The likelihood of obtaining the desired functional diversity for small molecules is also correlated to the nature of the biological targets of interest. Thus, it is more likely to find enzyme inhibitors than small molecules that block protein–protein interactions ***(8)***.

Functional diversity of small molecules for chemical genomics experiments can therefore be defined in terms of obtaining maximum affinity to a given target protein by a minimum of molecular volume in order to minimize unwanted binding to other proteins. The realization of this Max–Min concept in a library of small-molecule compounds poses a serious challenge to organic chemists. As the molecular volume of small molecules is rather similar and does not leave enough room for major variations, the only way to obtain both affinity and selectivity is a maximum diverse distribution of atoms in this small volume. This can not be achieved by varying a small number of building blocks like the 20 amino acids for proteins, but requires the whole repertoire of organic chemistry to assemble novel chemical scaffolds. This recent understanding has led not only to a reappreciation of natural-product chemistry but also to a series of initiatives that aim at the development of novel chemistries to obtain chemical diversity.

2.1. Drug-Like Compound Libraries

A large range of small-molecule physico-chemical properties can be computed and used for diversity selection and drug likeliness. Thus, drug-like molecules should follow the "rules of five" that requires a logP <= 5, molecular weight <= 500, number of hydrogen bond acceptors <= 10, and number of hydrogen bond donors <= 5. Molecules violating more than one of these rules may have problems with bioavailability ***(9)*** or the likelihood of having good drug properties ***(10)***. A polar surface area of <120 Å^2 has been shown to correlate with bioavailability and can be computed in a straightforward way ***(11)***. A new method based on quantum chemistry performs the rapid, automated fragment-wise construction of an approximate charge density (σ-profiles) on the molecular surface—describing better the surface of the ligand a target protein will interact with ***(12)***.

An optimally diverse compound library of drug-like molecules for chemical genomics experiments would be a collection of representatives from a variety of compound clusters of similar molecules. A comparison between various simple and more complex structural descriptors and clustering methods has been given by Brown and Martin ***(13)***, using large sets of different molecules. Interestingly, the simplest method, counting 153 different substructure keys using a priori knowledge about substructure elements from a fragment dictionary, provided a better similarity measure than using more complex two- or three-dimensional criteria. This points to the difficulty of generating general representations of molecules, and it is probably impossible to hope to characterize a large, diverse data set by a small set of representatives. Calculated octanol-water partition coefficients, molar refractivities, or dipole moments can be used as additional criteria to assemble a useful library for biological testing ***(14)***. Exclusion criteria should also include knowledge on promiscuous or reactive substructures such as, for example, excluding compounds that exhibit thiourea, nitro, or α-haloketo groups such as defined by Rishton ***(15)***.

The above principles have been applied with success to select compound libraries for chemical genomic experiments. We would like to give two representative examples for such a successful selection.

Monastrol was identified as a novel antimitotic compound from a selected library of 16,320 commercially available compounds ***(16)***. The library was screened by applying a series of phenotype-based screens, selecting 139 compounds that induced phosphonucleolin levels, than subsequently eliminating 52 compounds from these 139 that targeted tubulin and a further 42 antimitotic compounds that target the interphase cytoskeleton, leaving overall only 5 compounds with the desired cellular effect. From these, monastrol arrests mammalian cells in mitosis with mono-astral spindles surrounded by a ring of chromosomes. The target of monastrol was found to be the mitotic motor protein Eg5, a bipolar kinesin known to be required for spindle bipolarity. This cellular activity of monastrol is very sensitive to structural changes: the chemically similar DHP2 compound is already completely inactive (**Fig. 1**).

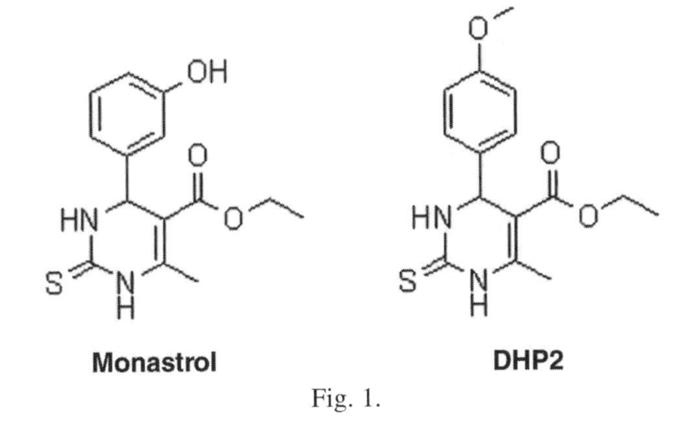

Fig. 1.

In a conceptually similar experiment, compounds were selected from commercial vendors to give a screening library of 73,400 diverse, drug-like molecules that comply with the rules of five. Screening of this library in a permeable yeast two-hybrid system was accomplished to select molecules that block the protein-protein interaction of cI-DBD-H-Ras(G186) with AD-Raf-1, but not of LexA-DBD-hsRPB7 and AD-hsRPB4. From this library, 38 active compounds were selected for further screening in mammalian cells to inhibit serum-induced transcriptional activation through SRE and AP-1 sites from the c-fos promoter. MCP1 was than selected by performing additional cell-based screening to remove unselective or toxic compounds *(17)*. Interestingly, the chemical structure of MCP1 resembles other structures that are able to inhibit protein–protein interactions, such as, for example, FKB-001 *(18)*, featuring flexible side chains that enable the molecule to fold into cavities at the protein surface. On the other hand, both molecules (**Fig. 2**) are accessible via fast and parallel synthesis methods *(19)*.

As opposed to selecting libraries of general molecules from historic compound collections, for combinatorial libraries the problem of a diversity metric is simplified due to the limited number of building blocks that are used for their synthesis. Using volume, lipophilicity, charge, and H-bond donor or acceptor descriptors, it was shown that combinatorial libraries may be designed to exhibit the same diversity as commercial drugs with respect to the density of these structural fragments *(20)*. Subsets from a large virtual library can then be chosen by D-optimal design to assemble an information-rich diverse library. More advanced methods (CATS and TOPAS) have been developed recently to select and synthesize compounds for screening based on two-dimensional criteria *(21,22)*.

2.2. Target-Family Compound Libraries

Proteins of a certain family usually exhibit structurally similar binding sites; a prominent example is the adenosine triphosphate (ATP)-binding site of the

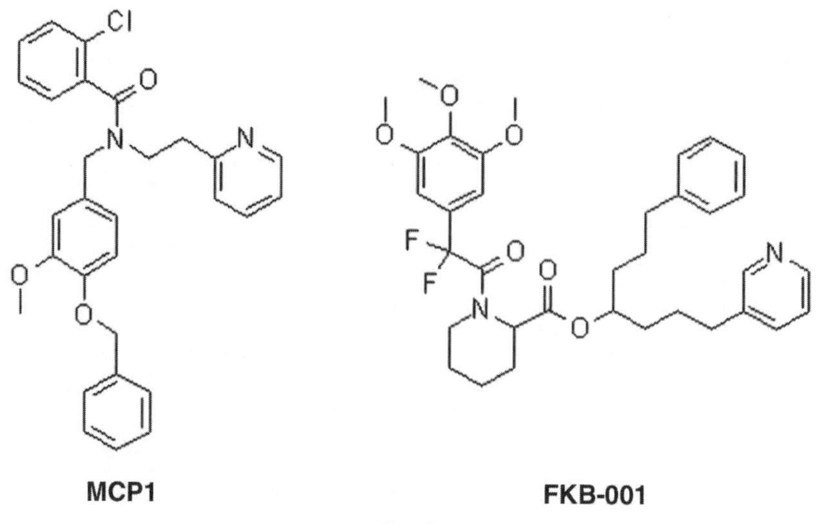

Fig. 2.

approx 518 kinases identified so far, also called the human kinome *(23)*. Not surprisingly, compounds that mimic the structural features of ATP are used with success to synthesize more or less selective kinase inhibitors. These inhibitors can be used both as tool compounds to study the effect of a particular kinase in cell signaling events or as potential drug candidates. A great deal of chemical, biological, and biostructural information on kinase inhibitors is available and provides a straightforward starting point for designing compound libraries that target kinases. In a recent example, biostructural information was used to classify kinases according to the similarity of their small-molecule ligands, and develop a fast virtual screening method—structural interaction fingerprint (SIFt)—that allowed rapid identification of novel and selective transforming growth factor-β kinase inhibitors. This method, based on experimental data, combines for the first time protein sequence homology and the chemical structure information of the ligand into one truly chemo-genomic method *(24)*.

In another example of combinatorial parallel chemistry, we have recently used the Ugi three-component reactions (Ugi 3-CR) to construct a library of 16,840 protease inhibitors *(25)*. It has been demonstrated previously that the Ugi-3CR reaction provides a useful chemical scaffold for the design of serine protease inhibitors: N-substituted 2-substituted-glycine *N*-aryl/alkyl-amides have been identified that are potent factor Xa, factor VIIa, or thrombin inhibitors. The three variable substituents of this scaffold, provided by the amine, aldehyde, and isonitrile starting materials, span a favorable pyramidal pharmacophoric scaffold that can fill the S1, S2, and S3 pockets of the respective protease. This library was screened against five proteases (factor Xa, trypsin, urokinase, tryptase, and chymotrypsin) at different concentrations to create an exhaustive structure–activity relationship (SAR) data set. Using this SAR, both selective and potent inhibitors could be identified against all five proteases, such as the four prototypic factor Xa inhibitors shown in **Fig. 3**.

2.3. Natural Products

By intuition, natural products are thought to exhibit a higher degree of complexity than traditional synthetic drugs *(26)*. This hypothesis was verified by analyzing the chemical core structures of natural products *(27)*. In addition, natural products tend to have a higher molecular weight, more oxygen atoms, and fewer nitrogen atoms than synthetic drug molecules. A quantitative index of complexity was developed that uses the number and size of rings and the connectivity of each atom *(28)*. Alternatively, complexity can be defined as the number of domains or substructures contained in a molecule that are available for interacting with the target. A molecule with low complexity has fewer interaction sites than a molecule with greater complexity, that can therefore be more selective than simple compounds, also yielding fewer hits in primary screens *(29)*. Libraries of natural products can be obtained either as purified compounds or as extracts from microbes or plants. Although most natural products do not follow the rules of five, finding natural-product drug candidates has been a successful enterprise in the past—almost half of the small-molecule drugs on the market are either directly natural products or chemically derived from those.

However, the total synthesis of these complex structures poses a serious challenge to organic chemistry, resulting often in the inability to follow up first hits quickly by resynthesis and derivatization. Thus, the synthesis of complex natural products using efficient strategies, often accompanied by the development of novel reagents and reactions, was considered the "royal" discipline of organic chemists in the past. Today, we observe that those natural-product-oriented organic chemists are broadly using the developed know-how to bring together the pharmacophores of structurally diverse natural products with methods of combinatorial chemistry.

As a representative example, the rather complex natural-product dysidiolide was synthesized by methods amenable to combinatorial synthesis *(30)*. Using a cycloaddition-based approach to the dysidiolide core, a solid-phase synthesis of epi-dysidiolide (**Fig. 4**) and analogs thereof was developed.

The total synthesis proceeds in >10 steps on solid phase and includes various transformations, including an asymmetric Diels-Alder reaction, oxidation with singlet oxygen, and olefin metathesis. This synthesis sequence is among the most advanced and demanding solid-phase syntheses developed so far for chemical genomics experiments. It demonstrates that the total synthesis of complex natural products in multi-step sequences on solid phase is feasible.

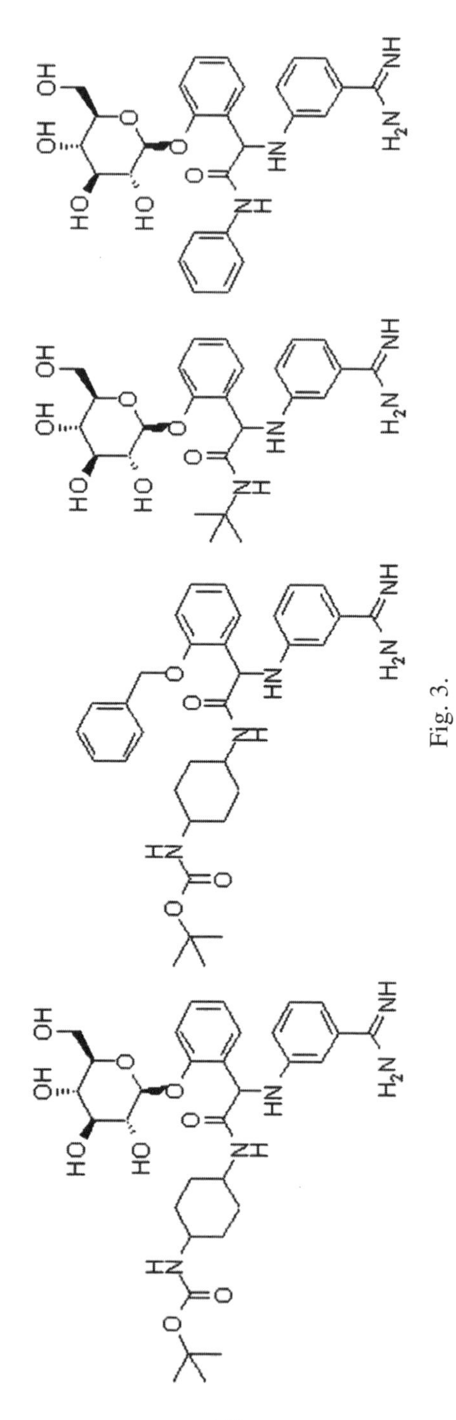

Fig. 3.

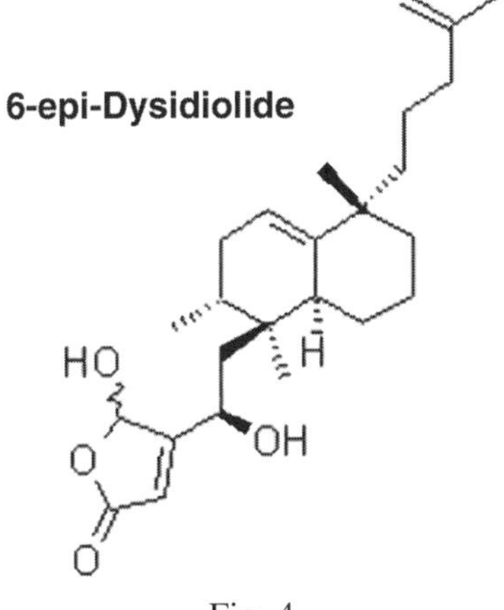

Fig. 4.

Moreover, the biological investigation of dysidiolide analogs has yielded selective Cdc25 inhibitors with high potency in enzymatic and cellular assays.

3. Diversity-Oriented Synthesis

Many natural products have been directly developed in the past as drugs—without the possibility for a significant synthesis of potentially better analogs. This observation has inspired a new strategy of synthesizing natural-product-like compounds using combinatorial, diversity-oriented synthesis (DOS) *(31)*. Thereby, the generation of molecular complexity by reaction sequences that are amenable to combinatorial chemistry has become an important theme of recent research *(32)*. Optimized reactions have been utilized to create compound libraries that are comparable in terms of structural diversity with natural products. In one example, the alkaloid martinellic acid has been the target for transition metal–catalyzed hetero-Diels-Alder reactions. Surprisingly, a reaction was found that yielded the desired scaffold in one step by a novel three-component coupling. In another example, a hetero-Diels-Alder scaffold was used to synthesize peptidomimetics, aiming to arrive at conformationally and proteolytically stable products derived from biologically active peptides *(33)*.

In a different approach, one may synthesize molecular scaffolds that have not been found in nature. These templates are often also called *new chemotypes*, expressing the hope of finding interesting and unexpected biological activities. For example, cycloadditions, imine formation, and Michael reactions are used to generate such novel backbones *(34,35)*.

Poly-substituted aromatic rings are difficult to synthesize—in a recent work, inhibitors of the cholesterine ester transfer protein (CETP) were synthesized, using a highly sophisticated assembly of basic organic reactions and comprising a three-component reaction *(36)*. This work certainly represents one of the best examples of how combinatorial chemistry has merged with the synthesis of high-diversity, complex compounds.

Multi-component reactions, first misinterpreted as rather rare exceptions in organic chemistry, are now commonly regarded as useful tools to generate highly diverse compound libraries. Only four synthetic steps are needed to generate complex, almost baroque structures *(37)* using an Ugi-type MCR, a Diels-Alder reaction, and a final olefin metathesis. Other examples of similar diversity-generating reaction sequences have been published, using oxidative bond formations or macrocyclization strategies by building up peptide scaffolds that are than cyclized via olefin metathesis *(38)*.

In most recent developments, it has been possible to devise synthetic strategies whereby a variety of chemical scaffolds can be obtained in one scheme *(39)*. Whereas in former examples of combinatorial synthesis a large number of compounds were synthesized that all shared the same chemical backbone, this new technology is opening new ways toward a diversity that can be used for chemical genomics in a designed strategy *(40)*.

4. Summary

The success of merging genomics and chemistry will depend on finding small molecules that can interact specifically with a given protein target molecule. This goal might be especially difficult to achieve when dealing with proteins that are similar in their sequence, three-dimensional structure, or function, as are kinases. One way, as outlined in this chapter, is to improve chemical methods to arrive at suitable small molecules. However, the concept of chemical genomics might also inspire and call for completely new technologies that integrate chemistry and genomics even better.

As an illustration, we would like to cite the work of Shokat *(41)*. To allow for the fast selection of suitable molecules, the three-dimensional structure of the v-Src kinase was changed through site-directed mutagenesis, creating a new binding pocket in the vicinity of the active site while retaining its catalytic properties. It was then shown that an ATP analog, created through introducing a side chain by chemical means into ATP, resulted in a highly specific substrate that is converted only by the mutated v-Src kinase and not by other members of this large enzyme family. It was thereby possible to study the pharmacological role of v-Src kinase without affecting other kinases. This method is the first experimental merger of genomics and small-molecule chemistry, allowing the creation of tools for chemical genomics studies. Other ideas along these lines may include variants of dynamic combinatorial libraries *(42)* or click chemistry *(43)*, where the target protein assists the synthesis of the small molecule by selecting a binding molecule out of a large pool of possible molecules.

The examples described here in any case illustrate a new quality of investigating the interaction of small molecules within complex biological systems. How can one use the chemical-genomics answers obtained to design better drugs, more rapidly and more efficiently? Typically, the biological responses towards a chemical entity are complex and multi-dimensional, such as gene-expression patterns, protein binding, phenotype reversals, and phenotype inductions. Multi-dimensional optimization algorithms will be required to translate such biological feedback into the creation of better drugs. In low-dimensional optimization spaces, traditional structure-activity relationships work well. Higher-dimensional search spaces, as provided by chemical genomics, need heuristic optimization procedures, as exemplified by neuronal networks or genetic algorithms *(42)*. Genetic algorithms utilize biological genomics information in combination with a recently introduced, DNA-like description of small synthetic molecules, to provide an opportunity for optimizing small molecules more efficiently using nature's evolutionary principles. Ultimately, one could envisage a co-evolutionary drug-discovery process that would enable the discovery of novel drug targets by using small molecules, which are then improved by respective biological feedback loops into highly potent and selective new drug candidates.

References

1. Schreiber, S. L. (1998) Chemical genetics resulting from a passion for synthetic organic chemistry. *Bio. Med. Chem.* **6,** 1127–1152.
2. Kubinyi, H. and Müller, G. (eds.). (2004) Chemogenomics in drug discovery, a medicinal chemistry perspective. In *Methods and Principles in Medicinal Chemistry*, Wiley-VCH, Weinheim, Germany.
3. Ferenc Darvas, F., Guttman, A., and Dormann, G. (2004) *Chemical Genomics.* Marcel Dekker, New York.
4. Salemme, F. R. (2003) Chemical genomics as an emerging paradigm for post-genomic drug discovery. *Pharmacogenomics* **4,** 1–11.
5. Crews, C. M. and Splittgerber, U. (1999) Chemical genetics: exploring and controlling cellular processes with chemical probes. *TIBS* **5,** 317–320.
6. Morphy, R., Kay, C., and Rankovic, Z. (2004) From magic bullets to designed multiple ligands. *Drug Discovery Today* **9,** 641–651.
7. Kuntz, I. D., Chen, K., Sharp, K. A., and Kollmann, P. A. (1999) The maximal affinity of ligands. *Proc. Natl. Acad. Sci. USA* **96,** 9997–10,002.
8. Spencer, R. (1998) High-throughput screening of historic collections: observations on file size, biological targets, and file diversity. *Biotechnol. Bioeng.* **61,** 61–67.
9. Lipinski, C. A., Lombardo, F., Dominy, B. W., and Feeney, P. J. (1997) Experimental and computational approaches to estimate solubility and permeability in drug discovery and development settings. *Adv. Drug. Delivery Rev.* **23,** 4–25.
10. Proudfoot, J. R. (2002) Drugs, leads, and drug-likeness: an analysis of some recently launched drugs. *Bioorg. Med. Chem. Lett.* **12,** 1647–1650.

11. Ertl, P., Rohde, B., and Selzer, P. (2000) Fast calculation of molecular polar surface area as a sum of fragment-based contributions and its application to the prediction of drug transport properties. *J. Med. Chem.* **43,** 3714–3717.
12. www.cosmologic.de. Accessed on March 30, 2005.
13. Brown, R. D. and Martin, Y. C. (1996) Use of structure-activity data to compare structure-based clustering methods and descriptors for use in compound selection. *J. Chem. Inf. Comput. Sci.* **36,** 572–584.
14. Shemetulskis, N. A., Dunbar, J. B., Dunbar, B. W., Moreland, D. W., and Humblet, C. (1995) Enhancing the diversity of a corporate database using chemical clustering and analysis. *J. Comp. Aided Mol. Design* **9,** 407–416.
15. Rishton, G. M. (1997) Reactive compounds and in vitro false positives in HTS. *Drug Discovery Today* **2,** 382–384.
16. Mayer, T. U., Kapoor, T. M., Haggarty, S. J., King, R. W., Schreiber, S. L., and Mitchtison, T. J. (1999) Small molecule inhibitor of spindle bipolarity identified in a phenotype-based screen. *Science* **286,** 971–974.
17. Kato-Stankiewicz, J., Hakimi, I., Zhi, G., et al. (2002) Inhibitors of Ras-Raf-1 interaction identified by two-hybrid screening revert Ras-dependent transformation phenotypes in human cancer cells. *Proc. Nat. Acad. Sci. USA* **99,** 14,398–14,403.
18. Dubowchik, G. M., Vrudhula, V. M., Dasgupta, B., et al. (2001) 2-Aryl-2,2-difluoroacetamide FKBP12 ligands: synthesis and X-ray structural studies. *Org. Lett.* **3,** 3987–3990.
19. Lu, Y., Sakamuri, S., Chen, Q.-Z., et al. (2004) Solution phase parallel synthesis and evaluation of MAPK inhibitory activities of close structural analogues of a Ras pathway modulator. *Bioorg. Med. Chem. Lett.* **14,** 3957–3962.
20. Martin, E. J., Blaney, J. M., Siani, M. A., Spellmeyer, D. C., Wong, A. K., and Moos, W. H. (1995) Measuring diversity: experimental design of combinatorial libraries for drug discovery. *J. Med. Chem.* **38,** 1431–1436.
21. Schneider, G., Chomienne-Clement, O., Hilfiger, L., et al. (2000) Virtual screening for bioactive molecules by evolutionary de novo design. *Angew. Chemie Int. Ed.* **39,** 4130–4133.
22. Schneider, G., Lee, M.-L., Stahl, M., and Schneider, P. (2000) De novo design of molecular architectures by evolutionary assembly of drug-derived building blocks. *J. Comput. Aided Mol. Des.* **14,** 487–494.
23. Manning, G., Whyte, D. B., Martinez, R., Hunter, T., and Sudarsanam, S. (2002) The protein kinase complement of the human genome. *Science* **298,** 1912–1934.
24. Deng, Z., Chuaqui, C., and Singh, J. (2004) Structural Interaction Fingerprint (SIFt): a novel method for analyzing three-dimensional protein-ligand binding interactions. *J. Med. Chem.* **47,** 337–344.
25. Weber, L. Fractal theory applied to structure-activity relationships. Euro-QSAR 2004, Istanbul, Turkey, September 5–10, 2005.
26. Verdine, L. G. (1996) The combinatorial chemistry of nature. *Nature* **384,** 11–13.

27. Lee, M.-L. and Schneider, G. (2001) Scaffold architecture and pharmacophoric properties of trade drugs and natural products. *J. Comb. Chem.* **3,** 284–289.
28. Barone, R. and Chanon, M. (2001) A new and simple approach to chemical complexity. Application to the synthesis of natural products. *J. Chem. Inf. Comput. Sci.* **41,** 269–272.
29. Hann, M. M., Leach, A. R., and Harper, G. (2001) Molecular complexity and its impact on the probability of finding leads for drug discovery. *J. Chem. Inf. Comput. Sci.* **41,** 856–864.
30. Brohm, D., Metzger, S., Bhargava, A., Müller, O., Lieb, F., and Waldmann, H. (2002) Natural products are biologically validated starting points in structural space for compound library development: solid phase synthesis of dysidiolide-derived phosphatase inhibitors. *Angew. Chem. Int. Ed.* **41,** 307–311.
31. Schreiber, S. L. (2000) Target-oriented and diversity-oriented organic synthesis in drug discovery. *Science* **287,** 1964–1969.
32. Weber, L. (2000) High-diversity combinatorial libraries. *Curr. Opin. Chem. Biol.* **4,** 295–302.
33. Creighton, C. J., Zapf, C. W., Bu, J. H., and Goodman, M. (1999) Solid-phase synthesis of pyridones and pyridopyrazines as peptidomimetic scaffolds. *Org. Lett.* **1,** 1647–1649.
34. Peng, G., Sohn, A., and Gallop, M. A. (1999) Stereoselective solid-phase synthesis of a triaza tricyclic ring system: a new chemotype for lead discovery. *J. Org. Chem.* **64,** 8342–8349.
35. Brooking, P., Crawshaw, M., Hird, N. W., et al. (1999) The development of a solid-phase tsuge reaction and its application in high throughput robotic synthesis. *Synthesis* **11,** 1986–1992.
36. Paulsen, H., Antons, S., Brandes, A., et al. (1999) Stereoselective Mukaiyama-Michael/Michael/Aldol domino cyclization as the key step in the synthesis of penta-substituted arenes: an efficient access to highly active inhibitors of cholesteryl ester transfer protein (CETP). *Angew. Chem. Int. Ed.* **38,** 3373–3375.
37. Lee, D., Sello, J. K., and Schreiber, S. L. (2000) Pairwise use of complexity-generating reactions in diversity-oriented organic synthesis. *Org. Lett.* **2,** 709–712.
38. Reichwein, J. F., Wels, B., Kruijtzer, J. A. W., Versluis, C., Liskamp, R. M. J. (1999) Rolling loop scan: an approach featuring ring-closing metathesis for generating libraries of peptides with molecular shapes mimicking bioactive conformations or local folding of peptides and proteins. *Angew. Chem. Int. Ed.* **38,** 3684–3687.
39. Taylor, S. J., Taylor, A. M., and Schreiber, S. L. (2004) Synthetic strategy toward skeletal diversity via solid-supported, otherwise unstable reactive intermediates. *Angewandte Chemie* **43,** 1681–1685.
40. Burke, M. D. and Schreiber, S. L. (2004) A planning strategy for diversity-oriented synthesis. *Angewandte Chemie* **43,** 46–58.
41. Shah, K., Liu, Y., Deirmengian, C., and Shokat, K. M. (1997) Engineering unnatural nucleotide specificity for Rous sarcoma virus tyrosine kinase to uniquely label its direct substrates. *Proc. Nat. Acad. Sci. USA* **94,** 3565–3570.

42. Huc, I. and Lehn, J.-M. (1997) Virtual combinatorial libraries: dynamic generation of molecular and supramolecular diversity by self-assembly. *Proc. Natl. Acad. Sci. USA* **94,** 2106–2110.
43. Lewis, W. G., Green, L. G., Grynszpan, F., et al. (2002) Click chemistry in situ: acetylcholinesterase as a reaction vessel for the selective assembly of a femtomolar inhibitor from an array of building blocks. *Angew. Chem. Int. Ed.* **41,** 1053–1057
42. Illgen, K., Enderle, T., Broger, C., and Weber, L. (2000) Simulated molecular evolution in a full combinatorial library. *Chem. Biol.* **7,** 433–441.

3

Computer-Aided Design of Small Molecules for Chemical Genomics

Philip M. Dean

Summary

De novo design provides an *in silico* toolkit for the design of novel molecular structures to a set of specified structural constraints, and is thus ideally suited for creating molecules for chemical genomics. The design process involves manipulation of the input, modification of structural constraints, and further processing of the *de novo*-generated molecules using various modular toolkits. The development of a theoretical framework for each of these stages will provide novel practical solutions to the problem of creating compounds with maximal chemical diversity. This chapter describes the fundamental problems encountered in the application of novel chemical design technologies to chemical genomics by means of a formal representation. Formal representations help to outline and clarify ideas and hypotheses that can then be explored using mathematical algorithms. It is only by developing this rigorous foundation, that *in silico* design can progress in a rational way.

Key Words: Chemical genomics; *de novo* drug design; drug metabolism; gene expression; pharmacogenomics; polymorphism; drug selectivity; drug promiscuity.

1. Introduction

1.1. The Post-Genomics Challenge

The involvement of politicians on both sides of the Atlantic in the announcement of the completion of the first draft of the human genome heralded a lot of speculation about the use and abuse of the information that had been produced, or could be produced, when the genome was completed. As always with major developments of this kind, more problems are thrown up by the data than can be immediately solved. After all, the sequence is merely a string of letters representing one-dimensional information of the genes. How this information is

From: *Methods in Molecular Biology, vol. 310: Chemical Genomics: Reviews and Protocols*
Edited by: E. D. Zanders © Humana Press Inc., Totowa, NJ

to be translated into more useful knowledge was a more complex task and has led to an explosion of interest in bioinformatics. Many financially advanced nations have devoted considerable fractions of their science research budgets to new initiatives in bioinformatics; no one wants to be left behind, because the perceived commercial value of genomics is believed to be massive.

Bioinformatics is an exciting young discipline aimed mainly at sorting and categorizing data. The big challenges are twofold: the development of novel methods for automated linkage of disparate data silos, and the shift from one-dimensional genome sequence data into three-dimensional (3D) protein structure data. Automated linkage methods seek to find relationships such as possible protein–protein interactions, that can then be resolved by physical experiments. Once those transitions have been overcome, the genomic data will have been manipulated into knowledge that can be used as input for rational and automated methods for drug design. Primarily, we are hoping to develop new methods that will be exquisitely selective for specified proteins. Drug design on the genome scale is a huge challenge. However, if we can foresee a way through the challenge, even though the computing power needed is massive, then that way will surely be found.

1.2. Data Avalanches

Although the size of the human genome is not as large as initially envisaged (some 30–40,000 genes rather than the 140,000 genes predicted before the sequence of the first draft) there are still about 3 billion nucleotide bases in the sequence. The conversion of this gene information into protein sequences is more problematic, because there are many splice variants that lead to different proteins. The size of the human proteome may therefore be an order of magnitude more complex than the genome. Furthermore, the old dogma of one gene/one enzyme has been shown to be untrue, some proteins being multifunctional depending on where in the body they are expressed. For example, phosphoglucose isomerase is a multifunctional protein; in the cell it plays a part as an enzyme in both glycolysis and gluconeogenesis; however, outside the cell it acts as a cytokine with autocrine motility factor activity *(1,2)*. Similarly, the m-RNA splicing factor Prp19p also functions as a ubiquitin ligase *(3)*.

A subset of the proteome contains the targets for drugs and can be categorized into three principal groups:

- Receptors, where the drug modulates the activity of the receptor without the necessity of another small ligand being involved,
- Enzymes, where the drug interferes with the activity of the enzyme in turnover of substrate to product, and
- Protein–protein interactions, where the drug interferes with the formation of a protein complex necessary for the propagation of a biological effect.

If the size of the proteome is P, there will be no more than P^n possible n-mer interactions. However, as far as drug design is concerned, the size of the interacting surfaces that could be interfered with by a small molecule is small and unlikely to include three proteins; this limits the potential number of interactions to P^2. If the human proteome contains 10^5 proteins, the number of homodimer interactions is potentially 10^{10}.

Chemical space is thought to be massive: 10^{180} different structures with molecular weights <500; the number of drug-like molecules is thought to be much smaller at 10^{60}, whereas the number of chemotypes (novel core chemical scaffolds) may be 10^{40} *(4)*. These numbers can be put into a context of the mass of the visible universe, 10^{54}g. Placing these numbers within a computing perspective, a dedicated cluster of 10,000 teraflop machines would provide only 10^{23} flops per year. Clearly, if computing methods are to be used either to generate or to search these large chemical spaces, they would have to be efficient combinatorial optimizers to complete the job in a reasonable time.

The task for drug discovery is to reduce these numbers to practice by advanced computational optimization techniques and identify specific small molecules as starting material in the hit-to-lead process. In chemical genomics, we seek ultimately to develop methods to generate selective compounds from a structural genomics understanding of the proteome. Furthermore, it should be possible to predict potential lack of selectivity of current drugs through protein-binding promiscuity.

1.3. Personalized Medicines

Pharmacogenomics offers a way to study the effect of small changes in gene sequence on drug action; pharmacogenetics studies the effect of these changes on human populations. There are strong relationships between the two disciplines. The International Union of Basic and Clinical Pharmacology (IUPHAR) has recognized the importance of this link in the establishment of a section on the field of pharmacogenomics and pharmacogenetics. The general points for pharmacogenomic influences on drug action are at the steps of:

- Absorption
- Distribution
- Site of action
- Metabolism
- Elimination

At each of these general steps, there are many molecular mechanisms that can be involved in drug handling that have a significant bearing on chemical genomics. For example, absorption should not be thought of as a passive process but as a dynamic flux, forwards and backwards, across a set of molecular barriers

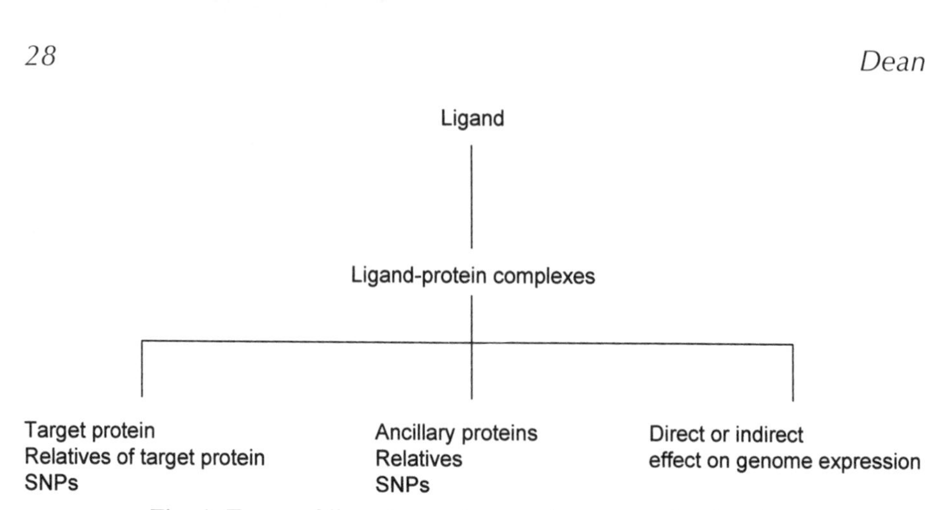

Fig. 1. Types of ligand–protein complex classed by function.

from the site of entry into the body to the target cell. Any change in the molecular structure of the barriers, pump mechanisms, or the rate of production of barrier molecules will have significant effects on the concentration of the drug at the therapeutic molecular target. Furthermore, the target cell may have induction mechanisms that can be triggered by the drug to start up a cycle of expulsion of the drug from the cell, leading to drug resistance. Historically, pharmacogenetics has focused on mutational changes in drug-metabolizing enzymes, the reason being that the enzymes have been simple to identify, isolate, and compare with the wild type.

The determination of the human genome sequence has potentially provided a much wider spectrum of points for interference in drug action. In principle, these genomic differences can be identified, and therefore diagnostic kits could be developed to identify problems in specific drug handling by individuals. A goal has thus been set in chemical genomics for personalizing medical treatment.

Figure 1 gives an overview of the complexity in the search for selective action of a ligand. It is more than likely that the ligand will bind to more than one protein type to form multiple ligand–protein complexes. These proteins may be related to the intended protein target—for example, sub-types of the same receptor where the binding site shows strong three-dimensional similarities—or the sites may differ only by single nucleotide polymorphisms (SNPs). Ligand complexes may form with ancillary proteins of unrelated function—for example, metabolic enzyme active sites, sites on transport proteins, allosteric sites, or plasma protein binding sites; there may be SNP variants to these as well. Other classes of site for ligand binding may be those that bind to functional elements in gene expression. In theory, all possible protein binding sites could be discovered by ligand–protein virtual docking experiments against known protein structures. If docking experiments on this scale were to be performed to identify all ligand-binding sites, it would provide powerful insights into ligand selectivity.

1.4. The Problems of Converting Genomic Visions Into Reality

Clearly, the completion of the human genome sequence offers medicine the possibility of huge steps forward in the conquest of disease. Probably most scientists in drug discovery have thought through the implications of this new knowledge for our own specific research area and can now envision a rational path towards possible solutions. The problems we face in converting vision into reality are largely problems of scale; hence the drive to industrialize the drug-discovery process using large computational efforts, micro-miniaturization of experiments, and robotic procedures to enhance the throughput of experiments.

It is well-known that drugs can have an effect on the up- and down-regulation of genes. The question that arises is: how do we relate gene expression in a tissue(s) to the molecular structure of drugs acting there? If there are differences in responses between individual patients, how do we handle genetic polymorphisms to small drug structures? Would this provide scientists with a handle for approaching the problem of personalized medicine through advances in chemical genomics?

2. A Formal Representation of Chemical Genomics Problems

2.1. Pharmacogenomics of Gene Expression

The chemical structure of the active drug molecules has to be related to their activity on all expressed genes, or gene products, within the body, whether that activity is viewed as a positive effect or one with drawbacks. Although in totality this is an exceptionally complex problem, it can be broken down into logical components for a more structured analysis.

The general problem can be expressed by considering a class of drug molecules, A, composed of a set of structurally similar members $\{A_1, A_2, A_3, A_k \ldots A_m\}$. Each drug molecule, A_k, of the set can be specified by a set of atomic coordinates for each atom a $\{x_{1a}, x_{2a}, x_{3a}\}$ and the molecule is composed of a set of molecular features F $\{F_1, F_2, F_3, F_k \ldots F_n\}$. Within a set of active compounds, there are likely to be substructural similarities; these similarities can be quantified so that similar molecular substructures in the set A can be found. A simple quantification system for molecular similarity is provided by the Tanimoto index of a pair of compounds. The index works by identifying common substructures within the molecules. Conversely, structural differences can also be identified. Correspondences in substructures and features can be determined by experiment and subsets of features can be related to activity.

Consider a set of organ tissues, O $\{O_1, O_2, O_3, O_k...O_p\}$, on which gene expression can be measured, then for each organ tissue there is a subset of genes G $\{G_1, G_2, G_3, G_k \ldots G_q\}$ that may exhibit modified expression. This expression of gene, G_k, may show a dynamic pattern of expression, E_{Gk}, where E_{Gk} is a time-dependent relationship over the period of stimulus, t, so

$$E_{Gk} = f(t).$$

This formal representation helps to provide a framework for investigating two questions:

1. Can the structural differences in a set of molecules, A, be correlated with changes in the genes, G, expressed in a particular organ, O_k?
2. If we consider a superset of drug molecules composed of M structurally different classes, M $\{A_{1 \ldots n}, B_{1 \ldots n}, C_{1 \ldots n}, \ldots Z_{1 \ldots n}\}$, can a particular subset of molecular features, F_k, be correlated with changes in the genes, $G_{1 \ldots n}$, expressed in organ, O_k? If the answer to this question is yes, we would discover a general set of substructure and feature markers that regulate a gene expression pattern, E_{Gk}.

Note that this formal representation says nothing about the mechanism of expression. It is still a superficial representation that is useful only for a top-level classification. What is needed is a further layer of sophistication that relates structural molecular features in the set of drug molecules with 3D interactions with their binding site. There is also the strong possibility that a structural molecular feature may also interact with more than one binding site, as is frequently the case with closely related gene families. A further layer of complexity needs to be added to consider gene expression networks with feedback to give a more realistic representation of the chemical genomics problem, but that is beyond the scope of this chapter. Useful reviews of the theory are given by Zhou et al. *(5)* and Hartemink et al. *(6)*.

2.2. Polymorphisms in Drug Design

Polymorphisms commonly affect chemical genomics in two ways: modifications to the active site for the drug or modifications to downstream handling of the drug molecule, e.g., metabolism or cell permeability. In many cases these differences are caused by SNPs.

2.2.1. Polymorphism in an Active Site

Consider a polymorphic site, S, for which there are a set of positions P $\{P_1, P_2, P_3, P_k, \ldots P_r\}$ where protein polymorphisms are possible, at each position, P_k, there is a set of possible residues R $\{R_1, R_2, R_3, R_k, \ldots R_s\}$; if a particular residue, R_k, at position, P_k, is absent or replaced within a patient and that residue is believed to contribute significantly to the activity of the current drug molecule, then that patient will show a marked difference in affinity for the drug. In a crude sense, the interaction energy between a drug and its binding site can be expressed as a summation of all pair-wise atomic interactions between the binding components. Replacement of atoms in a polymorphism therefore has a significant effect on affinity.

The drug designer then has two choices for discovering active compounds: one choice would be to design to the sub-set of site points that are common, with the hope that all patients could be treated effectively by the same drug; the other would be to design a different drug for each sub-population of individuals. The latter option is the personalized medicine approach. In many respects, this problem is identical to that encountered in practice with designing either promiscuity or selectivity into the drug–receptor interaction (*see* **Subheadings 4.4.** and **4.5.**). A good example of SNP problems in practice for drug designers is the case of treatment of chronic myeloid leukemia through the enzyme Abelson tyrosine kinase (ABL). Gleevec binds to ABL in its inactive form and keeps it inactive. However, many SNP mutations have arisen that prevent Gleevec binding *(7)*. The promising new compound BMS-354825 is able to bind to the mutated ABL *(8)*.

2.2.2. Polymorphism in a Metabolic Site

Polymorphisms in drug metabolism have a profound effect in chemical genomics and could also be considered formally. Drug metabolism is governed by the recognition of sub-structural molecular fragments by specific drug-metabolizing enzymes. The problem can be formally expressed: drug molecule A is composed of a set of sub-structural fragments F $\{F_1, F_2, F_3, F_k, \ldots F_r\}$; each fragment, F_k, is recognized by an enzyme Z_k, where Z_k is a member of the set, Z, of all drug-metabolizing enzymes. Recognition of the fragment by Z_k leads to the metabolism of A to metabolite M_k, where M_k is a member of the set of metabolites M. If the drug-metabolizing enzyme Z_k, for a particular fragment F_k on drug A, shows a set of polymorphisms where there are a set of positions P $\{P_1, P_2, P_3, P_k, \ldots P_r\}$ on the enzyme, and at each position P_k, there is a set of possible residues R $\{R_1, R_2, R_3, R_k, \ldots R_s\}$ for the site S, then different individuals will metabolize the drug less predictably. In turn, the drug metabolites may then create different gene-expression patterns. The questions to be addressed are:

1. What are the predicted routes of metabolism for the drug, given the set of polymorphisms?
2. How do we redesign the drug to eliminate the polymorphism problem?

2.3. Similarities in Binding-Site Structures

The evolution of protein structures shows conservation in recognition features for common binding sites. For example, the adenosine triphosphate (ATP)/guanosine triphosphate (GTP) binding site contains an amino acid motif A (P-loop) of [AG]-x(4)-G-K-[ST] *(9)*. Within this motif, ample variation is possible, with consequent differences in 3D structure in the binding sites. These differences will affect the binding of purine analogs to the binding sites, and a representation of this effect is quite similar to that given under **Subheading 2.2.** for polymorphisms.

The problem for chemical genomics is twofold:

1. Many compounds will cross-react by binding to variants of the site.
2. Different target proteins may well contain the same binding motif even though the function of the protein may be different; consequently, the drug will bind to different proteins and may interfere in different signaling pathways. Although in this case, the compound would be perfectly selective for the site, its action would be nonspecific, and that target should be avoided for drug design.

Progress in chemical genomics has been made towards linking ligand structural motifs to protein structural motifs by Deng, Chuaqui, and Singh ***(10)***; they used a method for fingerprinting the ligand and obtaining the complementary fingerprint of the binding site.

This formal representation of problems in chemical genomics identifies the areas where there will be considerable impact on drug design in the future. Currently, progress has been limited to very few examples in practice. However, chemical genomics is a rapidly expanding field, and these problems are ripe for exploration. The rest of this chapter focuses on the application of computational methods used in drug design to genomically derived information.

3. Computational Methods for Macromolecular Structure Determination

Therapeutic target identification and validation is a key problem for drug discovery. Protein interactions and pathway analysis provide clues for identifying proteins that may be key control points within a pathology. Where protein–protein interactions are involved, there is potentially an all-against-all problem to be resolved to pinpoint the interacting proteins.

If we are to industrialize chemical genomics and drug discovery efficiently, we need the 3D structures of all proteins, both for man and pathogens. This is the current challenge for structural genomics. Although there have been huge advances in the application of X-ray crystallography to protein structure determination using many thousands of crystallization attempts per day, the number of structures determined is still small: the total number of structures in the Protein Databank was 26,500 in July 2004. The structural genomics initiatives in North America, Europe, and Japan are tasked with providing high-resolution structures, either by X-ray crystallography or by nuclear magnetic resonance. However, experimental progress is slow, and in parallel to this large-scale computational structure prediction, methods are constantly being developed to automate the generation of new structures rapidly.

3.1. Homology Methods

The raw data for protein-structure generation is the protein sequence. In principle this data string encodes the 3D structure of the protein. How the protein folds is still unclear. Nevertheless, certain folding patterns are associated with specific sequence patterns, so it is possible to develop expert systems to provide 3D models of new folded proteins. Commercial software for homology modeling is widely available.

Automated homology methods can be used to model significant numbers of proteins on a genome-wide scale. However, the important issue is how closely the model represents structural reality. The Structural Genomics initiative has selected representatives from around 20,000 groups of proteins that could be used in a large-scale homology modeling process for most protein sequences ***(11)***. It is estimated that this represents 150,000 proteins. Active sites are quite closely conserved in sequence and structure, so it may be possible to use crystal data to represent the core elements of the site while modeling with molecular dynamics the more peripheral elements of the site.

4. Computational Methods for Designing Small Molecules

If genome-wide drug-design methods are to be developed, they have to be fast and automated so that the algorithm can design a good selection of molecular structures that can be synthesized readily. Synthetic tractability has always been a problem for *de novo* design methods, because by their nature the algorithms are geared to creating novel structures. However, if the molecules are to be built from small molecular fragments, the fragment assembly process can follow a limited set of rules that try to ensure synthetic tractability.

There are two general approaches to *de novo* drug design. First, where 3D coordinate data for the site are available, e.g., X-ray crystal data, nuclear magnetic resonance data, or good homology model data. Second, where this type of site data is missing, *de novo* design can proceed from the use of active ligands to build pharmacophore models of the site.

4.1. Molecular Design Criteria and Ligand Structural Variability

Consider a set of atoms, S, in the site, together with a set of tolerances, T, for the fit of a ligand(s), and a set of bounds, B, that limits the design region within the site; these constraints govern the local fit between the designed ligand and the site constraints. If the *de novo* structures are to be built from a complete set of molecular fragments, F, then in principle, all possible molecular structures could be considered by the generator, although not all would be acceptable. In the extreme case where the tolerances are set to zero, only perfectly fitting structures could be designed. These structures would be few if the bounds for design are restricted; they would be essentially isosteres.

As the tolerances are relaxed by increasing T, different local molecular substructures will be generated that satisfy the set of constraints imposed by the site. Hence, molecular-scaffold diversity will be introduced into *de novo* ligand design. For example, suppose there is a local steric constraint within a binding pocket for a six-membered aromatic ring; if the tolerance criteria are relaxed, then other ring systems may be accommodated. If the set of tolerances is uniformly relaxed, then by analogy, a larger structural variation could be accommodated across the site, but within the limits of B. This generation problem has been studied in some detail for the methotrexate binding site of dihydrofolate reductase, where 10,000 runs were performed; 26 solutions had the same connectivity as methotrexate but with slightly different orientations within the site; the principal known binding mode was mimicked ***(12)***. Each solution had a slightly different pose in the site.

If the site bounds are increased, ligand structural variation will be increased. To explore how much variation could be quantified, consider a simple case. Let there be five site points, with required features, that are needed for producing activity, and say *n* structures could be generated within a particular constraint set. If the limits of B are expanded twofold to include 10 possible site points, then the number of combinations of 5 from 10 is 252. It is more than likely that >252*n* structures are possible that fit the design criteria.

4.2. Structure-Based Methods

In this case, detailed site coordinate data are used to control *de novo* structure generation. The quality of the data is important; crystal-structure data with R < 2.1Å are ideal. Furthermore, it is often found that bound water molecules are associated with hydrogen bonding to the site and ligand; these clearly have an effect on ligand binding and consequently on drug-design strategies ***(13,14)***.

Two general strategies are often used. Interacting fragments are positioned in the site and a design mechanism tries to link the fragments together by other linker fragments. Alternatively, a single fragment is selected and positioned in the site, and structural elaboration ensues from that fragment, with other fragments being pieced together stepwise.

Automated *de novo* design methods seek to explore different molecular structures that could fit and bind to the site. They have general features in common:

1. A description of the site, usually the coordinates of the protein with a bounding box for the site; design is confined within the bounding box.

Table 1
Comparison Between the Numbers of Known Chemotypes for Four Targets With the Number of Chemotypes Reproduced by *De Novo* Structure Generation Using SkelGen Within 600 Generations Per Target

Target protein	Known chemotypes/motifs	Chemotypes/motifs reproduced by SkelGen
COX2	5 chemotype classes	5 chemotype classes
Estrogen receptor	10 chemotype classes	6 chemotype classes
CDK2	9 binding motifs	9 binding motifs
MMP3	5 binding motifs	5 binding motifs

2. A description of the strategy to be adopted for design, i.e., which site points to use to force certain interactions to be utilized.
3. A file of chemical fragments from which the ligand structures are to be built.
4. Rules for assembly of fragments within the site to ensure that appropriate chemistries are used in linking fragments together.
5. A penalty function, which enables a suggested structure to be assessed in the site for goodness of fit.
6. An optimization procedure, which controls the evolution of novel structures.
7. A stochastic process, which enables the algorithm to provide alternative solutions.

Two fundamental questions have to be asked about *de novo* design algorithms:

1. Can they reproduce known chemotypes for a range of targets without the algorithm having any knowledge of the known solutions?
2. How many novel structures can be grown for a target?

Stahl et al. ***(15)*** have attempted to answer the first question using four widely different targets: cyclo-oxygenase-2 (COX2), estrogen receptor (ER), cyclin-dependant kinase-2 (CDK2), and matrix metalloproteinase-3 (MMP3). In each case, the algorithm was run 600 times, and the output of each generated structure was compared with the known active chemotypes. To make this comparison more stringent, the output structures were compared with the known structures using a combined two-dimensional and 3D metric. The 3D metric took the known structure in the site as a reference structure and compared this with the coordinates of the output structure generated in the site. In this way, it is possible to superpose the two structures as bound in the site and generated in the site. **Table 1** summarizes the results for the four target sites. In each target, close structural chemotypes were found to the known active compounds. For COX2, CDK2, and MMP3, all known chemotypes were found. With ER, 6 of the 10 chemotypes were found; known *cis*-stilbene scaffolds could not be modeled into ER (1err Protein Databank structure) in a strain-free manner, so it is not surprising that the chemotype could not be generated. Many different alternative chemotypes were also grown.

Compute farms are ideally suited to molecular-scaffold generation in *de novo* design, because the process can be split into complete runs that can be distributed onto different processors. It scales linearly and is embarrassingly parallel.

4.3. Ligand-Based Methods

In the absence of a 3D structure for the target protein, and where there is little confidence in a homology model for the binding site, it may be possible to use compounds discovered to be active as the basis for further *de novo* structure generation. The compounds may be derived from either a high-throughput screen or, in the worst case, from a single active compound. This procedure uses *de novo* structure generation for a scaffold-hopping exercise. Because the information derived by this process is less precise, the method can be used to generate focused libraries that could be of value in screening with a novel target protein within a related gene family. In practical terms, this approach could be very useful, since most novel targets do not have 3D structural data. Lloyd et al. ***(16)*** have shown that at the estrogen receptor binding site, coordinate data of the site can be ignored and active ligands used as the training set to create a negative-image model of the site in which *de novo* structure generation can be performed. The structures generated can then be compared with known chemotypes. Once again, many known active chemotypes were generated for the ER, but this time without structural site information. Even with only one active ligand, a similar procedure can be used to create an image of the site and hence perform *de novo* structure generation. The big commercial value of this approach is in a scaffold-hopping exercise to explore intellectual property that is unencumbered by structures from the original training set or high-throughput screening hit(s).

4.4. Small-Molecule Selectivity

One aim of chemical genomics is to design potential drug structures that are selective enough for individual patient therapy. Suppose we are confronted with a target for which there are documented SNP variations within the binding site together with a 3D structure of the wild type; assume that design to the wild type generates an active compound, but let us say that the compound is inactive on the SNP variants. What approaches should we use to help ensure that the designed ligand is active in all known pharmacogenetic cases?

The first step would be to build all models of the SNP variations compared with the structurally known wild type. In the tailored therapy approach, for individual patients, different effective compounds could be designed and made. However, this is not likely to be a commercially viable strategy. A more attractive approach would be to establish a single design strategy that avoids interference by the SNP regions within the site. The altered SNP residues become a further constraint set that has to be avoided as *de novo* design proceeds. Alternatively, a less elegant solution would be to design ligands separately to all variant site structures within the SNP set. Each structure could then be cross-checked for binding to each of the other sites.

4.5. Small-Molecule Promiscuity

The opposite of selectivity in drug design is promiscuity; in this situation, a single compound acts on a number of sites. In chemical genomics, this could be useful as a tool for inhibiting all members of a gene family. It would provide pharmacologists with a chemical toolkit to ascertain whether a particular gene family could be important in an experimental or pathological scenario. The design strategy here would be very similar to the case of tailoring design to create active structures against a SNP set.

4.6. Molecular Scaffold Design and the Metabolome

Many problems in drug toxicity are related to how the drug is degraded. The inclusion of early-stage absorption, distribution, metabolism, and excretion properties in drug design has long been established as an important goal in drug discovery. The output of *de novo* structure generators is so varied that different scaffolds and substituent groups can be generated easily. Thus, the number of choices available for drug discovery is massive, and *in silico* predictors of metabolic pathways could be linked to the output of *de novo* generators. Post-processing the output could be performed such that structures that could be metabolized by undesired metabolic pathways could be eliminated from the generated set for consideration.

5. Conclusion

Figure 2 illustrates the flow of genomic information that will be available for selective drug discovery. A key process will be selection of an appropriate drug-design strategy, taking into account similarities in the binding site with other similar protein sites. By careful scripting, it should be possible to cycle through all the related sites to optimize strategies for selective design. The advantage of *de novo* structure generators is that they can design a wide variety of different chemical structures to fulfill the initial design strategies. Each scaffold then has to be assessed for appropriateness against the selected target. *In silico* filtering of each scaffold against absorption, distribution, metabolism, and excretion genomic flags should provide a method for ranking the scaffolds for minimal predicted adverse effects. Thus, the computing procedures for design can be linked by scripting to genomic databases to optimize drug discovery *in silico*.

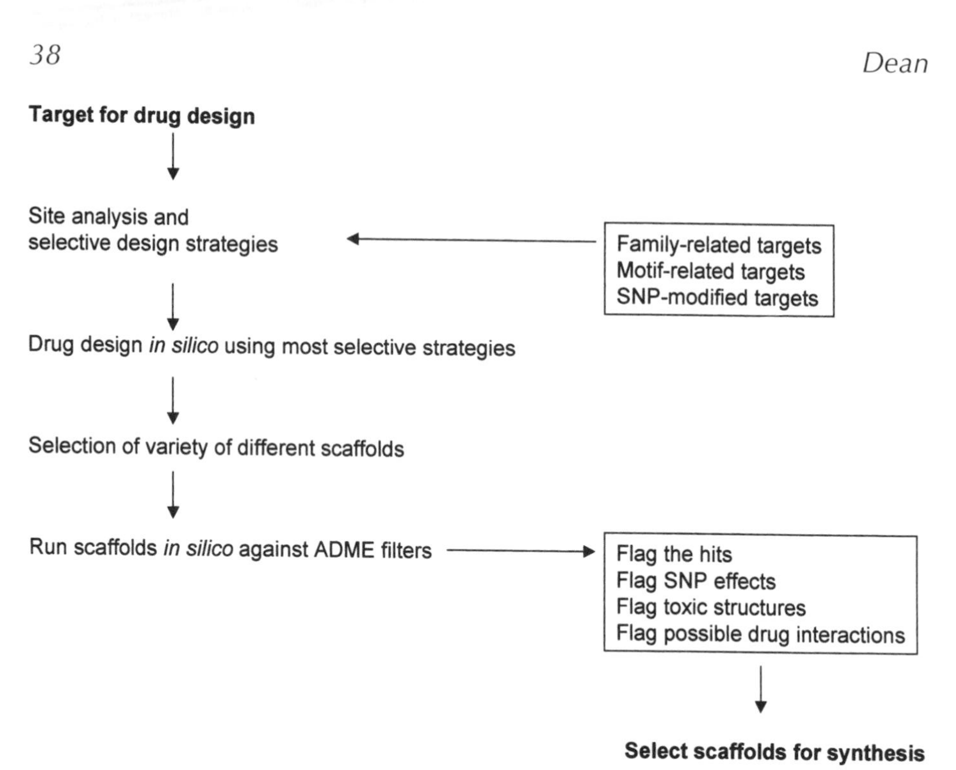

Fig. 2. Scheme for designing selective ligands.

Automated *de novo* drug design is a powerful tool for chemical genomics and is now advanced enough as a new technology to be applied to many problems revealed by genomics. A formal analysis of chemical genomics suggests that there could be a wide utility for *de novo* design techniques. Scaffold hopping by *de novo* design has provided many potential *in silico* ligands. It could also be used to handle specific requirements created by the action of SNPs, not just within the drug binding site but further afield in the metabolome. Chemical genomics by *in silico* methods is ripe for exploration on a massive scale.

References

1. Read, J., Pearce, J., Li, X., Muirhead, H., Chirgwin, J., and Davies, C. (2001) The crystal structure of human phosphoglucose isomerase at 1.6Å resolution: implications for catalytic mechanism, cytokine activity and haemolytic anaemia. *J. Mol. Biol.* **309,** 447–463.
2. Tsutsumi, S., Yanagawa, T., Shimura, T., et al. (2003) Regulation of cell proliferation by autocrine motility factor/phospoglucose isomerase signaling. *J. Biol. Chem.* **278,** 32,165–32,172.
3. Ohi, M. D., Vander Kooi, C. W., Rosenberg, J. A., Chazin, W. J., and Gould, K. L. (2003) Structural insights into the U-box, a domain associated with multi-ubiquitination. *Nat. Struct. Biol.* **10,** 250–255.

4. Bohacek, R. S., Martin, C., and Guida, W. C. (1996) The art and practice of structure-based drug design: a molecular modeling perspective. *Med. Res. Rev.* **16,** 3–50.
5. Zhou, X., Wang, X., and Dougherty, E. R. (2003) Construction of genomic networks using mutual-information clustering and reversible-jump Markov chain Monte-Carlo predictor design. *Signal Processing* **83,** 745–761.
6. Hartemink, A. J., Gifford, D. K., Jaakola, T. S., and Young, R. A. (2001) Using graphical models and genomic expression data to statistically validate models of genetic regulatory networks. *Pac. Symp. Biocomput.* 422–433.
7. Shah, N. P., Nicoll, J., Nagar, B., et al. (2002) Multiple BRC-ABL kinase domain mutations confer polyclonal resistance to the tyrosine kinase inhibitor imatinib (ST1571) in chronic phase and blast crisis chronic myeloid leukemia. *Cancer Cell* **2,** 117–125.
8. Shah, N. P., Tran, C., Lee, F. Y., Chen, P., Norris, D., and Sawyers, C. L. (2004) Overriding imatinib resistance with a novel ABL kinase inhibitor. *Science* **305,** 399–4001.
9. Koonin, E. V. (1999) http://kr.expasy.org/cgi-bin/nicedoc.pl?PDOC0017, last revised July 1999.
10. Deng, Z., Chuaqui, C., and Singh, J. (2004). Structural interaction fingerprint (SIFt): a novel method for analyzing three-dimensional protein-ligand binding interactions. *J. Med. Chem.* **47,** 337–344.
11. O'Toole, N., Raymond, S., and Cygler, M. (2003) Coverage of protein sequence space by current structural genomics. *J. Struct. Func. Genomics* **4,** 47–55.
12. Todorov, N. P. and Dean, P. M. (1997) An evaluation of a method for controlling molecular scaffold diversity in de novo ligand design. *J. Comput. Aided Mol. Des.* **11,** 175–192.
13. Lloyd, G. G., Garcia-Sosa, A., Alberts, I. L., Todorov, N. P., and Mancera, R. L. (2004) The effect of tightly bound water molecules on the structural interpretation of ligand-derived pharmacophore models. *J. Comput. Aided Mol. Des.* **18,** 89–100.
14. Mancera, R. L. (2002) *De novo* ligand design with explicit water molecules: an application to bacterial neuraminidase. *J. Comput. Aided Mol. Des.* **16,** 479–499.
15. Stahl, M., Todorov, N. P., Tames, T., Mauser, H., Boehm, H.-J., and Dean, P. M. (2002) A validation study on the practical use of *de novo* design. *J. Comput. Aided Mol. Des.* **16,** 459–478.
16. Lloyd, D. G., Buenemann, C. L., Todorov, N. P., Manallack, D. T., and Dean, P. M. (2004) Scaffold hopping in *de novo* drug design—ligand generation in the absence of receptor information. *J. Med. Chem.* **47,** 493–496.

II

Protocols

4

Design, Synthesis, and Screening of Biomimetic Ligands for Affinity Chromatography

Ana Cecília A. Roque, Geeta Gupta, and Christopher R. Lowe

Summary

Affinity chromatography is ideally suited to the purification of pharmaceutical proteins due to its unique bio-specificity characteristics. Tailor-made affinity ligands that represent a promising class of synthetic affinity ligands have been developed to target specific proteins and designed to mimic peptidal templates, natural biological recognition motifs, or complementary surface-exposed residues. These biomimetic ligands have been generated by a combination of rational design, combinatorial library synthesis, and subsequent screening of potential leads against target proteins. Small ligands based on a triazine scaffold also present exceptional selectivity and stability, which allows their use in harsh manufacturing environments.

Key Words: Affinity; biomimetic; ligands; synthetic; proteins; purification; design; combinatorial synthesis; screening.

1. Introduction

The cloning, expression, and production of proteins with significant biotechnological and therapeutic applications has to address crucial factors of post-translational modifications, folding, and stability of the final product. Furthermore, the mapping of the human genome and the explosion of proteomics have led to a glut of therapeutic targets and an increased interest in high-throughput screening for the identification of new proteins with therapeutic potential. However, a common concern is the design of inexpensive purification strategies for high-yield recovery of protein products that need to be compliant with the safety guidelines imposed by regulatory agencies, such as the Food and Drug Administration. Traditional methods to purify proteins from complex samples require a combination of steps. The initial enrichment usually involves precipitation and filtration

From: *Methods in Molecular Biology, vol. 310: Chemical Genomics: Reviews and Protocols*
Edited by: E. D. Zanders © Humana Press Inc., Totowa, NJ

techniques, whereas one or more chromatographic steps are employed in the subsequent intermediary and final purification stages. The tendency, however, has been to apply crude stocks directly onto chromatographic matrices, which places much greater demands on these materials to cope with cells, debris, and other foulants, and also on the development of matrices with high affinity and specificity for the protein of interest. *Affinity chromatography*, a term that was coined in 1968 by Cuatrecasas, is the most efficient method for tackling key issues in high-throughput proteomics and scale-up *(1)*. This technique exploits natural biological recognition and the predictive character of reversible, noncovalent binding between a protein and its complementary ligand, thereby granting this purification method the ability to reduce nonspecific interactions, increase operational yields, and eliminate undesirable contaminants. The adsorption is efficient even from diluted crude extracts, which makes it possible to achieve one-step purification of the target protein. Association constants between the ligand and the protein in the range 10^3–10^8 (M^{-1}) are normally most suitable for purification purposes.

Affinity ligands can be referred to as biospecific (natural ligands, such as antigen-antibody; lectin-glycoprotein; SpA-IgG) or as pseudo-biospecific, which include traditional synthetic ligands (dyes and metal-chelators) and biomimetics (biological, e.g., peptides, or synthetic, e.g., triazine-based ligands). Biomimetics are tailor-made molecules that mimic biological recognition between a target protein and a natural ligand; these have emerged as an alternative to biospecific ligands. The disadvantages to using natural ligands are that they require production and isolation, usually incurring high costs and problems in product characterization; have low yields at the coupling step; can leach during harsh elution steps; and are prone to degradation by conventional sterilization and cleaning-in-place schedules. The drawbacks of pseudo-biospecific affinity chromatography can be more easily surmounted, but they are still dependent on the biological or synthetic source of the ligand. Genetic engineering, combinatorial chemistry (alone or allied with molecular modeling), and high-throughput screening methodologies have contributed enormously to the development of highly resistant and specific biomimetic affinity ligands.

In this chapter, we describe the process of generating low-molecular-weight synthetic biomimetics for the purification of different target biomolecules, through the *de novo* design and synthesis of affinity ligands pioneered by Lowe and co-workers *(2)*. The roots of this strategy lay in the fact that textile dyes based on substituted triazines mimic the binding of natural anionic heterocyclic substrates, such as nucleotides, vitamins, and coenzymes, to proteins, and that rational changes in the structures of the dyes could enormously improve their selectivity *(3)*. The availability of crystallographic structures of proteins and protein–ligand complexes, together with the development of computer-based molecular modeling techniques, prompted the synthesis of "intelligent" triazine-scaffolded affinity ligands, with improved characteristics over their native counterparts. The methodology represents an integrated approach that combines structure-based ligand design, generating combinatorial ligand libraries, and screening in parallel for affinity against target proteins (**Table 1**). The general research methodology is schematically represented in **Fig. 1**, and comprises distinct stages: (a) design of a complementary ligand using the biological interaction or the target protein as templates; (b) synthesis of a near-neighbor combinatorial library of first-generation ligands, either in solid or solution phases; (c) evaluation of the affinity of the first-generation ligands; (d) solution-phase synthesis and characterization of the key lead ligands; and (e) immobilization, optimization, and chromatographic evaluation of the final adsorbent.

2. Materials

2.1. Design

1. Computer-aided molecular modeling. There is a wide range of commercially available software packages to perform molecular modeling, such as Quanta2000 and InsightII from Accelrys, which can run on a IRIX®6.5 Silicon Graphics® Octane® workstation from Silicon Graphics, Inc. Protein X-ray and nuclear magnetic resonance crystallographic structures are available from the Brookhaven database (www.rcsb.org/pdb/), which contains over 24,400 entries.

2.2. Synthesis

1. Cyanuric chloride (2,4,6-trichloro-sym-1,3,5-triazine; chloro-triazine; trichlorocyanidine). This is widely available. A high-purity (99%) compound should be used. It is a very reactive compound, and must be stored at 2–8°C in an anhydrous environment. It should be recrystallized in petroleum ether (*see* **Note 1**). Hazards: Poison, lachrymator, and irritant to eyes, skin and respiratory system. May be harmful if swallowed. Toxicity data: LD50 485 mg/kg oral, rat. *Should be handled in a fume hood with safety glasses and gloves, and treated as a possible cancer hazard.*
2. Epichlorohydrin (1-chloro-2,3-epoxypropane). Widely available. Used to epoxy activate the Sepharose® CL-6B beads or other surfaces. The extent of epoxy activation of beads may be determined (*see* **Note 2**). A high-purity (+99%) or equivalent should be used. It is a very unstable compound and must be stored at 0–4°C in an anhydrous environment. Hazards: Flammable, poison, toxic by inhalation or contact with skin, and if swallowed may cause cancer. Toxicity data: LD50 90 mg/kg oral, rat. *Should be handled in a fume hood with safety glasses and gloves, and treated as a possible cancer hazard.*
3. Ammonia aqueous solution (35% v/v). Widely available chemical. Used to introduce free amino groups in the epoxy-activated beads, which can be quantified by the 2,4,6-trinitrobenzenesulphonic acid (TNBS) test (*see* **Note 3**). Hazards: Poison,

Table 1
Strategies for the *De Novo* Design of Biomimetic Ligands for the Affinity Purification of Proteins

	Strategy			
Target protein	Design	Synthesis	Screening	References
Kallikrein	Complex kallikrein/kininogen (enzyme/substrate): mimicking Phe-Arg dipeptide on the substrate	Solution-phase synthesis	Affinity chromatography	***9***
Elastase inhibitor (enzyme	Complex elastase/turkey ovomucoid /inhibitor): mimicking the natural inhibitor	Solid-phase combinatorial chemistry (12 ligands)	Affinity chromatography	***10***
IgG (Fab fragment)	Complex PpL-IgG (ligand/protein): mimicking the natural ligand binding interfaces to IgG	Solid-phase combinatorial chemistry (169 ligands)	Affinity chromatography; quantitative ELISA; FITC-screening	***11,12***
IgG	Complex SpA-IgG (ligand/protein): mimicking the Phe132-Tyr133 dipeptide on the natural ligand	Solution-phase synthesis	Affinity chromatography	***4***
IgG	Refinement of SpA mimicking ligand ***(4)***	Solid-phase combinatorial chemistry (88 ligands)	Affinity chromatography	***13,14***
Human recombinant Factor VIIa	Complex tissue factor/factor VIIa; ligands designed to bind to the Gla-domain in Factor VIIa.	Solid-phase combinatorial chemistry; solution-phase synthesis of a sub-library	Affinity chromatography; SPR	***7,15***
Sugar moieties on glycoproteins	Protein-carbohydrate complexes: identification and mimicking of key residues determining monosaccharide specificity	Solid-phase combinatorial chemistry (80 ligands)	Affinity chromatography; 1H-NMR studies	***8,16,17***
Recombinant insulin precursor MI3	Study of the target protein per se and selection of appropriate binding sites	Solid-phase combinatorial chemistry (64 ligands); solution-phase synthesis of a sub-library	Affinity chromatography; SPR	***15,18,19***
Human α1-antitrypsin	Study of the target protein per se and selection of appropriate binding sites	Solid-phase combinatorial chemistry	Affinity chromatography	Unpublished
Prion Protein	Study of the target protein per se and selection of appropriate binding sites	Solid-phase combinatorial chemistry (49 ligands)	Affinity chromatography; ELISA on beads	***20***

SpA, Staphylococcal protein A; PpL, protein L from *Peptostreptococcus magnus*; IgG, immunoglobulin G; Gla, γ-carboxyglutamic acid; SPR, surface plasmon resonance; ELISA, enzyme-linked immunosorbent assay; FITC, fluorescein isothiocyanate; NMR, nuclear magnetic resonance

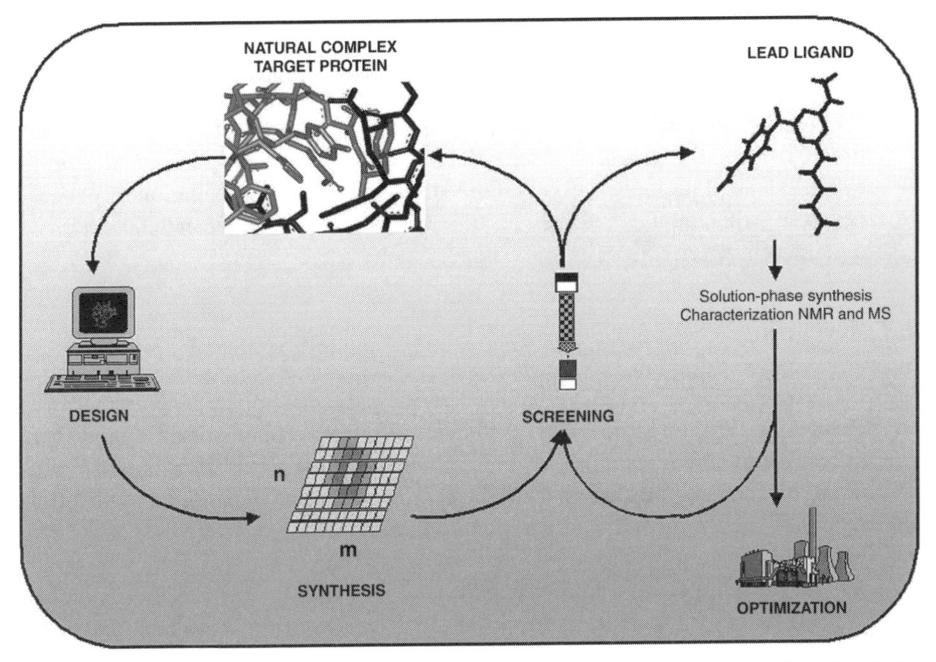

Fig. 1. Research strategy for the development of biomimetic affinity ligands. Relevant stages include 1. *de novo* design of ligands based on features of the complex ligand/protein or the protein *per se*; 2. synthesis of ligands in solid or solution phases; 3. screening for binding the target protein by affinity chromatography, enzyme-linked immunosorbent assay (ELISA), fluorescein isothiocyanate (FITC) system, or surface plasmon resonance (SPR); 4. selection and characterization of the lead ligand; 5. optimization of the affinity chromatography process.

corrosive alkaline solution, causes burns, harmful if swallowed, inhaled, or absorbed through skin. Toxicity data: LD50 3500 mg/kg oral, rat. *Should be handled in a fume hood with safety glasses and gloves.*

4. Ninhydrin (1,2,3-triketohydrindene monohydrate). Widely available chemical, light-sensitive. Ninhydrin reacts with free amines (2:1 molar ratio), giving a purple product (Ruhemann's purple resonance structure). Used for a qualitative test to determine the presence of aliphatic amines on the agarose beads as a 0.2% w/v solution in ethanol (*see* **Note 4**). Hazards: Harmful if swallowed; skin, eye, and respiratory irritant. Toxicity data: LD50 78 mg/kg intraperitoneal, mouse. *Should be handled in a fume hood with safety glasses and gloves.*
5. Amines. The amines used to sequentially substitute the chlorines of cyanuric chloride free or immobilized onto agarose vary from case to case. In general, the amines should be dissolved in an appropriate buffer, either an aqueous solution or an organic solvent such as dimethylformamide. Take precautions related to the reac-tivity of each compound, which in some cases may involve the protection of

reactive groups (*see* **Note 5**). Hazards and toxicity data should be checked individually for each compound.

6. Sepharose CL-6B. Product available from Amersham Biosciences (Piscataway, NJ), can be obtained as a suspension of beads in a 20% v/v aqueous ethanol solution. Must be stored at 4°C. Take special care during the handling of this gel to avoid extended periods of dryness and strong shear forces. Agitation of gel suspensions, when required, should be done with an orbital movement and not using a magnetic stirrer.

2.3. Screening

2.3.1. Affinity Chromatography

1. Binding and elution buffers. The buffers used for the screening of the ligands vary from case to case, depending on the type of protein studied, the standard conditions recommended for its use, and the type of interactions exploited in the affinity purification. Generally, the equilibration/binding/washing buffer is a buffer with pH 7.0–8.0 (e.g., phosphate-buffered solution [PBS]: 10 m*M* sodium phosphate, 150 m*M* NaCl, pH 7.4). Elution may be achieved by different methods: change in pH (lowering or increasing the pH), ionic strength, chaotropic agents and denaturants (e.g., 6 *M* urea and 3 *M* KSCN), organic solvents and detergents (e.g., ethylene glycol 50% v/v, CHAPS 1% v/v), increase or decrease of temperature, competitive elution, or complex elution buffers (including more than one agent or effect). Unfortunately, there is no general recipe for the binding and elution buffers, and it is therefore necessary to optimize the conditions for each specific case (*see* **Note 6**).
2. Regeneration buffer. Usually 0.1 *M* NaOH in 30% v/v isopropanol in water. This buffer is used to remove any physically adsorbed (not covalently attached to the matrix) ligand prior to screening, and after the screening procedure to remove retained protein. Special care should be taken when using isopropanol. (Hazards: flammable, irritant to eyes, respiratory system and skin. Toxicity data: LD50 10g/kg oral, human. *Should be handled with gloves, safety glasses, and avoid vapors.*)
3. Affinity chromatography. Disposable empty columns, e.g., Bond Elut TCA® (4-mL propylene columns with 20-µm frits) from Varian, Inc., can be used. Alternatively, if choosing an automatic system of sample/buffer loading and sample collection, e.g., the FPLC system from Amersham Biosciences, the affinity resins must be properly packed in columns recommended by the supplier.

2.3.2. Fluorescein-5-Isothiocyanate-Based Screening

1. The target protein must be conjugated with fluorescein-5-isothiocyanate (FITC) isomer I (F). Conjugation occurs through free amino groups of proteins or peptides, forming a stable thiourea bond. Conjugated proteins can be bought from most suppliers of biochemical products. Otherwise, the protein (P) can be conjugated in house, for example, using the FluoroTag™ FITC conjugation kit (Sigma, St. Louis, MO), with which different conjugation ratios can be obtained (molar F/P

of 2 is recommended). The conjugates may then be purified using pre-packed PD-10 columns (Amersham Biosciences) and characterized in terms of the F/P ratio:

$$Molar\ \frac{F}{P} = \frac{MW_{protein}}{389} \times \frac{A_{495} / 195}{A_{280} - [(0.35 \times A_{495})] / \varepsilon^{0.1\%}_{280}}$$

where $\varepsilon^{0.1\%}$ is the absorption at 280 nm of a protein at 1 mg/mL; A_{280} is the absorbance measured at 280 nm.

2. Glass slides.
3. A fluorescence microscope with appropriate filters for the fluorophore used.

2.3.3. Enzyme-Linked Immunosorbent Assay on Beads

1. 96-well Millipore MultiScreen®-HV clear sterile plates, with 0.45-µm pore size, hydrophilic, low-protein-binding Durapore® membrane (Fisher Scientific).
2. 96-well microtiter plates.
3. Antibody against the target protein (unlabeled or labeled, usually with an enzyme). The enzyme-linked immunosorbent assay (ELISA) system needs to be designed according to each situation.
4. Solution of the appropriate substrate for the enzyme chosen (e.g., 1 mg/mL *p*-nitrophenyl phosphate in 0.1 *M* diethanolamine-HCl, pH 9.8, for alkaline phosphatase conjugates; 3,3',5,5'-tetramethylbenzidine (TMB) or 3,3'-diaminobenzidine (DAB) for horseradish peroxidase conjugates).
5. ELISA plate reader equipment.
6. Centrifuge with an appropriate 96-well plate rotor (e.g., PK130 centrifuge from ALC, Italy, and rotor from Jencons-PLS).
7. Washing buffer: PBS Tween-20 (0.02% v/v).
8. Blocking buffer: casein 0.025% w/v in washing buffer.

2.3.4. Characterization of Affinity Interactions by Partition Equilibrium Studies

1. Eppendorf tubes containing solutions of the target protein (0.1–5 mg/mL in equilibration buffer).
2. Agarose-immobilized ligand.

2.3.5. Imaging Surface Plasmon Resonance (SPR) System

Several steps must be considered, each one with specific requirements:

1. Cleaning of SPR glass slides (optical grade SF2, refractive index 1.65, from UQG Optical Components). Sonicator (e.g., Lucas Dawe sonication bath) and different solutions: hydrochloric acid/nitric acid 3:1 v/v (solution A); ammonium cerium nitrate (1 *M*) (solution B); sulfuric acid/hydrogen peroxide 1:1 (Piranha solution); isopropanol. *Use caution in handling these solutions; harmful and corrosive.*
2. Deposition of gold films (thermal evaporation technique). Evaporator with deposition chamber (e.g., Edwards Auto306B from Bewhay), chromium-plated tung-

sten rods from Megatech, and gold powder Puratronic 20 mesh (99.9995% pure) from Alfa. *Harmful products, avoid contact with skin and inhalation.*

3. SPR sensor fabrication. 2-aminoethanothiol hydrochloride. Solution C: 11-mercapto-1-undecanol (5 m*M*) in water, ethanol, DMF, or 70% v/v DMF. Solution D: sodium hydroxide (0.4 *M*)/diglyme (0.4 *M*) (1:1) mixture. Solution E: DMF/0.5 *M* NaOH (70:30 v/v). Solution F: ethanol-water (20:80 v/v) with 11-mercapto-1-undecanol (5 m*M*). *Use care in handling these solutions: harmful and corrosive.*
4. SPR instrument. The instrument described in this chapter was built in house. The perspex flow cell was designed by Dr. Roger Millington (Cambridge, UK) and engineered by D. Wheatland Engineering (Surrey, UK). The flow cell consisted of a single channel surrounded by a viton O-ring (1.7 cm I.D. and a wall thickness of 1 mm; Nationwide Bearing, UK). A Gilson peristaltic pump (Villiers-Le Bel, France) was used to control the flow-rate throughout the system with PVC tubing (1 mm I.D.) delivering the fluidics. All lenses and prisms were purchased from Spindler and Hoyer.

3. Methods

3.1. Design

1. *De novo* design of biomimetic affinity ligands. The methodology varies with the type of strategy chosen (**Table 1**). For example, the design of SpA biomimetic ligands was based on the hydrophobic core dipeptide Phe132-Tyr133, which is located on the surface twist of a helix, and is found in the four highly conserved regions of SpA that interact with IgG. The aromatic side chains of the dipeptide protrude to contact a superficial cavity on IgG formed by residues Leu251, Ile253, His310, Gln311, Glu430, Leu432, Asn434, and His435. The distance between the α and ζ carbons of Phe132 and Tyr133 is 3.2 α and 8.5 ζ, respectively. By molecular modeling studies, it was observed that anilino and tyramino substitutents mimicked the side chains of Phe132 and Tyr133, and the triazine ring acted as the twist of helix, maintaining the two substituent groups in the optimal spatial geometry to mimic the natural side chains (**Fig. 2**). Using the same methodology, other ligands were designed, such as ligands containing tyramine as one substitutent (mirroring Tyr133) and *N*-*t*-butyl aniline or *N*-isopropyl aniline as the other substitutent (to mimic the side chain of Leu136, which is only 3.6 Å away from the Phe132 side chain) *(4)*.

3.2. Synthesis

3.2.1. Solid-Phase Combinatorial Synthesis of a Ligand Library (Fig. 3)

1. Epoxyactivation of agarose beads. The required amount of Sepharose CL-6B is washed with 40 mL of distilled water/g of gel on a sinter funnel (*see* **Note 7**). The washed agarose is transferred to a 1-L conical flask, and 1 mL of distilled water/g of gel is added. To this moist gel, 0.8 mL of 1 *M* NaOH/mL of gel and 1 mL of epichlorohydrin/mL of gel are added. The slurry is incubated for 10–12 h at 30°C on

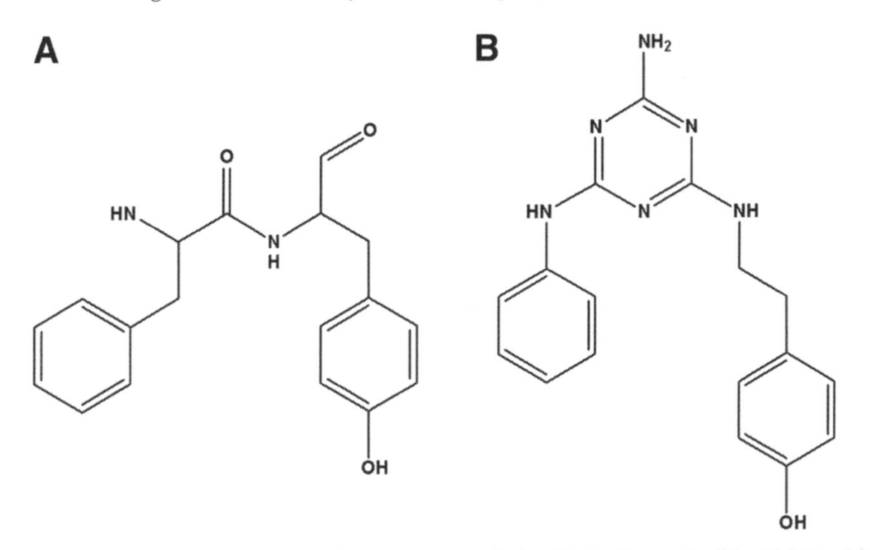

Fig. 2. Comparison between the structures of the SpA dipeptide Phe132-Tyr133 (**A**) and the biomimetic ligand (ApA) (**B**).

a rotary shaker. The epoxy-activated gel is washed with 40 mL of distilled water/g of gel on a sinter funnel and used directly for amination. The epoxy content is determined according to **Note 2**.

2. Amination of agarose beads. The washed epoxy-activated gel is suspended in 1 mL of distilled water/g of gel in a 1-L conical flask. About 1.5 mL of ammonia/g of gel are added and the gel incubated for 12 h at 30°C in a rotary shaker. Alternatively, 1,6-diaminohexanediamine (5 eq.) can be used to aminate the gel, yielding resins with a six-carbon spacer arm. The aminated gel is washed with 40 mL of distilled water/g of gel on a sinter funnel, and stored in 20% v/v ethanol at 0–4°C. The extent of amination is determined as described in **Note 3**. Aminated beads can also be purchased from Amersham Biosciences.
3. Cyanuric chloride activation. Aminated agarose is suspended in a 1-L conical flask, in a 50% v/v solution of acetone/water, using 1 mL of solution/g of gel. This mixture is maintained at 0°C in an ice bath on a shaker. Recrystallized cyanuric chloride (5 *M* excess to aminated gel) is dissolved in acetone (8.6 mL/g cyanuric chloride) and divided into four aliquots. The aliquots are added to the aminated gel, with constant shaking at 0°C and the pH maintained neutral by the addition of 1 *M* NaOH. Each aliquot is added after an interval of about 30 min, and samples of gel are taken in order to evaluate the presence of free amines (*see* **Note 4**). Usually, after the second aliquot, the ninhydrin test indicates that all the gel is activated. When the four aliquots are added, the gel is washed, with 1 L of each of the following acetone:water v/v mixtures-1:1, 1:3, 0:1, 1:1, 3:1, 1:0, and 0:1. Cyanuric chloride activated gel is not stored but used immediately for R_1 substitution.

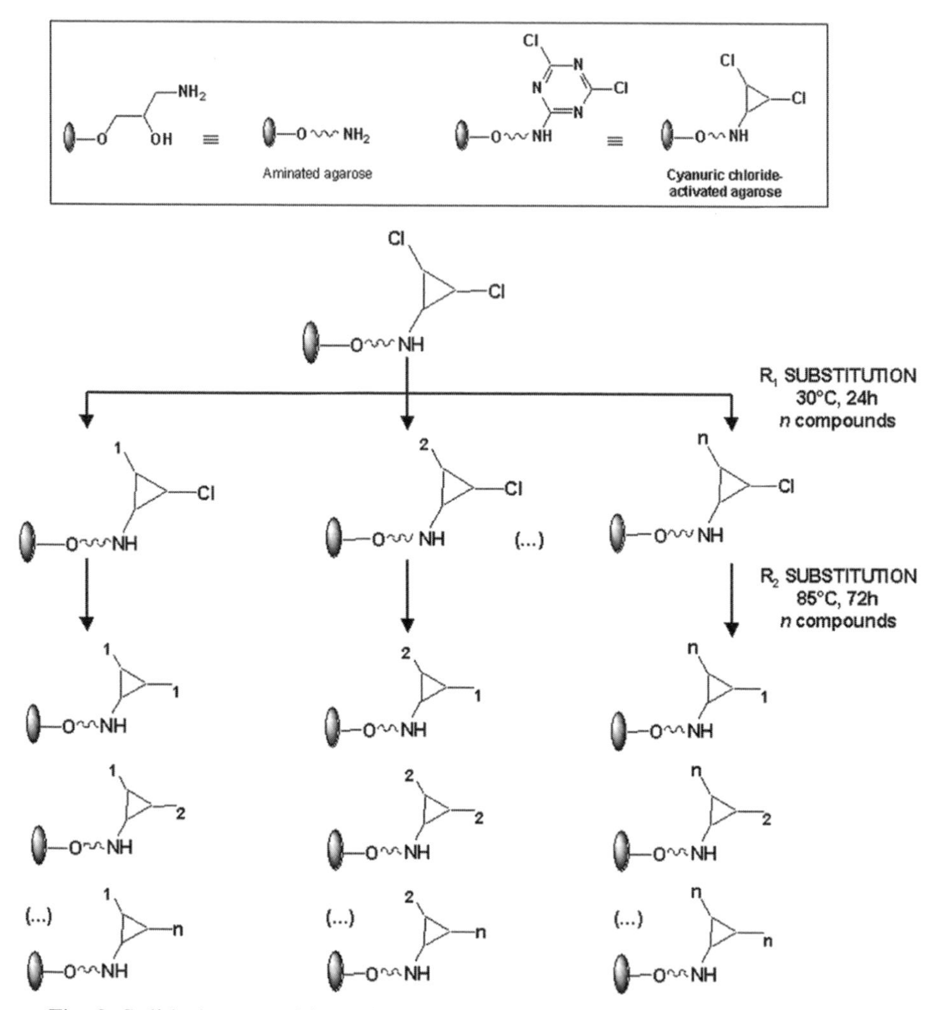

Fig. 3. Solid-phase combinatorial synthesis of triazine-scaffolded library of ligands using a modified "mix and split" strategy.

4. Nucleophilic substitution of R_1. Cyanuric chloride activated gel is divided into *n* aliquots, where *n* is the number of different amines used to synthesize the combinatorial library. A twofold molar excess (relative to the amount of amination of the gel) of each amine is dissolved in the appropriate solvent (1 mL/g gel) (*see* **Note 8**). The *n* aliquots are suspended in the previous mixture, and incubated at 30°C in a rotary shaker (200 rpm) for 24 h. After this period, each R_1-substituted gel is thoroughly washed on a sintered funnel with the appropriate buffer for each amine. The resulting gel is stored in 20% v/v ethanol at 0–4°C or used immediately for R_2 substitution (**Fig. 3**).

5. Nucleophilic substitution of R_2. The *n* amines selected are dissolved in 15 mL of appropriate solvent. Each amine is in 5 *M* excess to the amount of amination of the gel. Each aliquot of R_1-substituted gel is divided into 5-mL fractions, suspended in the previous mixture, and incubated at 85°C for 72 h. At the end of the synthesis, the gels are washed with appropriate solvent, weighed, and stored at 0–4°C in 20% v/v ethanol (**Fig. 3**).

3.2.2. Solution-Phase Synthesis of Triazine-Scaffolded Ligands

The conditions vary from case to case and need to be optimized accordingly. Solution-phase synthesized ligands are characterized by ^{1}H-NMR, ^{13}C-NMR, and mass spectroscopy, and further immobilized on a solid support (*see* **Note 9** or **step 7c**). As an example, we describe the solution-phase synthesis of ApA, 4-[2-(4-chloro-6-phenylamino-[1,3,5]-triazin-2-ylamino)-ethyl]-phenol (**Fig. 2B**).

1. Aniline (19.8 g, 212.6 mmol, 1.06 eq.) in acetone (50 mL) is added dropwise to a cyanuric chloride suspension (36.8 g, 200 mmol, 1 eq. in 200 mL acetone and 100 mL of ice:water), containing $NaHCO_3$ (5% w/v), while stirring in an ice bath. When aniline is not detected by thin-layer chromatography (silica, solvent dichloromethane [DCM]), the reaction is stopped by removing acetone in a rotary evaporator.
2. The product, 2-anilino-4, 6-dichloro-*s*-triazine (I) precipitated from aqueous solution, is then filtered and washed with water, 2 *M* HCl, and water again. Upon crystallization from DCM, white, needle-like crystals are obtained (36 g, 77%).
3. This product (I) (4.82 g, 20 mmol, 1 eq.) is then dissolved in acetone (100 mL), and a solution of tyramine (2.74 g, 20 mmol, 1 eq. in acetone:water [50 mL:10 mL[) added. The mixture is stirred at 50°C in an oil bath while a solution of $NaHCO_3$ (5% w/v) is added to maintain the pH between 6 and 7. The reaction is followed on silica TLC (solvent DCM:isopropanol; 96:4 v/v). When no tyramine is observed, the product is dried, redissolved in acetone (50 mL), and filtered.
4. The filtrate is precipitated with water (70 mL) and dried (II: 2-anilino-4-chloro-6-tyramino-*s*-triazine; 5.0 g, 75%).

3.3. Screening

3.3.1. Affinity Chromatography

Screening of affinity ligands is done by affinity chromatography (performed at room temperature) (**Fig. 4**).

1. The affinity ligands (1 g of moist gel) are packed into 4-mL columns (0.8 × 6 cm). Each matrix is washed with 2 × 3 mL regeneration buffer and then with distilled water to bring the pH to neutral. The resins are equilibrated with 10 mL of equilibration buffer.
2. Protein to be tested is reconstituted to 1 mg/mL in equilibration buffer and the absorbance at 280 nm measured. Protein solution (1 mL) is loaded onto each

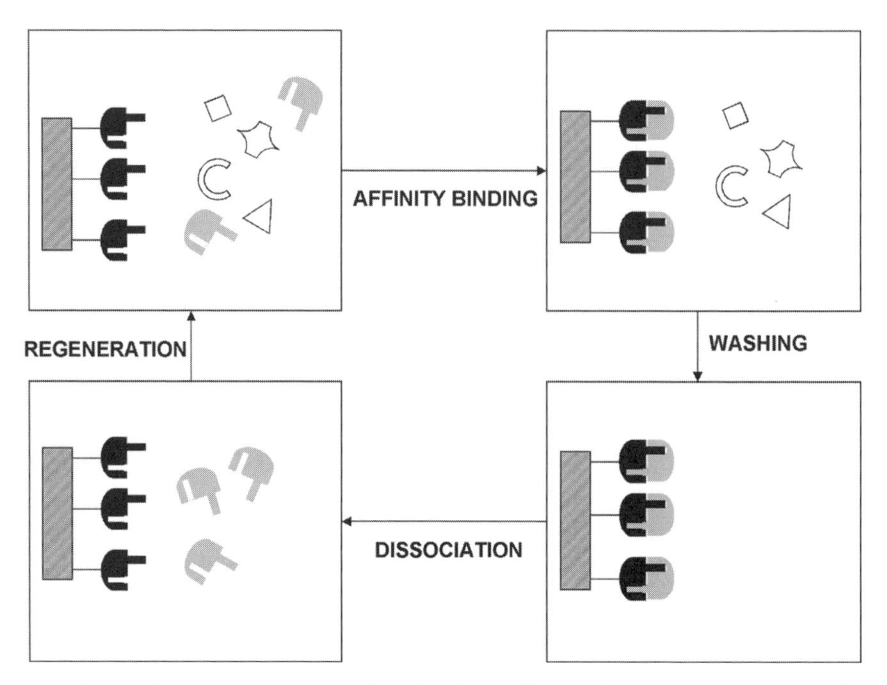

Fig. 4. The four principal steps involved in affinity chromatography. 1. Affinity binding: a solution containing the target protein and other contaminants is loaded onto the adsorbent holding the immobilized ligand; only the target protein will bind to the immobilized ligand. 2. Washing: the adsorbent is thoroughly washed with a suitable buffer to allow the flow-through of contaminants. 3. Dissociation: the target protein is dissociated from the immobilized ligand and recovered in the elution fractions. 4. Regeneration: the adsorbent is regenerated in order to remove any retained protein and then equilibrated for the next cycle of affinity purification.

column. The columns are washed with equilibration buffer until the absorbance of the samples at 280 nm reaches ≤ 0.005.

3. Bound protein is eluted with the elution buffer (1-mL fractions collected).
4. After elution, the columns are regenerated with regeneration buffer, followed by distilled water and equilibration buffer, and stored at 0–4°C in 20% v/v ethanol. Affinity chromatography is also utilized to test the capability of a lead ligand to purify the target protein from crude extracts (*see* **Note 10** and **Fig. 5**).

3.3.2. Screening With the FITC-Protein Conjugate

1. Each synthesized affinity matrix (50 µL) is mixed with 100 µL of distilled water in an Eppendorf tube, centrifuged for 2 min at 1430*g*, the supernatant discarded, and 2× 100 µL regeneration buffer added to the resin. The components are gently mixed and centrifuged for 2 min at 1430*g*, the supernatant discarded, and 2 × 100

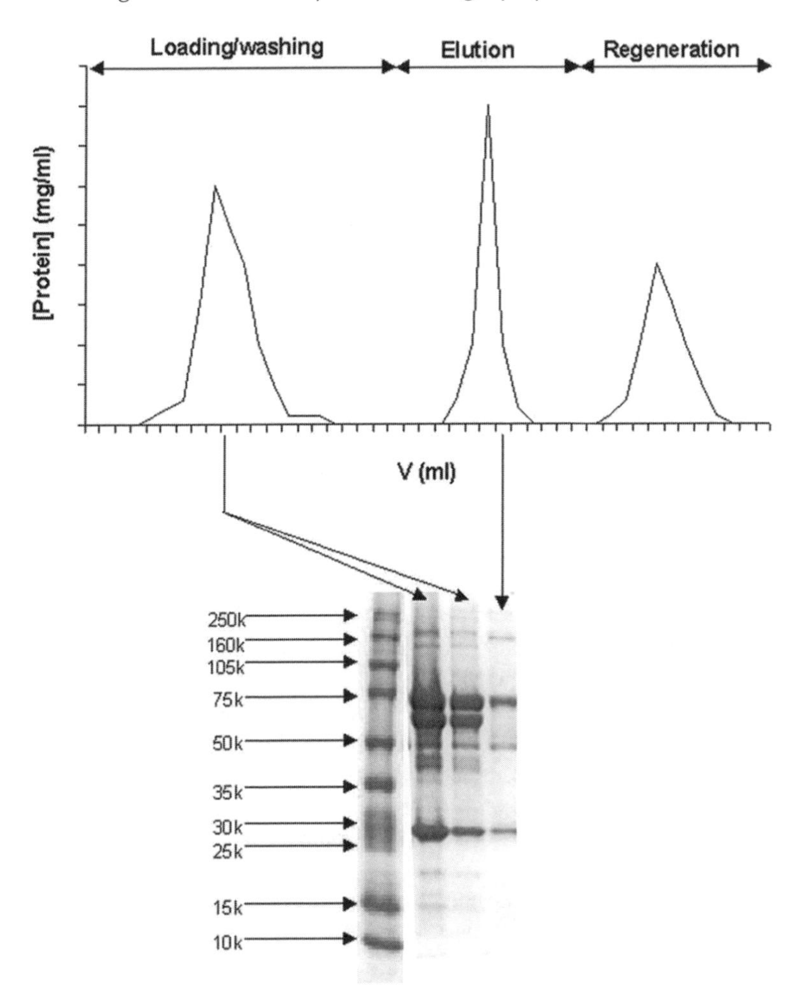

Fig. 5. Chromatogram (top) and corresponding sodium dodecyl sulfate-polyacrimide gel electrophoresis (SDS-PAGE) gel (bottom) for the one-step purification of human IgM from a crude extract (human plasma) by affinity chromatography. Lanes in the SDS-PAGE gel: first lane, molecular marker; second lane, loading solution (crude extract); third lane, washing of the adsorbent; fourth lane, elution fraction consisting of pure human immunoglobulin (Ig)M.

μL of distilled water added to the resin. The components are mixed and centrifuged for 2 min at 1430*g*, the supernatant discarded, and 2 × 100 μL of equilibration buffer added. The components are again mixed and centrifuged for 2 min at 1430*g*.

2. A FITC–target protein conjugate (50 μL; 1 mg/mL in equilibration buffer) is added to the resin, and the mixture incubated in the absence of light for 15 min with

orbital agitation. After this period, the resin is washed in the dark with 3 × 1 mL equilibration buffer (centrifuging the incubated resin with buffer at 1430*g* and then discarding the supernatant).

3. Each immobilized ligand matrix (1.5 μL) is placed on a microscope slide and observed under a fluorescence microscope (FITC-λ_{exc} = 495 nm; λ_{em} = 525 nm). The control experiments consist of repeating the procedure described above using Sepharose CL-6B, aminated agarose, and control ligand 0/0.

3.3.3. ELISA on Beads

1. Each immobilized ligand (50 μL moist gel) is placed in a well of the MultiScreen 96-well microtiter plate. Regeneration buffer (200 μL/well) is added and the wash solution removed by centrifuging the plates at 335*g* for 1 min. This procedure is repeated by washing the resins sequentially with distilled water, regeneration buffer, and equilibration buffer (2 × 200 μL/well). The washes are collected into a separate 96-well plate by stacking the MultiScreen plate above it, using a plate alignment frame.
2. After washing and equilibrating the ligand slurries, the target protein is added (e.g., 100 n*M* in equilibration buffer; 50 μL/well) and the plate incubated at 37°C or at room temperature, for 1 h with gentle shaking at 35*g*. The supernatant is then collected by centrifuging the MultiScreen plate above a standard ELISA plate at 335*g* for 1 min. The ligands are then washed three times with equilibration buffer (3 × 200 μL) and the unbound protein washes collected in separate ELISA plates (unbound protein can then be quantified by a quantitative ELISA, *see* **Note 11**).
3. The MultiScreen plate is then blocked with the blocking solution (200 μL/well) for 15 min at room temperature and then removed by centrifugation at 335*g* for 1 min.
4. A solution of an antibody recognizing the target protein, diluted in blocking solution (the dilution is specific for each antibody; check suppliers' recommendations), is added to each well (50 μL/well) and incubated for 1 h at 37°C.
5. The plate is then washed sequentially three times with washing buffer (200 μL/well) and the washes collected by centrifugation at 335*g* for 1 min. If the designed ELISA has a secondary antibody, the last two steps are repeated.
6. A solution of substrate (50 μL/well) is added and the plates incubated in the absence of light (time varies from case to case, but is usually 15–30 min). After incubation, the reacted solution is collected in an ELISA plate by centrifuging at 335*g* for 1 min, and the absorbance of the samples read at the characteristic wavelength (e.g., 405 nm for alkaline phosphatase assays).

3.3.4. Characterization of Affinity Interactions by Partition Equilibrium Experiments

1. The immobilized ligand to be studied is treated with regeneration buffer and then equilibrated in equilibration buffer.

2. A series of Eppendorf tubes are prepared with 1 mL of standard protein solutions in equilibration buffer (0.1–5 mg/mL; confirm concentration by A_{280} measurement).
3. Immobilized ligand (0.1 g of moist-weight gel previously dried under vacuum in a sintered funnel) is added to each Eppendorf tube and incubated for 24 h at room temperature and under orbital agitation.
4. After this period, the Eppendorf tubes are centrifuged (1 min; 1430*g*) to settle the matrix and the supernatant taken to measure the A_{280}. The control experiment involves incubating the partitioning solute with unmodified Sepharose CL-6B.
5. Data collected from these experiments can be fitted into Scatchard plots to determine the association constant (K_a).

3.3.5. Imaging SPR System for Screening Affinity Ligands

Several steps are involved and are as follows:

1. Cleaning of SPR glass slides and prisms (sonication at room temperature). For the removal of the gold and chromium–a. sonication in a bath with solution A for 5 min, followed by extensive washing with water; b. sonication with solution B for 5 min. For the cleaning of the glass surfaces–c. incubation in Piranha solution for 10 min, followed by extensive rinsing with water; d. sonication in isopropanol for 10 min, followed by 10–15 min oven drying at 50°C (until thoroughly dry).
2. Deposition of gold films on SPR glass slides and prisms. Cleaned glass slides or prisms are placed in a deposition chamber, where deposition of a 2-mm chromium adhesion layer is accomplished by reducing the chamber pressure below 3×10^{-6} Torr (deposition rate = 5Å/s; temperature = 25°C). Samples are left in the chamber for 1 h to cool, after which they can follow directly to step c or be stored in a vacuum dessicator under nitrogen.
3. SPR sensor fabrication (10-spot format). Ligands synthesized in solution phase (0.12 mmol) are dissolved in DMF (4 mL), and 2-aminoethanethiol hydrochloride (18.51 mg, 0.24 mmol) and $NaHCO_3$ (10 mg, 0.12 mmol in 1 mL of water) added. The reaction proceeds for 15 h at 90°C with vigorous stirring and followed by TLC until completion, when the product is precipitated with water, collected on 42-grade ashless filter paper, and dried in the oven at 50°C for 1 h. Gold-coated glass slides or prisms are immersed in solution C and allowed to react for 2–12 h at room temperature, after which they are washed and stored in the respective solvent systems.
3. The hydroxylated surfaces are reacted with epichlorohydrin (0.5 *M*) in solution D, for 4 h at room temperature. The surface is washed sequentially with water, ethanol, and water. Solution E (5 mL) containing each amino-ethanethiol-activated ligand (5 m*M*) and mercaptoethanol (5 m*M*) is hand spotted on the gold surface in predefined linear positions and reacts for 12 h at room temperature, after which the slide is immersed in DMF (50 mL) to remove unbound ligands.
4. Unreacted epoxide groups are quenched with solution E (5 mL) containing mercaptoethanol (5 m*M*), reacting 12 h at room temperature. The devices are then

washed in DMF and stored in ethanol at room temperature. Alternatively, gold-coated slides can also be covered with dextran (*see* **Note 12**).

5. SPR experiments. SPR measurements are made by tracking the change in the SPR minima over the time course of the experiment and converting the total minimal angle shift to protein surface loadings (ng/mm^2), because the device is calibrated previously (*see* **Note 13**). A run consists of loading a protein sample (50–150 μg/mL) in binding buffer at 0.25 mL/min for 15–30 min, rinsing with binding buffer for 2 min at 0.25 mL/min followed then by elution buffer at 1 mL/min for 4 min and regeneration buffer at 1 mL/min for 4 min. The concentration of protein is increased until signal saturation (concentration expresssed as maximum surface coverage concentration). Values for K_a (affinity constants) for the ligand–protein interaction can be calculated from these experiments.

4. Notes

1. Cyanuric chloride recrystallization. Cyanuric chloride (30 g, 0.16 mol) is dissolved in hot petroleum ether (500 mL) with constant stirring in an oil bath. Heated petroleum ether is poured over a fluted filter paper and the solution of cyanuric chloride filtered into a 1-L conical flask. The saturated solution of cyanuric chloride is left overnight, covered, to allow formation of crystals. The crystals are filtered and dried under reduced pressure. The dried crystals are stable at room temperature in an airtight container. The yield is about 95%.
2. Extent of epoxyactivation of agarose beads. Sodium thiosulphate (1.3 *M*, 3 mL) is added to 1 g of epoxy-activated gel and incubated at room temperature for 20 min. This mixture is neutralized with 0.1 *M* HCl and the amount of HCl used is recorded. The volume of 0.1 *M* HCl added corresponds to the number of OH^- moles released (10 μmol for each 100 μL added), which equals the μmol of epoxy groups/g gel. Therefore, the extent of epoxy activation is expressed as μL HCl used/10 μmol/g gel. The protocol usually results in 25 μmol epoxy groups/g moist weight gel.
3. Extent of amination on agarose beads with the TNBS test *(5)*. Aminated gel (0.1 g) is hydrolyzed with 500 μL of 5 *M* HCl at 50°C for 10 min. On cooling, the hydrolyzed sample is neutralized with 5 *M* NaOH and added to 1 mL of 0.1 *M* sodium tetraborate buffer (pH 9.3) and 25 μL of 0.03 *M* TNBS. Samples are incubated at room temperature for 30 min prior to measuring their absorbance at 420 nm. The negative control is 1 mL of distilled water to which sodium tetraborate buffer and TNBS solution (amounts cited above) are added. Calibration curves are constructed with 6-aminocaproic acid (0–2 μmol/mL). Usual values obtained are 20–25 μmol amine groups/g moist weight gel.
4. Qualitative test for aliphatic amines. A small amount of moist gel (approx 1 mL) is placed on a filter paper and ninhydrin in ethanol (0.2%, w/v) sprayed on it. The filter paper is heated with a hairdryer (very carefully to avoid burning) until development of color. Purple or brown coloration indicates, respectively, the presence or absence of free aliphatic amines. Alternatively, the sample of moist gel is placed

in a test tube, two or three drops of ninhydrin solution added, and the test tube heated until development of color (adapted from **ref. *6***).

5. Example of a protected amine. The protected amino acid N^{ε}-*tert*-butoxycarbonyl (Boc)-l-lysine, when used as the amine to synthesize the triazine-based ligands, is deprotected prior to screening with protein. Protected immobilized ligands, after washing with the appropriate solvent in a sintered funnel, are immersed in 99% v/v trifluoroacetic acid (TFA) and stirred for 1 h in the fume hood. Methanol is added to dilute the TFA, and the waste solution is carefully disposed of. The matrixes are washed sequentially with methanol and distilled water (3 × 10 gel volumes) and stored in 20% v/v ethanol at 4°C.
   a. Screening of ligands for rFVIIa (bind to the protein in a Ca^{2+}-dependent manner). Binding buffer–20 m*M* Tris-HCl, 50 m*M* NaCl, 2.5 m*M* $CaCl_2$, pH 8.0. Elution buffer–20 m*M* Tris-HCl, 50 m*M* NaCl, 5 m*M* ethylene diamine tetraacetic acid (EDTA), pH 7.5 *(7)*.
   b. Screening of ligands binding to sugars. Binding buffer–10 m*M* Tris, 200 m*M* NaCl, 20% v/v ethylene glycol, 1 m*M* Mn^{2+}, 1 m*M* Ca^{2+}, pH 7.0. Elution buffer–0.5 *M* α-D-methyl mannoside dissolved in equilibration buffer *(8)*.
7. The modified "mix-and-split" combinatorial method is used for the synthesis of the ligand library yields $n \times n$ members, where n represents the number of different amines chosen. Usually 5 g of each immobilized ligand are synthesized; however, this amount depends on the screening strategy preferred (for example, the FITC-based system or the ELISA on beads require less resin than the conventional affinity chromatography).
8. Common solvents. For hydrophilic amines such as 4-aminobutyric acid, distilled water is the preferred solvent. For hydrophobic amines such as 4-aminobenzamide, a DMF:water 50% v/v solution may be an option. In any case, usually 1 molar equivalent of $NaHCO_3$ is added in order to neutralize the HCl released during nucleophilic substitution. *Caution: DMF is harmful and considered a potential carcinogen. Should be handled in a fume hood.*
9. Coupling disubstituted triazinyl ligands to aminated agarose. To 1 g of moist aminated agarose (24 µmol/g) a solution is added containing 5 molar equivalents of the disubstituted triazinyl and 5 molar equivalents of $NaHCO_3$ in an appropriate solvent (usually 50% v/v DMF:H_2O). The coupling reaction is carried out at 85°C (30 rpm) for 72 h. Agarose beads are then sequentially washed with DMF:water (1:1; 1:0; 1:1; 0:1 v/v) and stored in a solution of 20% v/v ethanol at 0–4°C. The ligand concentration or density of immobilized ligands can be determined. Immobilized ligands are washed with regeneration buffer and then neutralized by washing with distilled water. Moist gel (30 mg) containing the immobilized ligand is hydrolyzed in 5 *M* HCl (0.3 mL) at 60°C for 10 min. On cooling, ethanol (3.7 mL) is added to the hydrolyzed ligand and its absorbance read at the characteristic wavelength estimated for each ligand, against a solution of unmodified agarose submitted to the same treatment. The determination of the extinction coefficient, ε_λ, for each ligand is made by constructing a standard curve with the measurements of the absorbance read at the characteristic wavelength for different free-ligand

concentration solutions. Repeating the above-described procedure with 30 mg of unmodified Sepharose-CL 6B constituted the control experiment.

10. Example of purification of a target protein from real feedstocks. Crude extracts are diluted in binding buffer and 1 mL of each loaded onto pre-equilibrated columns containing 1 g of immobilized ligand. The resins are washed extensively with washing buffer until the absorbance at 280 nm of the collected samples reaches ≤0.005. Bound protein is eluted and samples collected (1 mL). Loading, washing, and elution samples (10 µL of each) are denatured by adding 5 µL of sample buffer (2 mL glycerol, 2 mL 10% w/v sodium dodecyl sulfate (SDS), 0.25 mg bromophenol blue, 0.5 mL β-mercaptoethanol, and 2.5 mL of stacking gel 4X, for a 10-mL solution) and boiling for 5 min. These samples are then applied to wells of SDS-polyacrimide gel electrphoresis (PAGE) gels (e.g., 4–20%). The gels are run by applying constant voltage or current intensity (running buffer [1 L]: 3 g Tris-base, 14.4 g glycine, 1 g SDS) and then stained (Coomassie-blue or silver staining), after which they are dried, scanned, and the relative intensities of each band analyzed (for example, with Scion Image software, www.scioncorp.com). The bands corresponding to the target protein are then analyzed to evaluate the purity and the purification yield of the elution fractions.
11. Quantitative ELISA. This is an alternative method to quantify unbound and eluted protein from the immobilized ligands. It can start with ELISA plates containing unbound protein from **step 6** or ELISA plates coated with samples collected during the affinity chromatography assay (**step 4**). Protein samples are diluted 1:10 in coating buffer (0.05 *M* sodium carbonate-bicarbonate, pH 9.6) and 100 µL of each added to wells of the ELISA plate. The plates are incubated for 1 h at 37°C or at 0–4°C for 12 h. After the incubation period, the wells are washed three times with washing buffer (PBS Tween-20, 0.02% v/v). Then, washing buffer (200 µL) is added to each well and plates are incubated at room temperature for 1 h. The wells are washed three times, after which 100 µL of the appropriate enzyme-conjugated antibody, diluted in washing buffer, are added to each well. The plates are incubated for 2 h at room temperature, and the wells washed three times. The substrate solution is added (100 µL/well) and the plates incubated for the required time. The absorbance in each well is read at the specific wavelength. When necessary, some samples were further diluted (1:20, 1:100, 1:400) and the ELISA assay repeated. Calibration curves to correlate the concentration of the target protein and the absorbance at the specific wavelength are constructed.
12. Covering gold-coated slides with dextran. Gold-coated slides are incubated in solution F for 24 h at room temperature, and then reacted with epichlorohydrin (0.5 *M*) in solution D for 4 h at room temperature. After washing sequentially with water, ethanol, and water, the surface is treated with a basic dextran solution (3 g of dextran T-500 in 0.1 *M* NaOH [1 mL]) for 20 h at room temperature, after which the surface is washed with water. The dextran-coated surface is then treated with epichlorohydrin (0.5 *M*) in solution D for 4 h. The surface is washed sequentially with water, ethanol, and water, then immersed in solution E containing aminoetha-

nethiol-activated ligand (50 μ*M*) and mercaptoethanol (5 m*M*) for 12 h, and then washed thoroughly in DMF.

13. Calibration of the SPR device by adsorption of ^{125}I-labeled target protein. Protein solutions (100 μL; 1–150 μg/mL in binding buffer) comprising known amounts of labeled and unlabeled protein are added to the reaction chamber with a total adsorption area of 46 mm^2 (perspex immobilization block containing glass microscope slide pieces [10 × 10 mm^2] coated with a gold layer). The adsorption is left to proceed for 400 s, after which the gold-coated glass pieces are rinsed three times with binding buffer (100 μL), removed from the immobilization block, and counted on a ISO-Data multi-head 500 γ counter (ICN, Biomedicals). These data are used to calculate the mean surface coverage of protein with respect to the initial concentration added. *Radioactive isotopes are carcinogenic. Wear gloves, use shielding, absorbent carbon, and vented hoods when appropriate. Monitor contamination with a good-quality Geiger counter and wipe tests. Users of radioactive isotopes must have appropriate training.*

References

1. Lowe, C. R. (2001) Combinatorial approaches to affinity chromatography. *Curr. Opin. Chem. Biol.* **5,** 248–256.
2. Lowe, C. R., Burton, S. J., Burton, N. P., Alderton, W. K., Pitts, J. M., and Thomas, J. A. (1992) Designer dyes–biomimetic ligands for the purification of pharmaceutical proteins by affinity-chromatography. *Trends Biotechnol.* **10,** 442–448.
3. Ansell, R. J. and Lowe, C. R. (1999) Artificial redox coenzymes: biomimetic analogues of NAD(+). *Appl. Microbiol. Biotechnol.* **51,** 703–710.
4. Li, R. X., Dowd, V., Stewart, D. J., Burton, S. J., and Lowe, C. R. (1998) Design, synthesis, and application of a Protein A mimetic. *Nat. Biotechnol.* **16,** 190–195.
5. Snyder, S. L. and Sobocinski, P. Z. (1975) An improved 2,4,6-trinitrobenzene sulfonic acid method for the determination of amines. *Anal. Biochem.* **64,** 284–288.
6. Kaiser, E., Colescott, R., Bossinger, C. D., and Cook, P. I. (1970) Color test for detection of free terminal amino groups in the solid-phase synthesis of peptides. *Anal. Biochem.* **34,** 595–598.
7. Morrill, P. R., Gupta, G., Sproule, K., et al. (2002) Rational combinatorial chemistry-based selection, synthesis and evaluation of an affinity adsorbent for recombinant human clotting factor VII. *J. Chromatogr. B* **774,** 1–15.
8. Palanisamy, U. D., Hussain, A., Iqbal, S., Sproule, K., and Lowe, C. R. (1999) Design, synthesis and characterisation of affinity ligands for glycoproteins. *J. Mol. Recognit.* **12,** 57–66.
9. Burton, N. P. (1992) Design of novel affinity adsorbents for the purification of trypsin-like proteases. *J. Mol. Recognit.* **5,** 55–68.
10. Filippusson, H., Erlendsson, L. S., and Lowe, C. R. (2000) Design, synthesis and evaluation of biomimetic affinity ligands for elastases. *J. Mol. Recognit.* **13,** 370–381.

11. Lowe, C. R., Roque, A. C. A., and Taipa, M. A. (2003) Affinity Adsorbents for Immunoglobulins PCT/GB 03/04524.
12. Roque, A. C. A., Taipa, M. A., and Lowe, C. R. (2004) A new method for the screening of solid-phase combinatorial libraries for affinity chromatography. *J. Mol. Recognit.* **17(3),** 262–267.
13. Teng, S. F., Sproule, K., Hussain, A., and Lowe, C. R. (1999) A strategy for the generation of biomimetic ligands for affinity chromatography. Combinatorial synthesis and biological evaluation of an IgG binding ligand. *J. Mol. Recognit.* **12,** 67–75.
14. Teng, S. F., Sproule, K., Husain, A., and Lowe, C. R. (2000) Affinity chromatography on immobilized "biomimetic" ligands synthesis, immobilization and chromatographic assessment of an immunoglobulin G-binding ligand. *J. Chromatogr. B* **740,** 1–15.
15. Morrill, P. R., Millington, R. B., and Lowe, C. R. (2003) Imaging surface plasmon resonance system for screening affinity ligands. *J. Chromatogr. B* **793,** 229–251.
16. Palanisamy, U. D., Winzor, D. J., and Lowe, C. R. (2000) Synthesis and evaluation of affinity adsorbents for glycoproteins: an artificial lectin. *J. Chromatogr. B* **746,** 265–281.
17. Gupta, G. and Lowe, C. R. (2004) An artificial receptor for glycoproteins. *J. Mol. Recognit.* **17(3),** 218–235.
18. Disley, D. M., Morrill, P. R., Sproule, K., and Lowe, C. R. (1999) An optical biosensor for monitoring recombinant proteins in process media. *Biosens. Bioelectron.* **14,** 481–493.
19. Sproule, K., Morrill, P., Pearson, J. C., et al. (2000) New strategy for the design of ligands for the purification of pharmaceutical proteins by affinity chromatography. *J. Chromatogr. B* **740,** 17–33.
20. Soto Renou, E. N., Gupta, G., Young, D. S., Dear, D. V., and Lowe, C. R. (2004) The design, synthesis and evaluation of affinity ligands for prion proteins. *J. Mol. Recognit.* **17(3),** 248–261.

5

The Role and Application of *In Silico* Docking in Chemical Genomics Research

Aldo Jongejan, Chris de Graaf, Nico P. E. Vermeulen, Rob Leurs, and Iwan J. P. de Esch

Summary

In silico docking techniques are being used to investigate the complementarity at the molecular level of a ligand and a protein target. As such, docking studies can be used to identify the structural features that are important for binding and for *in silico* screening efforts in which suitable binding partners can be identified. Here we describe a practical approach for setting up docking simulations using different docking programs. We also cover the analysis and rescoring of the obtained docking poses. Possible pitfalls in the docking studies are discussed and hints are provided to resolve commonly occurring problems.

Key Words: *In silico* docking; AutoDock; FlexX; GOLD; virtual screening; automated docking; molecular docking.

1. Introduction

Chemical genomics provides us with an overwhelming amount of information on how small molecules influence cell function. The effects that these small compounds have are often initiated by direct interactions with the proteins. In computational chemistry, docking is being used to generate and investigate these interactions at the molecular level. A three-dimensional (3D) structure of the protein under study is necessary for these approaches. Fortunately, there is a wealth of structural target information available. Recent advances in structural biology, most notably nuclear magnetic resonance (NMR) and X-ray analysis, provide us with valuable 3D information about receptors and enzymes. Docking routines are a key tool with which to investigate protein–ligand interaction because they combine the molecular features of the protein targets with the

From: *Methods in Molecular Biology, vol. 310: Chemical Genomics: Reviews and Protocols*
Edited by: E. D. Zanders © Humana Press Inc., Totowa, NJ

molecular features of the interacting ligands. By combining algorithms to generate different poses (docking) and scoring functions that evaluate the protein–ligand interactions, one can use docking methods to predict energetically favorable conformations and orientations of ligands within the binding site of the protein. Automated docking tools are employed, for example, in *in silico* screening experiments, in which databases of (putative) ligands are docked into the binding site of a protein target. In this way, compounds that have a relatively high chance of binding to the target can be selected for further investigation. *In silico* docking simulations (or "virtual screening") can be considered the computational counterpart of high-throughput screening experiments, although they are devoid of the practical problems concerning assay automation or limitations in quality or quantity of the chemical compound library *(1)*. It is the role of *in silico* docking programs to design new compounds, select possible lead compounds from (in-house) databases, and provide insight into the molecular features responsible for binding and activity of compounds.

Docking simulations can generate an avalanche of data that must be carefully analyzed and evaluated. The accuracy of the docking simulations is highly dependent on the applied docking and scoring algorithms. Extensive overviews of docking algorithms *(2)* and docking scoring functions *(3)* have been published. Different combinations of docking and scoring algorithms have been applied to protein–ligand complexes with varying accuracy and success, depending on the protein system and the physico-chemistry of the target-ligand interactions *(1,4,5)*. This suggests that a docking-scoring strategy should be specifically optimized for the system under study. Other unresolved issues are the inclusion of both ligand and protein flexibility in the docking, and the inclusion (or omission) of explicit water molecules in the binding pocket *(6,7)*. The whole procedure can thus hardly be run as an "out-of-the-box procedure" and should, ideally, be tuned to the specific problem at hand.

This chapter aims to present a very practical description of how to set up *in silico* docking experiments. Many problems and pitfalls that can be encountered during these experiments will be discussed. Because a thorough discussion of the theoretical background of the algorithms is beyond the scope of this chapter, references to important papers describing fundamental aspects are specifically mentioned in the text.

2. Materials

Various software tools are available with which to build, handle, and investigate structures *in silico*, and there are also quite a few docking programs available. All have their own set of features, strengths, and weaknesses. It is impossible to describe a generic protocol for the performance of docking experiments using any combination of software modules. Fortunately, the general setup will be similar for alternative programs. Therefore, the following subsections describe the use of a combination of specified software. The described protocols are by no means meant to replace the manuals of the particular programs. For a thorough discussion of the theoretical background of the applied algorithms and the many fine-tuning options, the reader is referred to the manuals and the accompanying literature.

2.1. Software

- Molecular modeling programs (building, viewing); SYBYL6.8
- Automated docking programs (AutoDock v3.0.5, FlexX, GOLD v2.1)
- Scoring functions (SCORE, X-SCORE and CSCORE)

The software used runs on most flavors of Unix and Linux. The procedures described here are intended for use on Unix-based systems. Commands are highlighted in boldface. In SYBYL, > signifies a selection of a menu-option, [] indicates a button to be pressed.

3. Methods

The methods described as follows outline the different steps in setting up, running, and evaluating a typical docking run.

3.1. Target Preparation

3.1.1. Sources of Target Structures

One of the most important issues in *in silico* docking is the availability of a (high-resolution, i.e., <1.5 Å) protein structure. Publicly available protein structures are deposited in the Protein Data Bank (PDB) Brookhaven *(8)* (http://www.rcsb.org/pdb). Structures of interest can be selected, assessed, and downloaded in the so-called PDB format. Most structure files contain the 3D coordinates of the heavy atoms of the protein and the relative positions of water molecules (or rather, the positions of the oxygen atoms of the water molecules). Also included are ligands, co-factors, and any other co-crystallized molecules. In the PDB file, these co-crystallized molecules are easily identified, as they are either described in the header of the file and/or have residue names deviating from the naturally occurring amino acids. In general, most of these interacting molecules must be removed from the protein structure, either manually via use of a text editor or by means of the modeling software (e.g., SYBYL). If no suitable protein structure is at hand, one can resort to the construction of a suitable 3D model based on homology with related proteins. However, the accuracy of these models is limited, and this approach can seriously compromise the success of the docking exercise.

Once a protein model has been found (or constructed), the resolution of the model should be critically assessed. The header of the PDB file contains the resolution and other relevant information (like references to the original papers, the methods used, the unit cell symmetry, B factors, missing amino acids, secondary structure elements, and so on). Often the reliability of the structural data varies within the model, leaving room for speculation on, for example, the conformation of loops and positions of water molecules, small molecular fragments, or even residues. Certain programs (i.e., PROCHECK *[9]*, WhatCheck *[10]*, ERRAT *[11]*) that are often based on statistical data gleaned from the PDB structure databases can check the structure for errors or deviations from generally found (geometrical) values ***(12)***. In the case of oligomeric structures, the quality of the individual monomers should be carefully compared with respect to the site of interest.

3.1.2. The Positions of Hydrogen Atoms

A notable feature of most structures imported from the PDB is the absence of hydrogen atoms. Only very high resolution structures (<<1.0 Å) contain information on the positions of hydrogen atoms. Most modeling software tools allow protonation of any given protein structure. Often this is done using preconstructed residue fragments in which information regarding the pH-dependency of protonation state is included. However, the influence of the local environment is not taken into account, and this should be done manually. Programs exist with which to optimize the hydrogen-bonding network within a protein structure (WhatIf ***[13–15]***). In particular, the protonation of histidine residues should be considered carefully, as they can exist in different tautomeric states under physiological conditions. In conjunction with this, the side chains of histidine, asparagine, and glutamine might be flipped 180° around their χ_1, χ_2, and χ_3 torsion angles, respectively, because it is hard to differentiate between the electron densities of N and O ***(16)***. By flipping torsion angles within the local hydrogen-bonding network, Nielsen et al. have achieved a significant improvement of the electrostatic characteristics of active sites and binding pockets of structures present in the PDB ***(17)***.

Some force fields and docking procedures work with the concept of *essential* hydrogens, i.e., hydrogens bonded to heteroatoms or hydrogens that could be involved in formation of a hydrogen bond. It is important to adhere to the procedure used to calibrate the docking program and to apply the correct method of protonation.

As most programs add hydrogen atoms according to preset geometrical rules (slight differences between programs may occur), optimization of these positions using fast line-minimization algorithms is advisable. The impact of the position of hydrogen atoms on docking is significant, as hydrogen atoms form the main interaction surfaces between protein and ligands. Incorrect placement of a hydrogen atom may be detrimental to the formation of a hydrogen bond and may lead to a lower final score of the docking *(18)*. A logical work-around would be to repeat the docking simulation with structural snapshots from a minimization run (or even a molecular dynamics run, which would also include protein flexibility in the docking simulation; *see* **Subheading 3.3.3.**).

3.1.3. The Importance of Water Molecules

Many X-ray structures contain crystal water molecules, as they are restrained in their movement by the protein environment and can thus be detected by crystallographic methods. They provide important information regarding the possible interaction of polar functional groups with the protein. Traditionally, water molecules were not taken into account during docking runs. They were removed from the PDB structure, after which docking could be performed in a completely empty protein. The free energy of desolvation of the ligand and the ligand-binding site was incorporated into the scoring functions via an appropriate energy term *(19–22)*. The possibility of water-mediated interactions was mainly ignored. Recently, the interest in the role of water molecules in docking has increased, as it has been realized that, in real life, proteins are never empty and water molecules can play an essential role in ligand–protein binding *(23–25)*. The active site or binding pocket is either filled with solvent, a ligand, or a substrate molecule. The loss of (a number of) waters from the active site or binding pocket and their gain of translational and rotational freedom once they are part of the bulk solvent also has an important (entropic) contribution to the observed binding affinity. The number of waters present and their position in the binding site might even differ among different classes of ligand molecules, and should be carefully analyzed. Good examples of this have been thoroughly described for proteases by Babine and Bender *(26)*. Several programs exist for the calculation of suitable positions for water molecules within a protein structure (PASS *[27]*, Dowser *[28]*, GRID *[29]*).

Various approaches have become available to include water molecules in the automated docking process, albeit with varying success (GLIDE *[30]*, SLIDE *[31]*, AutoDock *[32]*, and the particle placement procedure using FlexX *[33]*). Nevertheless, we highly encourage the inclusion of (a different number of) water molecules during docking simulations.

3.1.4. Preparing the Target File

After the putative binding site of the target protein has been prepared as described in the previous subheading, the protein structure file needs to be specifically prepared for the docking programs. General modeling programs can be used for this purpose. (Commands given below refer to the SYBYL modeling program, but other programs can be used equally well.)

Different docking programs require different procedures for setting up the target and the ligands in a state suitable for docking. For example, AutoDock and FlexX require the presence of partial and formal charges, respectively, whereas GOLD uses polarizabilities, which are derived from the atom type of the particular atom. The atom type (*see* description of any general molecular force field) is a very important input parameter, as it sets the topology of the atom, the protonation state, and the (partial) charges of the atom (and in the case of GOLD, the polarizabilities). With AutoDock, solvation parameters must be included in the final file used for the macromolecule. One can achieve this by switching to a file format derived from the PDB format, which allows for the specification of partial charges and solvation parameters (PDBQS). The AutoDock utility "mol2topdbqs" takes care of this conversion.

Open the PDB file; in SYBYL:
File, Read, protein_target.pdb
In FlexX, save the protein as a PDB file:
File, Save As, [Brookhaven], protein_FLEXX.pdb
In AutoDock, add polar hydrogen atoms:
Biopolymer > Add Hydrogens . . . , [All], [OK], Essential Only, [OK]
Load partial charges:
Biopolymer > Load Charges . . . , [All], [OK], KOLL_UNI(Kollman United), [OK]
Save the protein:
File, Save As, protein_AUTODOCK.mol2
File, Save As, [Brookhaven], protein_AUTODOCK.pdb
mol2topdbqs protein_AUTODOCK.mol2 (creates protein_AUTODOCK.pdbqs)
In GOLD, add all hydrogen atoms:
Biopolymer > Add Hydrogens . . . , [All], [OK], All, [OK]
Load partial charges:
Biopolymer > Load Charges . . . , [All], [OK], KOLL_ALL(Kollman All), [OK]
Save the protein:
File, Save As, protein_GOLD.mol2

3.1.5. Including Active Site Water Molecules: An Example

Energetically favorable positions for water molecules can be identified using the program GRID ***(29)***. A rectangular box is put around the binding pocket

(21.75 × 21.75 × 21.75 Å^3, with grid distance 0.333 Å) and hydrated with 25 water molecules (using an energy cut-off value of 5 kcal/mol). The volume of the different ligands can be calculated using the AutoDock-tool "pdb-volume" to obtain the dimensions of a "minimal" box. After centering this minimal box in the binding pocket, water molecules within 4.5 Å or one-half the length of the largest ligand-box dimensions are excluded from the docking studies in order to permit docking of larger compounds as well. Hydrogen atom positions of both crystal and predicted waters are optimized using DOWSER *(28)*. The energy cut-off of DOWSER must be modified to perform optimization calculations on buried waters with a cut-off larger than 12 kcal/mol. The AutoGrid program (part of the AutoDock package) must be modified to enable docking in the presence of water; i.e., an oxygen atom bound to two hydrogen atoms should be identified as a potential hydrogen bond acceptor *(34)*. The above procedure for the positioning of water molecules in the ligand-binding site has recently been applied successfully in the binding mode prediction of cytochrome P450 and thymidine kinase protein–ligand complexes *(35)*.

3.2. Ligand Preparation

One can obtain 3D structures of ligands in several ways:

1. By building the ligand *de novo* within SYBYL (or any other program capable of building small molecules).
2. By extracting the structure from a (PDB) file of a ligand–protein complex deposited at the Protein Databank.
3. By loading the ligand (or ligand derivative/homolog) structure file from the publicly, commercially, or in-house available databases. Examples are Cambridge Structural Databank (CSD—http://www.ccdc.cam.ac.uk/products/csd/) and the National Cancer Institute database (http://cactus.nci.nih.gov/). Here is a short tutorial for preparing ligand molecules within SYBYL:

Build/Edit > Sketch Molecule, (M1:Empty),[OK]

Do not add hydrogen atoms yet. SYBYL works with a force field. In a force field, atom characteristics and intramolecular connections are defined by atom types.

Use:

View > Label > Atom Types . . . ,[All],[OK]

To view the atom types of the different atoms and change them:

Build/Edit > Modify > Atom . . . , Type, [OK], pick atom, [OK], choose atom type, [OK]

For a nice flow sheet to derive proper SYBYL atom types, *see* the SYBYL manual or http://www.sdsc.edu/CCMS/Packages/cambridge/pluto/atom_types.html. Specific attention should be paid to the protonation states of heteroatoms (for example, a protonated sp3 nitrogen atom has atom type N.4, and a deprotonated carboxylate oxygen atom has atom type O.co2).

Add hydrogen atoms and check the molecule:
Build/Edit > Add > Hydrogens
Calculate partial charges for all atoms:
Compute > Charges > Gasteiger
Add formal charges to each O.co2 oxygen (–0.50), each O.3 oxygen in PO_2 (–0.50), in PO_3 (–0.66), and in SO_3 (each –0.5), each N.4 nitrogen (+1.00), and each N.3 nitrogen in $C(NH_2)_2$ (+0.50).
An energy minimization of the ligand molecule should be performed to optimize the molecular conformation:
Compute > Minimize; Method: Conj Grad, Max. Iterations:1000, m1, [OK]
Save the molecule:
File, Save As, LIGAND_cryst_H.mol2
The actual docking can be commenced only if the files describing the macromole cule and the ligand(s) are prepared in the correct way and stored in the right format (*see* **Notes 1–5**).

3.3. Docking of the Ligands in the Protein-Binding Site

A docking program basically consists of two parts: a exhaustive, fast, and robust conformational search of the ligand, preferably exhaustive, fast, and robust, and a method of calculating and scoring the observed interactions, which, ideally, should be fast and accurate.

3.3.1. Docking Programs Used

The increasing interest in drug design has led to an explosion in the number of docking programs. In this chapter, the use of the following programs is described: FlexX ***(37)***, GOLD ***(38)***, and AutoDock ***(39)***. Both FlexX and GOLD are commercially available, whereas AutoDock is freeware. This selection is based purely on our experience with these programs and their availability to us.

3.3.1.1. FlexX

The docking program FlexX ***(37)*** is based on sophisticated physicochemical models. FlexX applies an incremental construction algorithm through which ligands can be placed into the active site. The binding pocket must either be identified manually or created through the specification of a radius surrounding a reference ligand structure already present in the binding pocket. The program uses a database for torsion angles, with which conformers can be created, and an interaction geometry database, allowing exact description of intermolecular interaction patterns. For scoring, the Bohm function (with minor adaptations necessary for docking) is applied ***(20,40)***. More recently, algorithms for the placement of discrete water molecules into the binding pocket ***(33)*** and the docking of hydrophobic ligands ***(41)*** have been added.

3.3.1.2. GOLD

The program GOLD ***(38)*** uses an energy function that is based partly on the conformational and nonbonded contact information extracted from a crystallographic database (CSD). It uses a genetic algorithm to find the best protein–ligand interaction. The ligand is completely flexible during the docking run, whereas the protein is partly flexible. The torsion angles of the hydroxyl groups in serine, threonine, and tyrosine are optimized by GOLD so that the optimal interaction with the ligands can be found. The NH_3^+ group of lysine is similarly optimized. The user must provide both the ligands and a point (Cartesian coordinates or atom) that is in contact with the proposed binding pocket. A binding pocket in which the ligands are to be fitted is identified from this point using a flood-fill algorithm.

3.3.1.3. AutoDock

The original version of the AutoDock ***(39)*** program used a simulated annealing approach, exploring the conformational space with the Metropolis algorithm. The energy evaluation is performed through the use of grid-based molecular affinity potentials. As of version 3.0 ***(19)***, a hybrid search technique that applies a modified genetic algorithm has been added. As this algorithm allows an update of the genotype via modification of the phenotype by a local search method, it has been named a Lamarckian genetic algorithm (LGA). The ligand is flexible, whereas the protein is kept rigid. The user must specify a box around the protein (or part of interest) in order to calculate the grid-based potentials, the rotatable bonds of the ligand, and a random or arbitrarily chosen starting conformation. The interaction energy of the ligand with every grid point is evaluated and added.

3.3.2. Identification of the Binding Pocket

Identification of the binding pocket is an important step in the docking procedure. Docking programs are generally incapable of identifying binding pockets. For example, extending the docking grid used by AutoDock to encapsulate the whole protein normally results in docking poses located at the outskirts of the protein. Here, the energy function returns the lowest value, as interaction with protein atoms is minimal and interaction with the vacuum surrounding has no contribution to the overall energy function. Generally, the residues delineating the binding pocket can be inferred from a crystal structure containing an inhibitor or ligand. Alternatively, programs capable of detecting cavities within protein structures can be used to locate possible binding pockets. The "Putative Active Site with Spheres" (PASS) algorithm ***(27)*** uses geometry to characterize binding regions, by evaluating buried volumes within protein structures and ranking them based on size, volume, and shape. The active site center determined by

PASS is taken as the AutoDock affinity grid center, the probe location for FlexX studies, and the starting position of the GOLD flood-fill algorithm. AutoDock affinity grid calculations were carried out with the same grid box as used for the prediction of positions of active-site water molecules. The FlexX binding pocket is by default defined as the amino acid residues within 6.5 Å from the "translated" ligand structure (described previously). GOLD has a default value of 10 Å for the radius used in the flood-fill algorithm.

3.3.3. Including Flexibility of the Protein

The inclusion of full protein flexibility in automated ligand–protein docking is a very difficult task, considering the enormous increase in conformational space owing to the introduction of new degrees of freedom associated with both protein backbone movement and flexibility of (binding) amino acid residues ***(2,6)***. Until now, no automated ligand–protein docking program that attempts to explicitly consider the full flexibility of both the ligand and protein structure has been validated successfully ***(7,42)***.

Nevertheless, methods have been implemented that allow partial treatment of protein flexibility ***(6,43)***. These approaches include: the use of averaged conserved coordinates ***(44)*** or protein–ligand interaction fields ***(32)***, the use of side-chain libraries ***(45)***, statistical corrections of scores of ligands docked across multiple active sites ***(46)***, mean field theory ***(31)*** or Monte-Carlo sampling ***(47)***, docking to a large ensemble of protein molecular dynamics (MD) snapshots ***(48)***, and even an aggressive (minima mining) optimization algorithm to allow any or all rotatable bonds in selected protein side chains to be treated as continuous degrees of freedom during the docking process ***(49)***. However, to be able to describe the docking of ligands to protein targets in an even more accurate and extensive way, we suggest focusing on the development of docking search algorithms that also optimize the positions of several water molecules during the docking process. To our knowledge, no docking tool that, in addition to the ligand and the protein, also considers flexible water molecules as a third partner in protein–ligand interactions has been properly validated.

3.3.4. Running the FLEXX Docking Program

FlexX can be interactively run within SYBYL. If only one ligand is to be docked:
Options > FlexX > Run-one-Ligand

A previously created receptor description file (RDF) can be used (**Fig. 1**), or a new *protein.rdf* file (**Fig. 2**) can be created by pressing "Create." The RDF file can be customized by enabling "Customize RDF File," which allows the addition of metals, cofactors, or water molecules and the determination of the protonation state of several amino acid residues (**Fig. 3**):

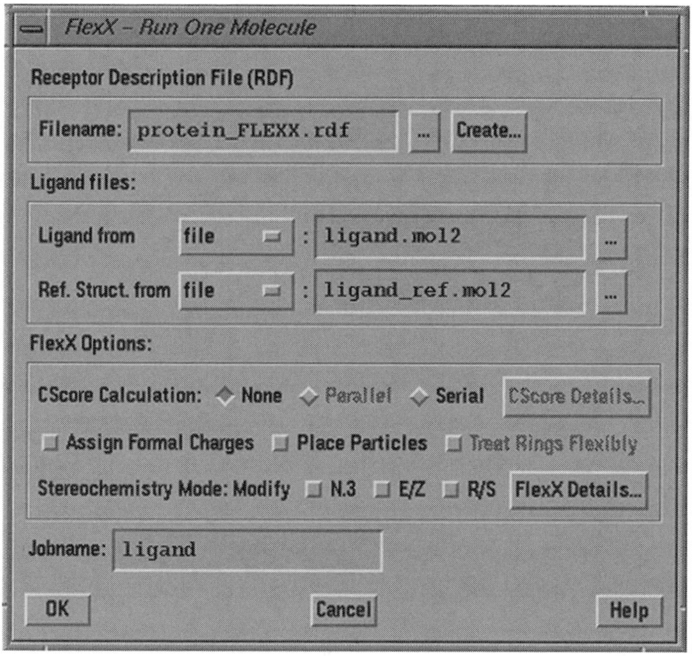

Fig. 1. Setting up a FlexX docking run for one ligand molecule. Receptor description file (RDF): *Filename*: specify a filename for the RDF to be created (or browse using a filebrowser). Use *Create* to create the RDF from the PDB structure (*see* **Fig. 2**). Define ligand starting and reference structures and FlexX docking parameters Ligands from: LIGAND.mol2. Ref. Structure from: LIGAND_cryst_H.mol2. FlexX options: *CScore calculation* allows calculation of docking scores according to the Cscore module. The *Assign Formal Charges* setting enables automated assignment of the formal charge. *Place particles* is meant for including virtual particles in the docking procedures *(33)* and is mainly used for including water molecules. *Treat ring flexibility* option can be used when a Confort license has been obtained; it allows the inclusion of ring conformers in the docking. Stereochemistry Mode: *Modify N.3* allows modification of N sp^3 stereo centers; *E/Z* refers to the conformation around double bonds; *R/S* to the conformation around chiral atoms. *FlexX Details* allows for specifying the number of desired answers (typically 50; *see* **Fig. 4**).

PDB Filename: protein_FLEXX.pdb
Active-Site File: protein_poc.pdb
Create from Ref. Struct. (7 Å)
Customize RDF File . . . : *Add specific metals, cofactors, and/or water molecules or determine protonation stat pf ASP, CYS, GLU, HIS, SER*

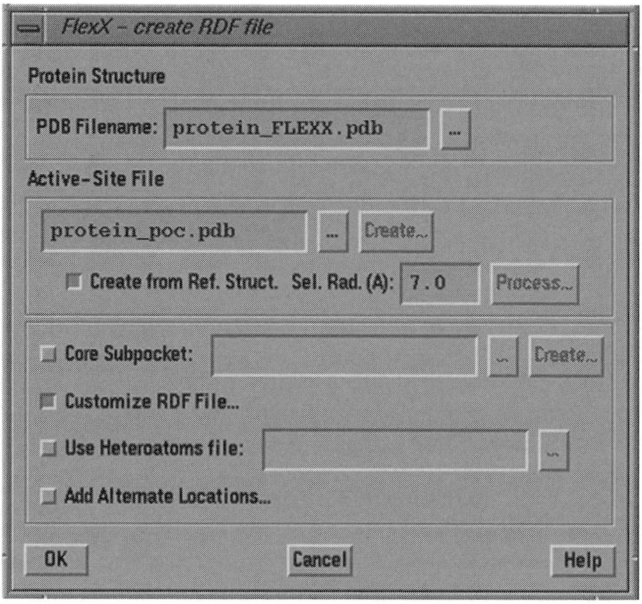

Fig. 2. Creation of the receptor description file (RDF) for use with FlexX. Protein Structure: Specify the correct PDB structure. Active-Site File: The FlexX binding pocket can be defined as the amino acid residues within 7 Å from a reference structure (e.g., a ligand structure "translated" to the protein active site midpoint as determined by PASS; *see* **Subheading 3.3.2.**). Customize RDF File: enables specification of metals, cofactors, protonation states, and torsions of residues (*see* **Fig. 3**).

[OK]

Define the ligand and reference structures and FlexX docking parameters (**Fig. 1**):

Ligand from: LIGAND.mol2
Ref. Structure from: LIGAND_REF.mol2
Assign Formal Charges
Stereochemistry Mode: Modify N.3;
Jobname: LIGAND

(Other) FlexX parameter values can of course be adapted with the "FlexX Details" function. Here, the appropriate scoring function can be chosen and the desired number of solutions and amount of output can be specified (**Fig. 4**).

When the FlexX job has finished, a table with results is presented. The results can be exported to a table:

Edit (in spreadsheet) **> Export, TSV(Tab Separated), Filename: flexx_ligand.tab**

To save the table:

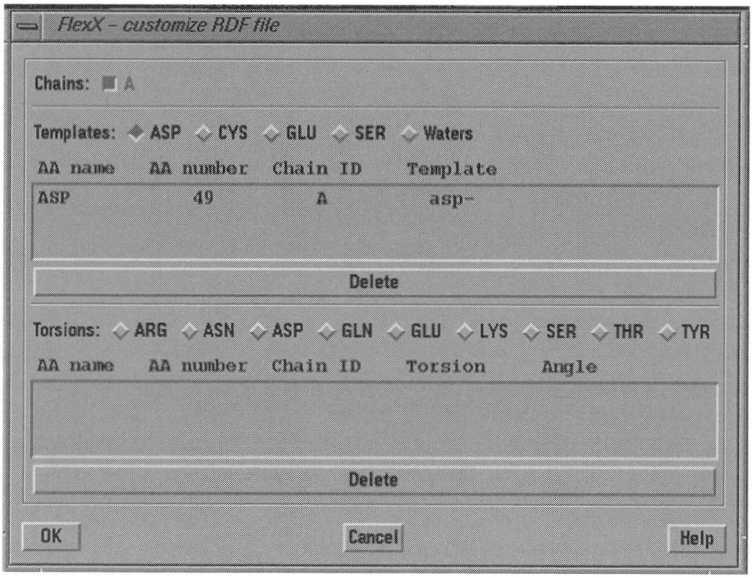

Fig. 3. Customization of the receptor description file. *Chains* allows specification of the chains present in the protein that have to be included (grayed out if only one chain is present); *Templates* to add specific metals, cofactors, and/or water molecules or determine protonation state of Asp, Cys, Glu, His, Ser; *Torsions* allows setting of the dihedral angle of the terminal hydrogen atom of the specified residue.

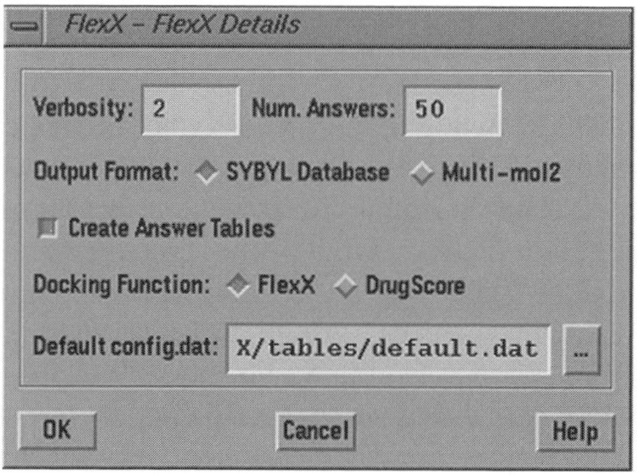

Fig. 4. Two different scoring functions can be used during the docking process: FlexX ***(20,50)*** and DrugScore ***(54)***. Verbosity: determines the amount of output generated. Num. Answers: the maximum number of docking solutions to be saved by FlexX. Output format: either a database of all conformations for all ligands is created or a single multimolecule file is created for each successfully docked ligand. No Answer Tables can be generated from such a Multi-Mol2 file. Create Answer Tables: allows easy browsing within SYBYL of all the answers, but requires additional disk space and processing time.

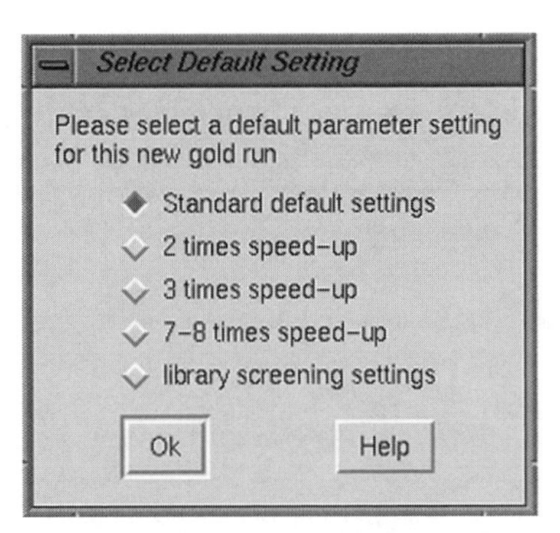

Fig. 5. First pop-up of GOLD graphical user interface. Selection of running mode (differences in scoring function and genetic algorithm settings).

Save (in spreadsheet)**, Filename: flexx_ligand.tbl**

For the docking of *multiple* ligands:

Options > FlexX > Run-multiple-Ligands

And follow the same procedure as for one ligand, choosing a multiple mol2 file containing all ligands (MLT.mol2, generated with cat LIGAND_1mol2 LIGAND_2.mol2 . . . LIGAND_N.mol2 > MLT.mol2:

Filename > MOL2 File: MLT.mol2

3.3.5. Running the GOLD Docking Program

GOLD comes with its own graphic user interface (GUI), allowing the program to run interactively. The user can choose among various running modes, each with its own demand of central processing unit (CPU) time and accuracy (**Fig. 5**). A configuration file specifying the necessary input parameters can be set up through the GUI (**Fig. 6**) and either saved or run interactively (however, *see* **Note 1**). The generated configuration file (**Fig. 7**) can be edited manually if desired and run from the command line as follows:

gold_auto protein.conf >& protein_GOLD.log &

The resulting conformations can be stored in a directory per ligand and a list of best hits is stored in the file "bestranking.lst" (*see* **Note 6**).

Fig. 6. Graphical user interface of GOLD. Variables and parameters can be set interactively. The resulting configuration file can be saved and edited manually at a later stage. Input parameters and files: the user has to provide a point (coordinates or atom—for example, the active site center determined by PASS; *see* **Subheading 3.3.2.**) in contact with the proposed binding pocket, and the ligands. A binding pocket in which the ligands are to be fitted is identified from this point using a flood-fill algorithm. GOLD

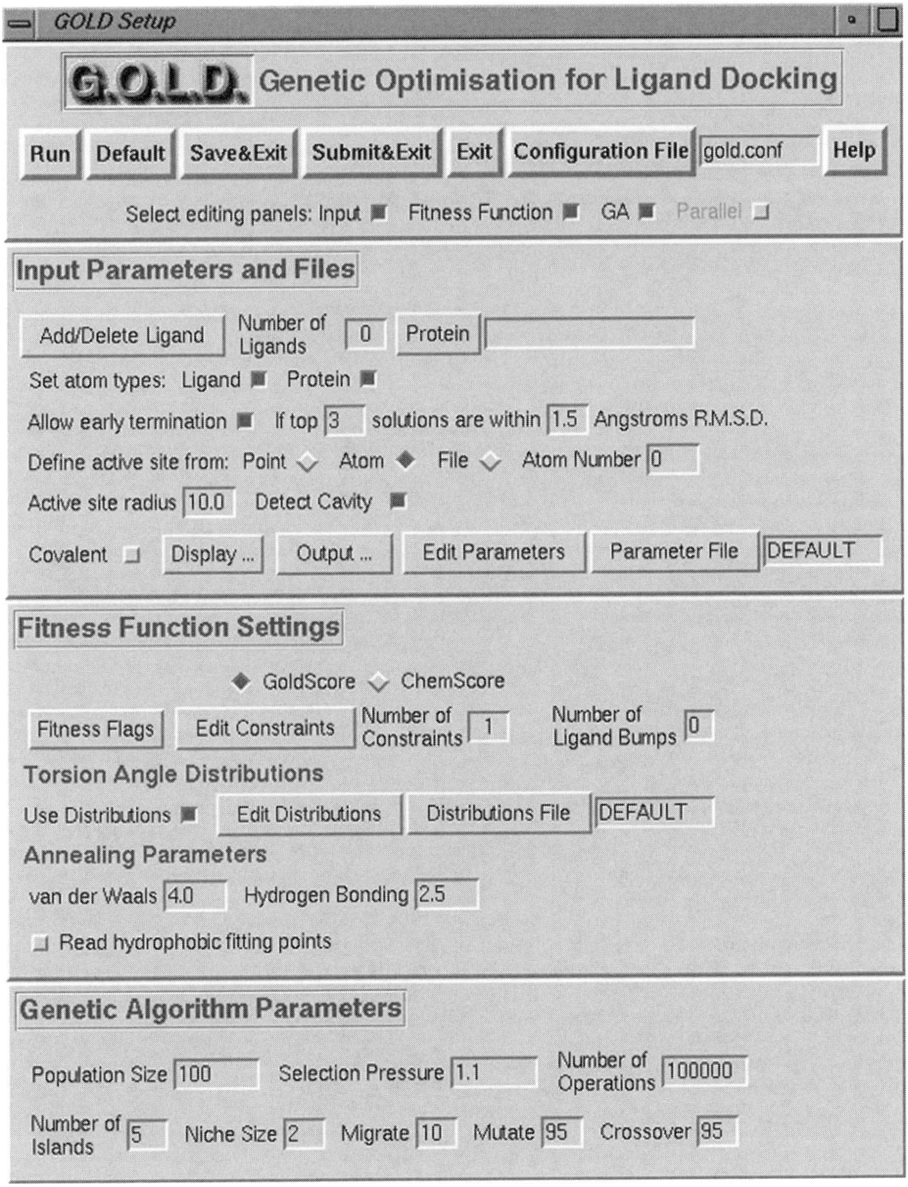

GOLD Setup

G.O.L.D. Genetic Optimisation for Ligand Docking

Run | Default | Save&Exit | Submit&Exit | Exit | Configuration File | gold.conf | Help

Select editing panels: Input Fitness Function GA Parallel

Input Parameters and Files

Add/Delete Ligand | Number of Ligands 0 | Protein

Set atom types: Ligand Protein

Allow early termination If top 3 solutions are within 1.5 Angstroms R.M.S.D.

Define active site from: Point Atom File Atom Number 0

Active site radius 10.0 Detect Cavity

Covalent | Display ... | Output ... | Edit Parameters | Parameter File | DEFAULT

Fitness Function Settings

GoldScore ChemScore

Fitness Flags | Edit Constraints | Number of Constraints 1 | Number of Ligand Bumps 0

Torsion Angle Distributions

Use Distributions | Edit Distributions | Distributions File | DEFAULT

Annealing Parameters

van der Waals 4.0 Hydrogen Bonding 2.5

Read hydrophobic fitting points

Genetic Algorithm Parameters

Population Size 100 Selection Pressure 1.1 Number of Operations 100000

Number of Islands 5 Niche Size 2 Migrate 10 Mutate 95 Crossover 95

Fig. 6 *(Continued)* can be instructed to terminate docking on a given ligand if the *n* top-ranked answers obtained so far are within *x* Å rms deviation of one another, where *n* and *x* are user-defined quantities. Fitness function settings: two different scoring functions can be used during the docking process—namely, GoldScore *(38)* and ChemScore *(53)*. A specific feature of the GOLD program is that distance, hydrogen bonding template, and similarity constraints can be included during docking.

```
  GOLD CONFIGURATION FILE

  generated by gold front end


  POPULATION
popsiz = 100
select_pressure = 1.100000
n_islands = 5
maxops = 100000
niche_siz = 2

  GENETIC OPERATORS
pt_crosswt = 95
allele_mutatewt = 95
migratewt = 10

  FLOOD FILL
radius = 10
origin = 0 0 0
do_cavity = 1
floodfill_atom_no = 1255
cavity_file = cavity.atoms
floodfill_center = atom

  DATA FILES
protein_datafile = /home/silly/jongejan/protein_GOLD.mol2
ligand_data_file = /home/silly/jongejan/ligand_1.mol2 10
ligand_data_file = /home/silly/jongejan/ligand_2.mol2 10
.
.
.
ligand_data_file = /home/silly/jongejan/ligand_N.mol2 10
param_file = DEFAULT
set_ligand_atom_types = 1
set_protein_atom_types = 0
directory = /home/silly/jongejan/GOLD_DOCK_run1
tordist_file = DEFAULT
make_subdirs = 0
save_lone_pairs = 1
fit_points_file = fit_pts.mol2
read_fitpts = 0

  FLAGS
display = 0
internal_ligand_h_bonds = 0
n_ligand_bumps = 0
flip_free_corners = 0
flip_amide_bonds = 0
flip_planar_n = 1
use_tordist = 1
start_vdw_linear_cutoff = 2.5
initial_virtual_pt_match_max = 4

  TERMINATION
early_termination = 1
n_top_solutions = 3
rms_tolerance = 1.5

  CONSTRAINTS

  COVALENT BONDING
covalent = 0

  SAVE OPTIONS
save_score_in_sdfile = 1
```

Fig. 7. Example of the configuration file used as input for GOLD.

3.3.6. Running the AutoDock Program

The scoring function of AutoDock uses a precalculated grid, on which at certain points in space the interaction energy between different atomic probes with the protein structure is considered. To calculate such a grid, the utility Auto Grid is used. First, a grid parameter file (GFP; **Fig. 8**), *protein_AUTODOCK.gpf,* must be generated, specifying the likely position of the ligand in the macromolecule (*see* **Note 7**):

mkgpf3 LIGAND.pdbq protein_AUTODOCK.pdbqs

Alternatively, instead of using mkgpf3 to create the GPF file, a previously generated template GPF file can be used.

The GPF is used to calculate the grid maps (the –l option specifying the grid logfile [GLG]):

autogrid3 -p protein_AUTODOCK.gpf -l protein.glg &

The configuration file containing the docking parameters (a docking parameter file [DPF], *LIGAND.protein_AUTODOCK.dpf*) is generated as follows:

mkdpf3 LIGAND.pdbq protein.pdbqs

The actual docking is performed through the use of this DPF file (the –l option specifying the output to be sent to the docking logfile [DLG]). This file should be inspected and, if necessary, edited (**Fig. 9**) according to the specific needs. The run is submitted by:

autodock3 -p LIGAND.protein_AUTODOCK.dpf -l LIGAND.protein.dlg &

Originally, AutoDock was devised to dock a single ligand per submitted job. However, it is possible to perform batch jobs to enable docking or screening of databases. The main issue here is to generate the correct grid maps, i.e., including halogens and other elements. Unfortunately, interaction potentials of these atoms are not always available (for a list of available parameters, *see* http://www.scripps.edu/pub/olson-web/doc/autodock/parameters.html). The user can generate his or her own template GPF including the desired parameters. To include these extra potential maps, it is necessary to recompile AutoGrid and AutoDock, and specify more atom types in the gpf3gen(.awk) utility must be specified. Through the use of simple scripts, a "multi-docking" version of AutoDock can be created, simplifying the process of database docking and screening.

3.4. Analysis

The analysis and interpretation of docking results can be both difficult and time-consuming. When one is working with only a limited number of ligand molecules, visual analysis of the found docking conformations is still feasible. However, in the screening of large databases, this is hardly an option. The fact

```
receptor protein_AUTODOCK.pdbqs              #macromolecule
gridfld  protein_AUTODOCK.maps.fld              #grid_data_file
npts      60 60 60                  #num.grid points in xyz
spacing  .375                           #spacing (Angstroms)
gridcenter -0.114 0.711 -0.706       #xyz-coordinates or "auto"
types CANOSH                           #atom type names
smooth 0.500                        #store minimum energy within radius (Angstroms)
map protein_AUTODOCK.C.map                          #filename of grid map
nbp_r_eps  4.00 0.0222750 12  6   #C-C lj
nbp_r_eps  3.75 0.0230026 12  6   #C-N lj
nbp_r_eps  3.60 0.0257202 12  6   #C-O lj
nbp_r_eps  4.00 0.0257202 12  6   #C-S lj
nbp_r_eps  3.00 0.0081378 12  6   #C-H lj
nbp_r_eps  3.75 0.0230026 12  6   #C-X lj
nbp_r_eps  2.79 0.0538015 12  6   #C-M lj
sol_par 12.77 0.6844                  #C atomic fragmental volume, solvation param.
constant 0.000                          #C grid map constant energy
map protein_AUTODOCK.A.map                          #filename of grid map
nbp_r_eps  4.00 0.0222750 12  6   #A-C lj
nbp_r_eps  3.75 0.0230026 12  6   #A-N lj
nbp_r_eps  3.60 0.0257202 12  6   #A-O lj
nbp_r_eps  4.00 0.0257202 12  6   #A-S lj
nbp_r_eps  3.00 0.0081378 12  6   #A-H lj
nbp_r_eps  3.75 0.0230026 12  6   #A-X lj
nbp_r_eps  0.00 0.0000000 12  6   #A-M lj
sol_par 10.80 0.1027                  #A atomic fragmental volume, solvation param.
constant 0.000                          #A grid map constant energy
map protein_AUTODOCK.N.map                          #filename of grid map
nbp_r_eps  3.75 0.0230026 12  6   #N-C lj
nbp_r_eps  3.50 0.0237600 12  6   #N-N lj
nbp_r_eps  3.35 0.0265667 12  6   #N-O lj
nbp_r_eps  3.75 0.0265667 12  6   #N-S lj
nbp_r_eps  2.75 0.0084051 12  6   #N-H lj
nbp_r_eps  3.50 0.0237600 12  6   #N-X lj
nbp_r_eps  2.54 0.0555687 12  6   #N-M lj
sol_par  0.00 0.0000                  #N atomic fragmental volume, solvation param.
constant 0.000                          #N grid map constant energy
map protein_AUTODOCK.O.map                          #filename of grid map
nbp_r_eps  3.60 0.0257202 12  6   #O-C lj
nbp_r_eps  3.35 0.0265667 12  6   #O-N lj
nbp_r_eps  3.20 0.0297000 12  6   #O-O lj
nbp_r_eps  3.60 0.0297000 12  6   #O-S lj
nbp_r_eps  1.90 0.3280000 12 10   #O-H hb
nbp_r_eps  3.35 0.0265667 12  6   #O-X lj
nbp_r_eps  2.39 0.0621176 12  6   #O-M lj
sol_par  0.00 0.0000                  #O atomic fragmental volume, solvation param.
constant 0.236                          #O grid map constant energy
map protein_AUTODOCK.S.map                          #filename of grid map
nbp_r_eps  4.00 0.0257202 12  6   #S-C lj
nbp_r_eps  3.75 0.0265667 12  6   #S-N lj
nbp_r_eps  3.60 0.0297000 12  6   #S-O lj
nbp_r_eps  4.00 0.0297000 12  6   #S-S lj
nbp_r_eps  2.50 0.0656000 12 10   #S-H hb
nbp_r_eps  3.75 0.0265667 12  6   #S-X lj
nbp_r_eps  2.79 0.0621176 12  6   #S-M lj
sol_par  0.00 0.0000                  #S atomic fragmental volume, solvation param.
constant 0.000                          #S grid map constant energy
map protein_AUTODOCK.H.map                          #filename of grid map
nbp_r_eps  3.00 0.0081378 12  6   #H-C lj
nbp_r_eps  2.75 0.0084051 12  6   #H-N lj
nbp_r_eps  1.90 0.3280000 12 10   #H-O hb
nbp_r_eps  2.50 0.0656000 12 10   #H-S hb
nbp_r_eps  2.00 0.0029700 12  6   #H-H lj
nbp_r_eps  2.75 0.0084051 12  6   #H-X lj
nbp_r_eps  1.79 0.0196465 12  6   #H-M lj
sol_par  0.00 0.0000                  #H atomic fragmental volume, solvation param.
constant 0.118                          #H grid map constant energy
elecmap protein_AUTODOCK.e.map                  #electrostatic potential map
dielectric -0.1146                  #<0,distance-dep.diel; >0,constant
#fmap protein_AUTODOCK.f.map                  #floating grid
# gpf3gen.awk 3.0.4 #
```

that these docking procedures are reasonably automated hardly makes them foolproof. Log and error files should be carefully checked not only for errors, but also for warnings, as sometimes in the process of docking, mismatched parameters have been replaced by default values in order to keep the program from crashing.

The actual score of the ligand is the primary result of the docking. However, the docking pose identified as the best solution by one program might not have been identified as the best solution by other docking programs (*see* **Subheading 3.5.**).

Experimental information from biophysical or biochemical methods are the main source of data for the validation of the found docking poses. Thus, docking poses should not only be ranked according their score, but also according to knowledge obtained from experiments:

- (Root mean square) deviation from a reference structure (what is the ligand binding mode compared with other [homologous] ligands?);
- Distance between the ligand (geometrical center) and the center of specific protein binding pockets (e.g., where in the protein does the ligand bind?);
- Distance between specific atoms of the ligand (e.g., preferred oxidation site) and parts of the protein target (e.g., catalytic center);
- Order of binding of ligands in a mutant protein.

3.4.1. Analyzing the Results Generated by FlexX

Results of a FlexX run can be analyzed within SYBYL:

Options > FlexX > Browse results

A window is opened in which the jobname (used earlier to submit the run) can be selected and the ligands with their corresponding FlexX score will be displayed. The "Show Failed Ligand" button gives ligands for which no docking solution has been obtained. The successfully docked ligands can be sorted according to their binding energy score by clicking the "Sorted" icon.

3.4.2. Analyzing the Results Generated by Gold

The docking poses found by GOLD are stored in MOL2 format. The ligand conformations are ranked according to their fitness score as calculated by the genetic algorithm. The final score is output in the mol2 file, and the best score for a particular ligand is transferred to a list (bestranking.lst). Results can be visualized using the usual software. A file called "fit_pts.mol2" gives an impression of the size of the detected cavity.

Fig. 8. *(Opposite page)* Example of grid parameter file for use in AutoDock. The file is largely self-explanatory. Reference is made to the grid maps that have been calculated using different molecular entities to probe the interaction with the protein.

```
seed      time pid  # for random number generator
types     CANOSH        # atom type names
fldprotein_AUTODOCK.maps.fld     # grid data file
mapprotein_AUTODOCK.C.map # C-atomic affinity map file
mapprotein_AUTODOCK.A.map # A-atomic affinity map file
mapprotein_AUTODOCK.N.map # N-atomic affinity map file
mapprotein_AUTODOCK.O.map # O-atomic affinity map file
mapprotein_AUTODOCK.S.map # S-atomic affinity map file
mapprotein_AUTODOCK.H.map # H-atomic affinity map file
mapprotein_AUTODOCK.e.map # electrostatics map file

move      LIGAND.pdb     # small molecule file
about     -0.114 0.711 -0.706    # small molecule center

# Initial Translation, Quaternion and Torsions
tran0     random        # initial coordinates/A or "random"
quat0     random        # initial quaternion or "random"
#ndihe    0             # number of initial torsions
#dihe0    random        # initial torsions

#torsdof 0 0.3113 # num. non-H tors.degrees of freedom & coeff.

# Initial Translation, Quaternion and Torsion Step Sizes and Reduction Factors
tstep     2.0           # translation step/A
qstep     50.0          # quaternion step/deg
dstep     50.0          # torsion step/deg
trnrf     1.            # trans reduction factor/per cycle
quarf     1.            # quat reduction factor/per cycle
dihrf     1.            # tors reduction factor/per cycle

# Hard Torsion Constraints
#hardtorcon 1 -180. 30.   # constrain torsion, num., angle(deg), range(deg)

# Internal Non-Bonded Parameters
intnbp_r_eps  4.00 0.0222750  12  6     #C-C lj
intnbp_r_eps  4.00 0.0222750  12  6     #C-A lj
intnbp_r_eps  3.75 0.0230026  12  6     #C-N lj
intnbp_r_eps  3.60 0.0257202  12  6     #C-O lj
intnbp_r_eps  4.00 0.0257202  12  6     #C-S lj
intnbp_r_eps  3.00 0.0081378  12  6     #C-H lj
intnbp_r_eps  4.00 0.0222750  12  6     #A-A lj
intnbp_r_eps  3.75 0.0230026  12  6     #A-N lj
intnbp_r_eps  3.60 0.0257202  12  6     #A-O lj
intnbp_r_eps  4.00 0.0257202  12  6     #A-S lj
intnbp_r_eps  3.00 0.0081378  12  6     #A-H lj
intnbp_r_eps  3.50 0.0237600  12  6     #N-N lj
intnbp_r_eps  3.35 0.0265667  12  6     #N-O lj
intnbp_r_eps  3.75 0.0265667  12  6     #N-S lj
intnbp_r_eps  2.75 0.0084051  12  6     #N-H lj
intnbp_r_eps  3.20 0.0297000  12  6     #O-O lj
intnbp_r_eps  3.60 0.0297000  12  6     #O-S lj
intnbp_r_eps  1.90 0.3280000  12 10     #O-H hb
intnbp_r_eps  4.00 0.0297000  12  6     #S-S lj
intnbp_r_eps  2.50 0.0656000  12 10     #S-H hb
intnbp_r_eps  2.00 0.0029700  12  6     #H-H lj

#intelec          # calculate internal electrostatic energy

# Simulated Annealing Parameters
#rt0 616.         # SA: initial RT
#rtrf 0.95        # SA: RT reduction factor/per cycle
#linear_schedule  # SA: do not use geometric cooling
#runs      10           # SA: number of runs
#cycles    50           # SA: cycles
#accs      100          # SA: steps accepted
#rejs      100          # SA: steps rejected
#select    m            # SA: minimum or last
```

```
# Trajectory Parameters (Simulated Annealing Only)
#trjfrq     100            # trajectory frequency
#trjbeg     1              # start trj output at cycle
#trjend     50             # end trj output at cycle
#trjout     protein_AUTODOCK.trj   # trajectory file
#trjsel     E              # A=acc only;E=either acc or rej

#watch      protein_AUTODOCK.watch.pdb    # real-time monitoring file

outlev      1              # diagnostic output level

# Docked Conformation Clustering Parameters for "analysis" command
rmstol      1.0            # cluster tolerance (Angstroms)
rmsref      protein_AUTODOCK.pdb   # reference structure file for RMS calc.
#rmsnosym          # do no symmetry checking in RMS calc.
write_all          # write all conformations in a cluster

extnrg      1000.          # external grid energy
e0max       0. 10000       # max. allowable initial energy, max. num. retries


# Genetic Algorithm (GA) and Lamarckian Genetic Algorithm Parameters (LGA)
ga_pop_size 50             # number of individuals in population
ga_num_evals 250000        # maximum number of energy evaluations
ga_num_generations 27000   # maximum number of generations
ga_elitism 1               # num. of top individuals that automatically survive
ga_mutation_rate 0.02      # rate of gene mutation
ga_crossover_rate 0.80     # rate of crossover
ga_window_size 10  # num. of generations for picking worst individual
ga_cauchy_alpha 0  # ~mean of Cauchy distribution for gene mutation
ga_cauchy_beta 1   # ~variance of Cauchy distribution for gene mutation
set_ga                     # set the above parameters for GA or LGA

# Local Search (Solis & Wets) Parameters (for LS alone and for LGA)
sw_max_its 300             # number of iterations of Solis & Wets local search
sw_max_succ 4              # number of consecutive successes before changing rho
sw_max_fail 4              # number of consecutive failures before changing rho
sw_rho 1.0         # size of local search space to sample
sw_lb_rho 0.01             # lower bound on rho
ls_search_freq 0.06        # probability of performing local search on an indiv.
set_psw1           # set the above pseudo-Solis & Wets parameters

# Perform Dockings
#do_local_only 50  # do only local search

#simanneal         # do as many SA runs as set by the "runs" command above

ga_run 10          # do this many GA or LGA runs

# Perform Cluster Analysis
analysis           # do cluster analysis on results
```

Fig. 9. Example of DPF for use in AutoDock. See the manual for the exact meaning of the variables, although the file is largely self-explanatory. “protein_AUTODOCK” and “LIGAND” need to be replaced by the appropriate labels.

3.4.3. Analyzing the Results Generated by AutoDock

The results of an AutoDock run are stored in PDB format. To view the final results, a PDB file containing the complex of all docked poses of the ligand in the protein can be generated using the “get-docked” utility (*see* **Notes 3** and **8**):

get-docked LIGAND.protein.dlg

Alternatively, the DLG file can be analyzed and the energies and the various energy components can be extracted. It is also possible to calculate the energy of a given ligand conformation with the protein structure or perform cluster analyses of the results of various runs (*see* AutoDock manual).

3.4.4. Verification of Generated Docking Solutions

The docking solutions can be compared with reference structures (i.e., an elucidated crystal structure of the protein–ligand complex) by the usual procedures:

3.4.4.1. FlexX in SYBYL

Load the protein active site:
File > Read > protein_poc.pdb > No
Do not center the molecule on screen.
Load the X-ray structure:
File > Read > *Xray.pdb* > No
Do not center the molecule on screen.
And load the docking solution:
File > Read >
Select "jobname," then "results" in the "Sub-directories" menu and *LIGAND.mol2* in the right "Files" menu.

3.4.4.2. GOLD

Display the X-ray structure, superimpose the empty protein structure, and read in the ligand file, LIGAND.mol2. Take care to select the empty protein structure as the frame of reference for the ligand structure. GOLD also supplies several utilities to display docking solutions "on the fly," calculate RMS differences with a reference structure, and perform hierarchical clustering ("grommet, smart_rms," and "rms_analysis" in the "sgi_utilities" directory).

3.4.4.3. AutoDock

Display the X-ray structure, superimpose the empty protein, and read in the PDB file containing all of the docking poses of the ligand obtained by the "getdocked" utility. Care should be taken to select the right frame of reference.

3.5. Rescoring

Scoring functions are an important factor in the performance of docking and screening tools. Many scoring functions have been developed and implemented in docking programs *(3)*. Docking solutions generated with the different automated docking programs can be ranked based on individual scoring functions or by consensus ranking of several scoring functions. Rescoring of docking solutions with other scoring functions is an important step in virtual screening and docking tools and helps to enhance the overall performance.

Docking poses generated by AutoDock, GOLD, and FlexX can be easily pooled and/or rescored with SCORE *(21)*, the three scoring functions available in the X-SCORE program *(22)*, and/or user-defined scoring functions. This can be achieved via simple scripts that extract the relevant information from the various output files (an example of such a script is available from the authors upon request). When the results are saved in a format readable to SYBYL (i.e., tab-separated columns), the energy scores can be listed and imported into a SYBYL spreadsheet:

File > Molecular Spreadsheet > New . . . , Datasource: Database, [OK], Database:LIGAND.mdb, [OPEN]

A spreadsheet is opened with all structures ranked in a column. Select all rows with conformations and add energy scores:

File (in spreadsheet) > Import, Format: TSV(Tab Separated), File:file.xls

The columns are filled with interaction energies and can be renamed accordingly:

Edit (in spreadsheet) > Rename, Columns

The conformations can be sorted by their energy value according to the different scoring functions:

View (in spreadsheet) > Sort

and visualized in the protein binding pocket. First open the protein structure.

File, Read, protein.mol2

and then import each conformer

Show Row Selection (in spreadsheet)

or

File (in spreadsheet) > Put Rows into Molecule Areas

Alternatively, if one has access to the CSCORE module of SYBYL, one can calculate scores of FlexX *(50)*, Gold *(38)*, PMF *(51)*, Dock *(52)*, and Chemscore *(53)* with a previously saved table of docked conformations. First, the table (corresponding to the database containing the conformations) is loaded:

File > Molecular Spreadsheet > New > Database (and select the database)

A spreadsheet with all the conformations is listed. By typing at the SYBYL prompt (in the shell) : cscore (followed by *Enter*); associated receptor mol2 file: protein.mol2 (followed by *Enter*); row expression : * (followed by *Enter*), one can calculate the scores for each row. The Cscore (in the sixth column) is a simple consensus scoring, specifying whether a score belongs to the top 5 score of each individual scoring list (0 if it is never found among the top 5 scores, 5 if it is always within the top 5).

3.6. Final Remarks

The exhaustive body of literature on virtual screening, docking, and scoring tools renders it impossible to describe every available program and method in detail. New docking algorithms and scoring functions are developed continually. Validation studies using various docking and scoring tools appear on a regular basis, enabling a more thorough comparison of the different methods. In this chapter, we have tried to give a practical guide to setting up a docking run, without going into too much detail on the scientific aspects. It is up to the reader to look at the technical aspects and merits of the various methods. The references herein will definitely provide a good starting point.

In silico docking tools remain indispensable for the analysis of large databases of chemical compounds with which to identify possible drug candidates. Information on the interactions responsible for binding can be extracted from the solutions generated by the docking programs and used to design even more successful lead compounds. The current advances in our understanding of the important factors in the docking of molecules and the quality of the docking programs will undoubtedly lead to even better predictions in the near future.

4. Notes

1. Ligands can be translated to the center of the active site as determined by the PASS algorithm ***(27)*** (*see* **Subheading 3.3.2.**). AutoDock simulations are sensitive to the exact position of the ligand with regard to the protein. Reliability improves if the initial position of the ligand is close to the expected docking site. Input ligand geometries and orientations ***(18,30)*** and the number of independent docking runs used (convergence of the docking simulation) ***(35)*** can affect the performance of the docking programs. Care should thus be taken to save both protein and ligand in the same frame of reference.
2. Most molecular docking programs use the MOL2 format to describe the ligand. It allows for the identification of molecular fragments and includes all necessary information regarding partial charges, molecular topology, and atom types. Several programs for conversion between molecular formats exist (*see* URLs given at the end of this chapter). The outcomes of these conversion programs should be checked carefully, as they are not always error-free.
3. It is important to adhere to the procedure prescribed by a specific docking program concerning the preparation of the ligand structure (especially on the partial charge scheme used), as the reliability of the empirical scoring scheme has been estimated using the same settings.
4. As in the case of the protein structure, the use of the correct protonation state is also important when preparing the ligand structures. Alternative structures should be constructed manually and added to the database as individual entities. Stereoisomers and tautomers should be handled similarly. For the construction of tau-

tomers, a fully automated procedure (AGENT2.0) has been developed by Posposil and Balmer ***(36)***.

5. When using AutoDock, one must specify the rotatable bonds of the ligand. Fixed parts of the ligand are designated as ROOT, from which BRANCHES sprout; TORSIONS are special branches with only two nearest neighbors. This assignment can be done manually, but a helpful utility called AutoTors is included in the distribution. The specification of the various parts in the ligand allows for some fine-tuning of the conformational space search.
6. The configuration file should be carefully inspected, as the values for the "start_vdw_linear_cutoff" and "initial_virtual_pt_match_max" parameters in the generated configuration file differ from those specified in the GOLD GUI ("Van der Waals" and "Hydrogen bonding," respectively). They also deviate from the settings advocated in the manual (version 2.1).
7. The parameters in the GPF file are optimized for the energy function applied within AutoDock, but can be replaced by user-defined interaction potentials. AutoDock makes a distinction between aromatic ("A") and nonaromatic carbons ("C"), using different parameters (*see* **Fig. 8**). For this purpose, the PDB file should be edited either manually (change every "C" of the atom name of an aromatic carbon into an "A") or automatically, using the -A option of the AutoTors utility, which uses the out-of-plane angle to look for aromatic substructures.

   The "mkgpf3" utility makes use of the "gpf3gen.awk" script, which in the version of AutoDock used by us contains a line of code setting "A" equal to "C." As a result, there is no difference between aromatic and non-aromatic carbons. To remedy this behavior, either manually edit the "solvation parameters" of "A" in the GPF file (change "12.77 0.6844" to "10.80 0.1027") or comment out (by putting "#" in front of the lines) the following lines (545–547) in "gpf3gen.awk" (in the "share" directory of the AutoDock distribution):

   if (atype_lig == "A") { # some lines for "A" are not here yet atype_lig = "C" }.
8. In version 3.0.5 of AutoDock, the "get-docked" script refers to "dockedtopdb," where it should use "dockedtopdbq," as the former does not exist.

Web Pages

Docking Programs

AutoDock:
http://www.scripps.edu/pub/olson-web/doc/autodock/documentation.html
FlexX:
http://www.biosolveit.de/software/flexx/tutorial/tutor/
GOLD:
http://www.ccdc.cam.ac.uk/support/prods_doc/gold/GOLDdocnIX.html

Scoring Functions

X-SCORE:
http://sw16.im.med.umich.edu/software/xtool/manual/intro.html

SCORE:
http://www.es.embnet.org/Doc/score/intro.html
CSORE:
http://www.tripos.com/sciTech/inSilicoDisc/virtualScreening/cscore.html

Error-Checking of Protein Structures

ERRAT
http://www.doe-mbi.ucla.edu/People/Yeates/Gallery/Errat.html
PROCHECK
http://www.biochem.ucl.ac.uk/~roman/procheck/procheck.html
WHAT_CHECK
http://www.cmbi.kun.nl/gv/whatcheck

Molecular Format Conversion Programs

Babel
http://smog.com/chem/babel
OpenBabel
http://openbabel.sourceforge.net/babel.shtml
Mol2Mol
http://web.interware.hu/frenzy/mol2mol/index.html

Miscellaneous

Putative Active Site with Spheres (PASS)
http://www.ccl.net/cca/software/UNIX/pass/overview.shtml
DOWSER
http://mtzweb.scs.uiuc.edu/programs/documentation/dowser/Dowser.htm
AGENT2.0
http://www.pharma.ethz.ch/pc/Agent2/
GRID
www.moldiscovery.com/docs/grid21

References

1. Hou, T. J. and Xu, X. J. (2004) Recent development and application of virtual screening in drug discovery: an overview. *Curr. Pharm. Des.* **10,** 1011–1033.
2. Taylor, R. D., Jewsbury, P. J., and Essex, J. W. (2002) A review of protein-small molecule docking methods. *J. Comput. Aided Mol. Des.* **16,** 151–166.
3. Bohm, H. J. and Stahl, M. (2002) The use of scoring functions in drug discovery applications. *Rev. Comp. Chem.* **18,** 41–87.
4. Bissantz, C., Folkers, G., and Rognan, D. (2000) Protein-based virtual screening of chemical databases. 1. Evaluation of different docking/scoring combinations. *J. Med. Chem.* **43,** 4759–4767.

5. Paul, N. and Rognan, D. (2002) ConsDock: a new program for the consensus analysis of protein-ligand interactions. *Proteins: Struct. Funct. Gen.* **47,** 521–533.
6. Carlson, H. A. and McCammon, J. A. (2000) Accommodating protein flexibility in computational drug design. *Mol. Pharmacol.* **57,** 213–218.
7. McConkey, B. J., Sobolev, V., and Edelman, M. (2002) The performance of current methods in ligand-protein docking. *Curr. Sci.* **83,** 845–856.
8. Bernstein, F. C., Koetzle, T. F., Williams, G. J. B., et al. (1997) The protein data bank: a computer-based archival for macromolecular structures. *J. Mol. Biol.* **112,** 535–542.
9. Laskowski, R. A., MacArthur, M. W., Moss, D. S., and Thornton, J. M. (1993) PROCHECK: a program to check the stereochemical quality of protein structures. *J. Appl. Crystallogr.* **26,** 283–291.
10. Hooft, R. W. W., Vriend, G., Sander, C., and Abola, E. E. (1996) Errors in protein structures. *Nature* **381,** 272–272.
11. Colovos, C. and Yeates, T. O. (1993) Verification of protein structures: patterns of nonbonded atomic interactions. *Protein Sci.* **2,** 1511–1519.
12. Engh, R. A. and Huber, R. (1991) Accurate bond and angle parameters for X-ray protein structure refinement. *Acta Cryst.* **A47,** 392–400.
13. Hooft, R. W. W., Sander, C., and Vriend, G. (1996) Positioning hydrogen atoms by optimizing hydrogen-bond networks in protein structures. *Proteins* **26,** 363–376.
14. Glick, M. and Goldblum, A. (2000) A novel energy-based stochastic method for positioning polar protons in protein structures from X-rays. *Proteins: Struct. Func. Gen.* **38,** 273–287.
15. Nielsen, J. E. and Vriend, G. (2001) Optimizing the hydrogen-bond network in Poisson-Boltzmann equation-based pKa calculations. *Proteins* **43,** 403–412.
16. Word, J. M., Lovell, S. C., Richardson, J. S., and Richardson, D. C. (1999) Asparagine and glutamine: using hydrogen atom contacts in the choice of side-chain amide orientation. *J. Mol. Biol.* **285,** 1735–1747.
17. Nielsen, J. E., Andersen, K. V., Honig, B., et al. (1999) Improving macromolecular electrostatics calculations. *Prot. Engin.* **12,** 657–662.
18. Kramer, B., Rarey, M., and Lengauer, T. (1999) Evaluation of the FLEXX incremental construction algorithm for protein-ligand docking. *Proteins* **37,** 228–241.
19. Morris, G. M., Goodsell, D. S., Halliday, R. S., et al. (1998) Automated docking using a Lamarckian genetic algorithm and an empirical binding free energy function. *J. Comp. Chem.* **19,** 1639–1662.
20. Bohm, H. J. (1994) The development of a simple empirical scoring function to estimate the binding constant for a protein-ligand complex of known three-dimensional structure. *J. Comput. Aided Mol. Des.* **8,** 243–256.
21. Wang, R. X., Liu, L., Lai, L. H., and Tang, Y. Q. (1998) SCORE: a new empirical method for estimating the binding affinity of a protein-ligand complex. *J. Mol. Mod.* **4,** 379–394.
22. Wang, R. X., Lai, L. H., and Wang, S. M. (2002) Further development and validation of empirical scoring functions for structure-based binding affinity prediction. *J. Comput. Aided Mol. Des.* **16,** 11–26.

23. Poornima, C. S. and Dean, P. M. (1995) Hydration in drug design. 1. Multiple hydrogen-bonding features of water molecules in mediating protein-ligand interactions. *J. Comput. Aided Mol. Des.* **9,** 500–512.
24. Poornima, C. S. and Dean, P. M. (1995) Hydration in drug design. 2. Influence of local site surface shape on water binding. *J. Comput. Aided Mol. Des.* **9,** 513–520.
25. Poornima, C. S. and Dean, P. M. (1995) Hydration in drug design. 3. Conserved water molecules at the ligand-binding sites of homologous proteins. *J. Comput. Aided Mol. Des.* **9,** 521–531.
26. Babine, R. E. and Bender, S. L. (1997) Molecular recognition of protein-ligand complexes: applications to drug design. *Chem. Rev.* **97,** 1359–1472.
27. Brady, G. P. and Stouten, P. F. W. (2000) Fast prediction and visualization of protein binding pockets with PASS. *J. Comput. Aided Mol. Des.* **14,** 383–401.
28. Zhang, L. and Hermans, J. (1996) Hydrophilicity of cavities in proteins. *Proteins: Struct. Funct. Gen.* **24,** 433–438.
29. Goodford, P. J. (1985) A computational procedure for determining energetically favorable binding sites on biologically important macromolecules. *J. Med. Chem.* **28,** 849–857.
30. Friesner, R. A., Banks, J. L., Murphy, R. B., et al. (2004) Glide: a new approach for rapid, accurate docking and scoring. 1. Method and assessment of docking accuracy. *J. Med. Chem.* **47,** 1739–1749.
31. Schnecke, V. and Kuhn, L. A. (2000) Virtual screening with solvation and ligand-induced complementarity. *Perspect. Drug Discovery Des.* **20,** 171–190.
32. Osterberg, F., Morris, G. M., Sanner, M. F., Olson, A. J., and Goodsell, D. S. (2002) Automated docking to multiple target structures: incorporation of protein mobility and structural water heterogeneity in AutoDock. *Proteins* **46,** 34–40.
33. Rarey, M., Kramer, B., and Lengauer, T. (1999) The particle concept: placing discrete water molecules during protein-ligand docking predictions. *Proteins* **34,** 17–28.
34. Minke, W. E., Diller, D. J., Hol, W. G. J., and Verlinde, C. (1999) The role of waters in docking strategies with incremental flexibility for carbohydrate derivatives: heat-labile enterotoxin, a multivalent test case. *J. Med. Chem.* **42,** 1778–1788.
35. de Graaf, C., Pospisil, P., Pos, W., Folkers, G., Vermeulen, N. P. E. (2005) Binding mode prediction of cytochrome P450 and thymidine kinase protein-ligand complexes by consideration of water and rescoring in automated docking. *J. Med. Chem.* **48,** 2308–2318.
36. Pospisil, P., Ballmer, P., Scapozza, L., and Folkers, G. (2003) Tautomerism in computer-aided drug design. *J. Rec. Signal Transduction* **23,** 361–371.
37. Rarey, M., Kramer, B., Lengauer, T., and Klebe, G. (1996) A Fast Flexible Docking Method using an Incremental Construction Algorithm. *J. Mol. Biol.* **261,** 470–489.
38. Jones, G., Willett, P., Glen, R. C., Leach, A. R., and Taylor, R. (1997) Development and validation of a genetic algorithm for flexible docking. *J. Mol. Biol.* **267,** 727–748.
39. Goodsell, D. S. and Olson, A. J. (1990) Automated docking of substrates to proteins using simulated annealing. *Proteins: Struct. Func. Gen.* **8,** 195–202.

40. Klebe, G. (1994) The use of composite crystal-field environments in molecular recognition and the de novo design of protein ligands. *J. Mol. Biol.* **237,** 221–235.
41. Rarey, M., Kramer, B., and Lengauer, T. (1999) Docking of hydrophobic ligands with interaction-based matching algorithms. *Bioinformatics* **15,** 243–250.
42. Birch, L., Murray, C. W., Hartshorn, M. J., Tickle, I. J., and Verdonk, M. L. (2002) Sensitivity of molecular docking to induced fit effects in influenza virus neuraminidase. *J. Comput. Aided Mol. Des.* **16,** 855–869.
43. Halperin, I., Ma, B., Wolfson, H., and Nussinov, R. (2002) Principles of docking: An overview of search algorithms and a guide to scoring functions. *Proteins* **47,** 409–443.
44. Claussen, H., Buning, C., Rarey, M., and Lengauer, T. (2001) FlexE: efficient molecular docking considering protein structure variations. *J. Mol. Biol.* **308,** 377–395.
45. Schaffer, L. and Verkhivker, G. M. (1998) Predicting structural effects in HIV-1 protease mutant complexes with flexible ligand docking and protein side-chain optimization. *Proteins: Struct. Funct. Gen.* **33,** 295–310.
46. Vigers, G. P. A. and Rizzi, J. P. (2004) Multiple active site corrections for docking and virtual screening. *J. Med. Chem.* **47,** 80–89.
47. Taylor, R. D., Jewsbury, P. J., and Essex, J. W. (2003) FDS: flexible ligand and receptor docking with a continuum solvent model and soft-core energy function. *J. Comput. Chem.* **24,** 1637–1656.
48. Lin, J. H., Perryman, A. L., Schames, J. R., and McCammon, J. A. (2003) The relaxed complex method: accommodating receptor flexibility for drug design with an improved scoring scheme. *Biopolymers* **68,** 47–62.
49. Kairys, V. and Gilson, M. K. (2002) Enhanced docking with the mining minima optimizer: acceleration and side-chain flexibility. *J. Comp. Chem.* **23,** 1656–1670.
50. Rarey, M., Kramer, B., Lengauer, T., and Klebe, G. (1996) A fast flexible docking method using an incremental construction algorithm. *J. Mol. Biol.* **261,** 470–489.
51. Muegge, I. and Martin, Y. C. (1999) A general and fast-scoring function for protein-ligand interactions: a simplified potential approach. *J. Med. Chem.* **42,** 791–804.
52. Ewing, T. J. A. and Kuntz, I. D. (1997) Critical evaluation of search algorithms for automated molecular docking and database screening. *J. Comput. Chem.* **18,** 1175–1189.
53. Eldridge, M. D., Murray, C. W., Auton, T. R., Paoloinine, G. V., and Mee, R. P. (1997) Empirical scoring functions: I. The development of a fast, fully empirical scoring function to estimate the binding affinity of ligands in receptor complexes. *J. Comput. Aided Mol. Design* **11,** 425–445.
54. Gohlke, H., Hendlich, M., and Klebe, G. (2000) Knowledge-based scoring function to predict protein-ligand interactions. *J. Mol. Biol.* **295,** 337–356.

6

Synthesis of Complex Carbohydrates and Glyconjugates

*Enzymatic Synthesis of Globotetraose Using β-1,3-*N*-Acetylgalactosaminyltransferase LgtD From* Haemophilus infuenzae *Strain Rd*

Kang Ryu, Steven Lin, Jun Shao, Jing Song, Min Chen, Wei Wang, Hanfen Li, Wen Yi, and Peng George Wang

Summary

The lipopolysaccharide of capsule-deficient *Haemophilus infuenzae* strain Rd contains an *N*-acetylgalactosamine residue attached to the terminal globotriose moiety in the Hex5 glycoform. Genome analysis identified an open reading frame, HI1578, referred to as LgtD, whose amino acid sequence shows a significant level of similarity to those of a number of bacterial glycosyltransferases involved in lipopolysaccharide biosynthesis. To investigate its function, overexpression and biochemical characterization were performed. Most of the protein was obtained in a highly soluble and active form. Standard glycosyltransferase assay, high-performance liquid chromatography (HPLC), and liquid chromatography (LC)/mass spectrometry (MS) show that LgtD is an *N*-acetylgalactosaminyltransferase with high donor substrate specificity, and globotriose is a highly preferred acceptor substrate for the enzyme.

Key Words: *Haemophilus infuenzae;* β-1,3-*N*-acetylgalactosaminyltransferase; enzymatic synthesis; globotetraose; LgtD; lipopolysaccharide; UDP-GalNAc; complex carbohydrates; glyconjugates.

1. Introduction

Haemophilus influenzae is a Gram-negative human pathogen that routinely colonizes the upper respiratory tract. Serotype b capsular strains are associated with invasive diseases such as meningitis, septicemia, epiglottises, pneumonia, and emphysema, particularly in infants ***(1,2)***. The presence of *N*-acetylglucosamine as a minor component of lipopolysaccharide (LPS) has been reported in *H. influenzae* type b strain A2 ***(3)*** and in the related species *Haemophilus ducreyi*

From: *Methods in Molecular Biology, vol. 310: Chemical Genomics: Reviews and Protocols*
Edited by: E. D. Zanders © Humana Press Inc., Totowa, NJ

(4). In a capsule-deficient *H. influenzae* strain Rd, structural analysis of LPS epitopes *(5,6)* has shown the *N*-acetylglucosamine to be attached to a terminal globotriose moiety in the Hex5 glycoform, giving the globotetraose unit, GalNAcβ1-3Galβ1-4Galβ1-4Glc, corresponding to the P antigen expressed in mammalian globo series glycolipid. Recently, a fingerprinting strategy was employed to establish the structure of LPS from strains with mutated putative glycosyltransferase genes *(5)*, one of which, the HI1578 gene, *lgtD*, appears to code for an *N*-acetylgalactosaminyltransferase involved in LPS extension.

To determine its enzymatic function and explore whether the *lgtD* gene product could be useful in the in vitro oligosaccharide synthesis, the *lgtD* gene from *H. influenzae* strain Rd was cloned and overexpressed in *Escherichia coli (7)*. Biochemical characterization of the recombinant protein confirms that the gene encodes a β-1,3-*N*-acetylgalactosaminyltransferase (GalNAcT) that is responsible for the addition of *N*-acetylgalactosamine to the terminal globotriose moiety in the LPS of *H. influenzae* strain Rd. Moreover, determination of its biochemical properties shows the potential application of this bacterial glycotransferase in large-scale synthesis of globo series oligosaccharides. Globoside is the most prominent neutral glycosphingolipid in human erythrocytes and is an essential structure of blood group P antigen. It has been suggested that globoside is an adhesion molecule on epithelial cells to various bacteria such as uropathogenic *E. coli (8)*, and a receptor for pig edema disease toxin *(9)*. Hence, the oligosaccharide structure of globoside, globotetraose, is a potential therapeutic agent that is difficult to make using traditional chemical methods. Compared with chemical synthesis, an enzymatic approach utilizing glycotransferases has proven to be more efficient in the production of complex carbohydrates. Recently, human β3GalNAcT, a key enzyme responsible for the synthesis of globoside, has been identified and expressed in mouse fibroblast L-cells *(10)*. However, eukaryotic host expression involves expensive tissue culture media and low yields of protein. Expression of mammalian glycosyltransferase in *E. coli*, on the other hand, normally results in the formation of inactive inclusion bodies. Because bacteria have been shown to synthesize oligosaccharide structures identical to those in mammals, bacterial glycosyltransferases would be better candidates for recombinant protein expression in *E. coli (11)*. This chapter gives an overview of the strategy for expression and purification of glycosyltransferase and synthesis of glyconjugates in vitro.

2. Materials

2.1. Cloning and Expression of β-1,3-N-Acetylgalactosaminyltransferase From H. infuenzae

1. Plasmid vector pET 15b (Novagen).
2. Host cell DH5α, BL21(DE3) (Novagen).

3. Oligonucleotide primers.
4. T4 ligase, restriction enzyme, DNA polymerase.
5. *Haemophilus infuenzae* type D strain RM118 (KW-20) chromosomal DNA (ATCC 5197D).
6. Polymerase chain reaction (PCR) purification kit, gel extraction kit, DNA mini-prep kit.
7. Ampicillin, isopropyl-β-D-thiogalactoside (IPTG).
8. Luria-Bertani (LB) media.
9. Cell washing buffer: 50 m*M* Tris-HCl buffer, pH 7.0.
10. Cell lysis buffer: 50 m*M* Tris-HCl, pH 7.0, 0.1 *M* NaCl, 0.5% Triton X-100 (w/v), 10% glycerol (w/v), and 10 m*M* 2-mercaptoethanol.
11. Sodium dodecyl sulfate (SDS)-polyacrylamide gel electrophoresis (PAGE) analysis and agarose gel electrophoresis equipment.
12. Sonicator.
13. PCR system.

2.2. Purification of β-1,3-N-Acetylgalactosaminyltransferase

2.2.1. Purification of GalNAcT Using Ni-NTA Column

1. Cleared lysate described under **Subheading 3.1.2.**
2. Ni-NTA superflow resin column, size exclusion column.
3. Lysis buffer (50 m*M* NaH_2PO_4, 300 m*M* NaCl, 10 m*M* imidazole; adjust pH to 8.0 using NaOH).
4. Wash buffer (50 m*M* NaH_2PO_4, 300 m*M* NaCl, 20 m*M* imidazole; adjust pH to 8.0 using NaOH).
5. Elution buffer (50 m*M* NaH_2PO_4, 300 m*M* NaCl, 250 m*M* imidazole; adjust pH to 8.0 using NaOH).
6. Fast protein liquid chromatography (FPLC) equipment.

2.2.2. Purification of GalNAcT Using Size Exclusion Column

1. Prepurified GalNAcT described under **Subheading 3.2.1.**
2. Hi-load 16/60 Superdex 200 column.
3. Equilibration and elution buffer: 50 m*M* Tris-HCl buffer, pH 7.0, containing 20% glycerol, 10 m*M* 2-mercaptoethanol, and 150 m*M* NaCl.
4. FPLC equipment.

2.3. Identification and Characterization of β-1,3-N-Acetylgalactosaminyltransferase Activity

1. Enzyme reaction buffer: 50 m*M* Tris-HCl, pH 7.5; 10 m*M* $MnCl_2$; 0.1% bovine serum albumin (BSA); 1 m*M* dithiothreitol (DTT); 0.3 m*M* UDP-d-[1-^{3}H]GalNAc as donor; 3 m*M* globotriose or lactose as acceptor; and various amounts of purified GalNAcT.
2. HPLC equipment.

3. NH_2 column.
4. Eluent mobile phase (H_2O:acetonitrile = 35%:65%).
5. Refractive index detector.
6. Liquid scintillation counter.
7. Dowex 1 × 8–200 chloride anion-exchange resin.
8. Reaction buffer for HPLC: 50 m*M* Tris-HCl, pH 7.5; 10 m*M* $MnCl_2$; 3 m*M* UDP-GalNAc as donor; 3 m*M* globotriose or lactose as acceptor; and purified GalNAcT.

2.4. Preparative Synthesis of Globotetraose and Structure Analysis

1. Si250F silica gel thin-layer chromatography (TLC) plate.
2. Fluorescent indicator.
3. Preparative synthesis reaction buffer (50 m*M* Tris-HCl, pH 7.5; 10 m*M* $MnCl_2$; 1 m*M* DTT; 10 m*M* UDP-GalNAc as donor; 12 m*M* globotriose as acceptor).
4. Bio-Gel P-2 gel filtration resin.
5. Mass spectrometer (LC/MS, electrospray ionization [ESI]).

3. Methods

The methods described below outline (1) cloning and expression of β-1,3-*N*-acetylgalactosaminyltransferase (GalNAcT), (2) purification of GalNAcT, (3) identification and characterization of GalNAcT, and (4) preparative synthesis of globotetraose and structure analysis

3.1. Cloning and Expression of β-1,3-N-Acetylgalactosaminyltransferase From H. infuenzae

3.1.1. pET Expression Vector

The pET system is the most powerful system yet developed for the cloning and expression of recombinant protein in *E. coli* ***(12–14)***. Target genes are cloned in pET plasmids under control of a strong bacteriophage T7 transcription signal; expression is induced by providing a source of T7 RNA polymerase in the host cell. T7 RNA polymerase is so selective and active that, when fully induced, almost all of the cell's resources are converted to target gene expression; the desired production can comprise more than 50% of total cell protein a few hours after induction. Although this system is extremely powerful, it is also possible to attenuate expression levels simply by lowering the concentration of inducer. Decreasing the expression level may enhance the soluble yield of some target proteins. Another important benefit of this system is its ability to maintain target genes transcriptionally silent in the uninduced state. Fusion tags can facilitate detection and purification of target protein, or may increase the possibility of biological activity by affecting solubility in the cytoplasm or export to the periplasm. The His-Tag sequence is very useful as a fusion partner for purification of proteins in general. It is especially useful for those proteins initially expressed as inclusion bodies, because affinity purification can be accomplished under totally denaturing conditions that solubilize the protein.

3.1.2. BL21 (DE3) Expression Host Cells

For protein production, a recombinant plasmid is transferred to an *E. coli* strain containing a chromosomal copy of the gene for T7 RNA polymerase. These hosts are lysogens of bacteriophage DE3, a lambda derivative that contains the immunity region of phage 21 and carries a DNA fragment containing the lacI gene, the lac UV5 promoter, and the gene for T7 RNA polymerase ***(12, 15)***. This fragment is inserted into the int gene, preventing DE3 from integrating into, or excising from, the chromosome without a helper phage. Once a DE3 lysogen is formed, the only promoter known to direct transcription of the T7 RNA polymerase gene is the lac UV5 promoter, which is inducible by IPTG.

***3.2. Cloning of GalNAcT From* H. infuenzae**

1. Prepare a reaction mixture for the appropriate number of samples to be amplified. Add the components in order while mixing gently. **Note 1** provides an example of a reaction mixture for the amplification of GalNAcT from *H. infuenzae*.
2. If the thermocycler is not equipped with a heated cover, overlay each reaction with DNase, RNase, and protease-free mineral oil.
3. Perform PCR using optimized cycling conditions (*see* **Note 2**).
4. Analyze the PCR amplification products on a 0.7% (w/v) agarose gel (*see* **Note 3**).
5. Purify the PCR amplification products using a PCR purification kit.
6. Digest the amplified PCR product and vector (pET-15b) with restriction enzymes (Nde I and XhoI).
7. Analyze the digested PCR product and vector on a 0.7% (w/v) agarose gel (*see* **Note 4**).
8. Purify the DNA using a gel extraction kit.
9. Assemble the components in a 1.5-mL microfuge tube (*see* **Note 5**). Add the ligase last.
10. Incubate at 4°C overnight. Also set up a control reaction.
11. Analyze the PCR amplification products on 0.7% (w/v) agarose gels (*see* **Note 6**).
12. Transform into subcloning host cell (DH5a) using heat shock or electroporation.
13. Plate cells on LB agar plates containing ampicillin (100 μg/mL) and incubate overnight at 37°C.
14. After transformation, analyze the transformant with colony PCR or plasmid miniprep procedures and check the restriction enzyme sites by digestion.
15. To confirm the correct inserts, sequence the purified plasmid.
16. Transform the correct recombinant plasmid into the expression host cell BL21 (DE3) using heat shock or electroporation. Analyze the transformants using the plasmid miniprep procedure and check the restriction enzyme sites by digestion.

3.3. Expression of Recombinant GalNAcT in E. coli BL21 (DE3)

1. Pick a single colony from a freshly streaked plate and inoculate 5 mL LB containing ampicillin into a test tube. Culture overnight at 37°C.
2. Inoculate 1 mL of culture into 50 mL LB containing ampicillin (100 μg/mL) in a 250-mL flask (*see* **Note 7**).
3. Incubate with shaking at 37°C until the OD_{600} reaches 0.8.
4. When the OD_{600} reaches 0.8, add IPTG from a 100 m*M* stock to a final concentration of 0.2 m*M* to induce protein expression.
5. Incubate at 30°C for 8 h.
6. Place the flasks on ice for 10 min and then harvest the cells by centrifugation at 5000*g* for 10 min at 4°C (save the supernatant for further analysis).
7. Wash the cells twice in one-half of their volume of cold cell washing buffer and resuspend in 0.25 times culture volume of chilled cell lysis buffer.
8. Disrupt the suspended cells using a suitable sonicator for the suspended volume. Sonicate on ice using a short burst (*see* **Note 8**).
9. Separate the supernatant by centrifugation at 12,000*g* for 30 min at 4°C (save the cell pellets for further analysis). Remove the cell debris in the supernatant using a syringe filter (0.2 μm). Perform the next purification step as soon as possible; the longer the sample remains at 4°C, the greater the risk of proteolysis.
10. Analyze each fractionated sample for expression level and solubility using SDS-PAGE (**Fig. 1** and **Table 1**).

3.4. Purification of β-1,3-N-Acetylgalactosaminyltransferase

3.4.1. Purification of GalNAcT Using Ni-NTA Column

1. Remove the top adaptor of the column and cap the bottom outlet (*see* **Note 9**).
2. Completely resuspend a 50% Ni-NTA Superflow slurry and pour the slurry into the column (*see* **Note 10**).
3. Allow the resin to settle (*see* **Note 11**).
4. Insert top adapter and adjust to top of bed (*see* **Note 12**).
5. Equilibrate column with five column bed volumes of lysis buffer (*see* **Note 13**).
6. Apply filtered lysate to the column described under **Subheading 3.1.2.** and wash with lysis buffer until the A_{280} is stable (*see* **Note 14**).
7. Wash with wash buffer until the A_{280} is stable (*see* **Note 15**).
8. Elute the protein with elution buffer (*see* **Note 16**) (**Fig. 1** and **Table 1**).

3.4.2. Purification of GalNAcT Using a Size-Exclusion Column

1. Assemble the column according to the manufacturer's instructions.
2. Equilibrate column with three column bed volumes of equilibration buffer until the A_{280} is stable.
3. Inject prepurified sample into column.
4. Elute the protein with elution buffer.
5. Determine the protein concentration using the Bradford method ***(16)***.

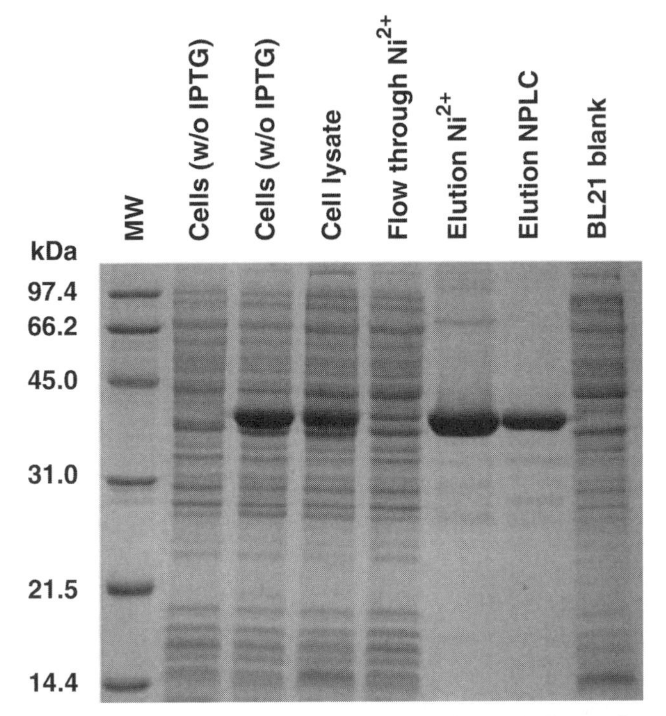

Fig. 1. Sodium dodecyl sulfate (SDS)-polyacrylamide gel electrophoresis (PAGE) analysis of expression and purification of recombinant protein. Ten-microliter aliquots were withdrawn at each step of the purification and loaded on a 12% SDS-PAGE gel in a Mini Protean III cell gel electrophoresis unit (Bio-Rad). The detection was performed with Coomassie blue staining. MW, low range (14–98 kDa) molecular weight marker (Bio-Rad).

Table 1
Purification of *β*-1,3-*N*-Acetylgalactosaminyltransferase From *Haemophilus influenzae* Strain Rd

Stages	Total protein (mg)	Recovery (%)	Total activity[a] (Units)	Specific activity (Units/mg)	Purification (fold)
Homogenate[b]	38.6	100	27	0.7	1
Cell lysate[b]	27.8	72	25	0.9	1.28
Nickel affinity	8.5	22	12.7	1.5	2.14
Gel filtration	5.8	15	10.4	1.8	2.57

[a]The reactions were performed using 10 µL cell or enzyme extract in a total of 100 µL under the standard assay conditions.

[b]Homogenates and cell lysates from *Escherichia coli* BL21 (DE3) strain harboring in the empty pET 15b plasmid were used as a control for these enzymatic activity assays.

Table 2
Preliminary Kinetic Parameters of GalNAcT from *Haemophilus influenzae* Strain

Substrate	K_m μM	K_{cat} s^{-1}	K_{cat}/K_m $(\mu M^{-1}s^{-1})$
UDP-GalNAc	58	7.8	1.34×10^{-3}
UDP-Gal	1.1×10^3	1.5	1.36×10^{-3}
Globotriose	8.6×10^3	18.6	2.16×10^{-3}
Lactose	3.3×10^3	1.7	5.15×10^{-5}

6. Analyze each fractionated sample for expression level and solubility using SDS-PAGE (**Fig. 1** and **Table 1**).

3.5. Identification and Characterization of β-1,3-N-Acetylgalactosaminyltransferase Activity

3.5.1. Enzymatic Assay

1. Make up the reaction mixture to a total volume of 0.1 mL in enzyme reaction buffer (*see* **Note 17**).
2. Incubate at 37°C for 20 min.
3. Add 0.1 mL of ice-cold 0.1 *M* ethylenediamine tetraacetic acid (EDTA) to terminate the reaction.
4. Add the Dowex-1 × 8–200 chloride anion exchange resin into working solution (0.8 mL, 1:1 v/v).
5. Separate the resin by centrifugation (10,000*g*, 5 min) and gently transfer 0.5 mL of the supernatant into a 20-mL plastic vial.
6. Add 5 mL of Scinti Verse BD into vial and vortex.
7. Count the radioactivity using a liquid scintillation counter.
8. Evaluate K_{cat} and K_m value in steady-state kinetics (**Table 2**).

3.5.2. HPLC Analysis

1. Make up the reaction mixture to a total volume of 0.3 mL in reaction buffer for HPLC.
2. Incubate at 37°C for 30 min.
3. Add 0.3 mL of acetonitrile.
4. Separate the precipitate by centrifugation (12,000*g*, 10 min) and filter using a syringe filter (0.2 μm).
5. Inject the column (Microsorb NH_2 100 Å, 4.6 × 250 mm, Varian, CA) with working solution to equilibrate.
6. Elute in the isocratic mode at a flow rate of 1.0 mL/min (acetonitrile:water = 65%: 35%).
7. Monitor eluted oligosaccharides using refractive index detector (at 30°C, integral mode).

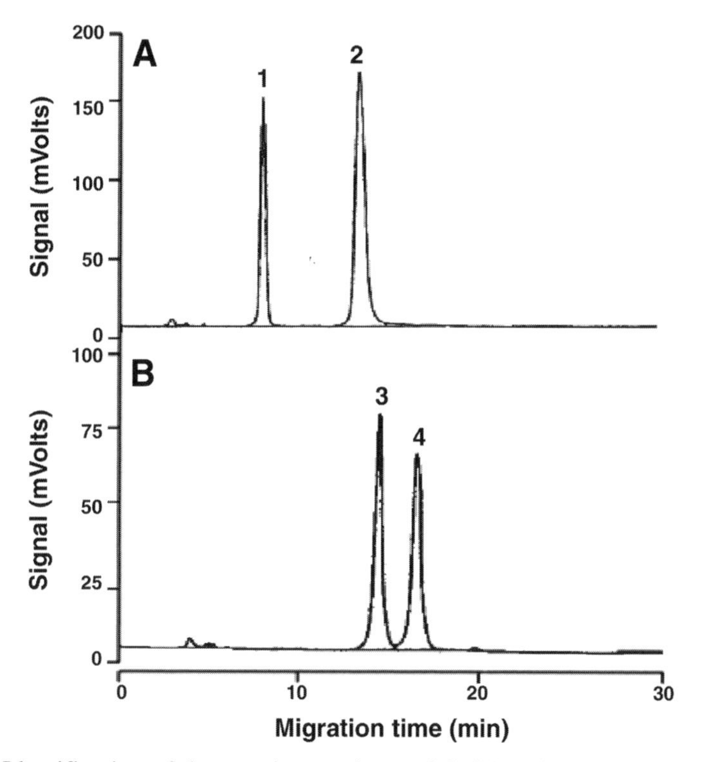

Fig. 2. Identification of the reaction products of GalNAc from *Haemophilus influenzae* Rd by high-performance liquid chromatography. The reactions were carried out as described in Methods. **(A)** Reaction with lactose as acceptor: 1, lactose; 2, GalNAcb 1,3Lac. **(B)** Reaction with globotriose as acceptor: 3, globotriose; 4, globotetraose.

8. Quantify the oligosaccharide by integrating the peak area. Obtain the chromo-phore peak calibration curve with pure oligosaccharide solution (**Fig. 2**).
9. Wash the column extensively with 65% acetonitrile in water.

3.6. Preparative Synthesis of Globotetraose and Structure Analysis

To obtain sufficient material for structural analysis, a preparative reaction was conducted with globotriose as acceptor. The reaction was optimized so that the substrate, UDP-GalNAc, was converted completely to the product. The high yield simplified the isolation of oligosaccharide product for further analysis.

3.6.1. Preparative Synthesis of Globotetraose

1. Make up the following reaction mixture to a total volume of 1.5 mL preparative synthesis buffer.
2. Add 15 mU of purified β-1,3-*N*-acetylgalactosaminyltransferase to initiate reaction.

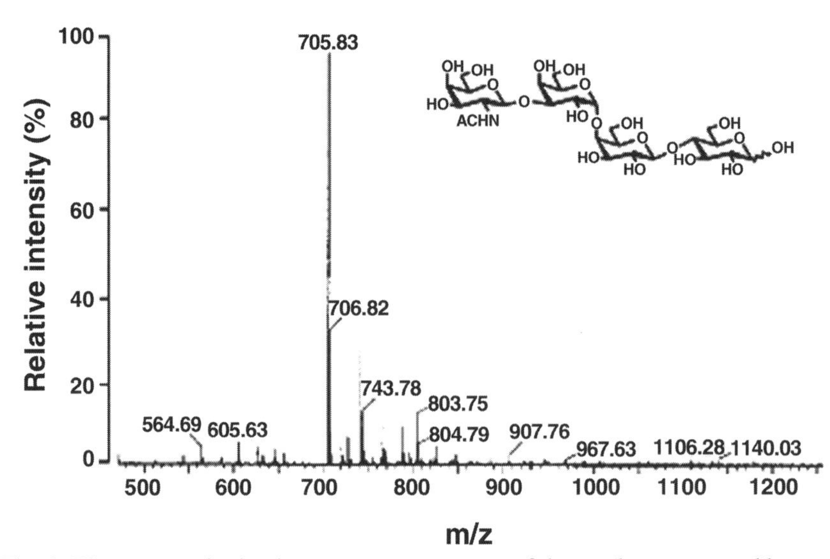

Fig. 3. Electrospray ionization mass spectrometry of the product generated by recombinant GalNAcT. Spectra were acquired in the negative mode. The molecular weight of globotetraose is 706.8. The ion peak at *m/z* 705.83 is attributed to the molecular ion $[M-H]^-$ of globotetraose.

3. Incubate at 37°C and monitor the progress of reaction by TLC (i-PrOH/H_2O/NH_4OH = 7:3:2 v/v/v) with a fluorescent indicator.
4. After complete conversion of donor substrate, boil the reaction solution at 95°C for 5 min.
5. Separate the precipitate by centrifugation (12,000*g*, 10 min) and filter using a syringe filter (0.2 μm).
6. Purify using Bio-Gel P-2 gel filtration (Bio-Rad, CA) with water as mobile phase.
7. Collect the desired fraction, lyophilize, and store at –20°C.

3.6.2. Structure Analysis of Synthesized Globotetraose LC/MS

MS has become an indispensable tool for the determination of carbohydrate structures. The information provided by this methodology ranges from accurate molecular weight determination to the complete primary structure, with picomole sensitivity. These remarkable advances have been possible because of the development of novel methods of ionization, such as fast atom bombardment (FAB) ionization, ESI, and matrix-assisted laser desorption/ionization (MALDI) (**Fig. 3**).

4. Notes

1. PCR reactions consist of 50 ng *H. influenzae* genomic DNA , 200 n*M* of each primer, 0.2 m*M* of each dNTP, 2.5 m*M* $MgCl_2$, 1 U *Pfu* polymerase, and a suitable buffer in a total of 50 μL.
2. After heating at 95°C for 5 min, 25 cycles were carried out, consisting of 30 s at 95°C, 1 min at 55°C, and 90 s at 72°C.
3. If the amplified PCR product is not the correct size, the PCR reaction must be optimized (components of PCR and reaction conditions). Deep *Vent* polymerase can be used, but both *Taq* and *Vent* polymerase frequently increase the error rate.
4. If DNA fragments are incompletely digested, add restriction enzyme (5 U) into digestion reaction solution or increase the incubation time.
5. Two microliters of 10X ligase buffer (200 m*M* Tris-HCl, pH 7.6; 100 m*M* $MgCl_2$; 250 μg/mL acetylated BSA), 2 μL of 100 m*M* DTT, 1 μL of 10 m*M* ATP, 2 μL of 50 ng/mL digested pET-15b vector and digested PCR product (0.3 pmol) in a total of 20 μL. The control reaction is single cut vector (NdeI or XhoI).
6. Generally, good ligation is shown when ligated DNA migrates at a higher position on agarose gels than the linear species.
7. For good aeration, add medium up to a maximum of 20% of the total flask volume.
8. Avoid extreme pressure and run times in this step (frequently, proteins are damaged by extreme conditions).
9. Assemble the column according to the manufacturer's instructions.
10. Avoid introducing air bubbles. Slowly pour slurry down a thin glass rod inserted into empty column. The column and bed sizes depend on the amount of His-tagged protein to be purified. Generally, the binding capacity of Ni-NTA superflow is 5–10 mg protein per mL resin; Ni-NTA superflow is supplied as 50% slurry.
11. The packing procedure can be accelerated by allowing the buffer to flow through by uncapping the bottom outlet. If desired, a peristaltic pump may be used, but do not exceed a flow rate of 2 mL/min. Do not allow resin to dry. If this should occur, resuspend resin in lysis buffer and repack the column. Before the bed volume has settled, more slurry may be added to increase bed volume.
12. Do not trap any air bubbles. The column can now be connected to the system.
13. The flow rate should not exceed 2 mL/min. Monitor elution at 280 nm; the baseline should be stable after washing with five column bed volumes.
14. Usually, 5–10 column volumes are sufficient. Monitor pressure at this step. If the lysate is very viscous, the pressure may exceed the recommended value. Reduce the flow rate accordingly. Start with a flow rate of 0.5–1 mL/min; if the His-tagged protein does not bind, the flow rate should be reduced. The flow rate may, however, be increased for protein elution. Collect fractions for SDS-PAGE analysis.
15. Usually, 5–10 column volumes are sufficient. Collect fractions for SDS-PAGE analysis.
16. If desired, a step gradient of elution buffer in wash buffer may be used to elute the protein. Five column volumes at each step are usually sufficient. The His-Tagged protein usually elutes in the second and third column volume. Imidazole absorbs

at 280 nm, which should be considered when monitoring protein elution. If small amounts of His-tagged proteins are purified, elution peaks may be poorly visible.

17. The globotriose of Acceptor was prepared using a whole-cell reaction system as previously described ***(17)***.

References

1. Campagnari, A. A., Gupta, M. R., Dudas, K. C., Murphy, T. F., and Apicella, M. A. (1987) Antigenic diversity of lipooligosaccharides of nontypable *Haemophilus influenzae*. *Infect. Immun.* **55,** 882–887.
2. Shan, F., Grattan, M., Germer, S., Linnsman, V., Hazlem, D., and Payne, R. (1984) Aetiology of pneumonia in children in Goroke hospital, Papua New Guinea. *Lancet* **ii,** 537–541.
3. Philips, N. J., Apicella, M. A., Griffiss, J. M., and Gibson, B. W. (1993) Structural studies of the lipopolysaccharides from *Haemophilus influenzae* type b strain A2. *Biochemistry* **32,** 2003–2012.
4. Schweda, E. K. H., Sundstrom, A. C., Eriksson, L. M., and Lindberg, A. A. (1994) Structural studies of cell envelope lipopolysaccharides from *Haemophilus ducreyi* strains ITM2665 and ITM4747. *J. Biol.Chem.* **16,** 12,040–12,048.
5. Hood, D. W., Cox, A. D., Wakarchuk, W. W., et al. (2001) Genetic basis for expression of the major globotetraose-containing lipopolysaccharide from *H. influenzae* strain Rd (RM118). *Glycobiology* **11,** 957–967.
6. Risberg, A., Masoud, H., Martin, A., Richards, J. C., Moxon, E. R., and Schweda, E. K. H. (1999) Structural analysis of the lipopolysaccharide oligosaccharide epitopes expressed by a capsule-deficient strain of *Haemophilus influenzae* Rd. *Eur. J. Biochem.* **261,** 171–180.
7. Shao, J., Zhang, J., Kowal, P., Lu, Y., and Wang, P. G. (2002) Overexpression and biochemical characterization of b-1,3-N-acetylgalactosaminyltransferase LgtD from *Haemophilus influenzae* strain Rd. *Biochem. Biophys. Res. Comm.* **295,** 1–8.
8. Lund, B., Lindberg, F. P., Baga, M., and Normark, S. (1985) Globoside specific adhesins of uropathogenic *Escherichia coli* are encoded by similar trans-complementable gene clusters. *J. Bacteriol.* **162,** 1293–1301.
9. Tyrrell, G. J., Ramotar, K., Toye, B., Boyd, B., Lingwood, C. A., and Brunton, J. L. (1992) Alteration of the carbohydrate binding specificity of verotoxins from Gal alpha 1-4Gal to GalNAcbeta 1-3Gal alpha 1-4Gal and vice versa by site-directed mutagenesis of the binding subunit. *Proc. Natl. Acad. Sci. USA* **89,** 524–528.
10. Okajima, T., Nakamura, Y., Uchikawa, M., et al. (2001) Expression cloning of human globoside synthase cDNAs. Identification of beta 3Gal-T3 as UDP-N-acetylgalactosamine: globotriaosylceramide beta 1,3-N-acetylgalactosaminyltransferase. *J. Biol. Chem.* **275,** 40,498–40,503.
11. Wakarchuk, W. W., Cunningham, A., Watson, D. C., and Young, N. M. (1998) Role of paired basic residues in the expression of active recombinant galactosyltransferases from the bacterial pathogen *Neisseria meningitidis*. *Protein Eng.* **11,** 295–302.

12. Studier, F. W. and Moffatt, B. A. (1986) Use of bacteriophage T7 polymerase to direct selective high-level expression of cloned genes. *J. Mol. Biol.* **189,** 113–130.
13. Rosenberg, A. H., Lade, B. N., Chui, D. S., Lin, S. W., Dunn, J. J., and Studier, F. W. (1987) Vectors for selective expression of cloned DNAs by T7 RNA polymerase. *Gene* **56,** 125–135.
14. Studier, F. W., Rosenberg, A. H., Dunn, J. J., and Dubendorff, J. W. (1990) Use of T7 RNA polymerase to direct expression of cloned genes. *Meth. Enzymol.* **185,** 60–89.
15. Grodberg, J. and Dunn, J. J. (1988) ompT encodes the *Escherichia coli* outer membrane protease that cleaves T7 RNA polymerase during purification. *J. Bacteriol.* **170,** 1245–1253.
16. Bradford, M. M. (1976) A rapid and sensitive method for the quantitation of microgram quantities of protein utilizing the principle of protein-dye binding. *Anal. Biochem.* **72,** 248–254.
17. Chen, X., Zhang, J.-B., Kowal, P., Andreana, P. R., and Wang, P. G. (2001) Transferring a biosynthetic cycle into a productive *E. coli* strain: large-scale synthesis of galactosides. *J. Am. Chem. Soc.* **123,** 8866–8867.

7

High-Throughput Cloning for Proteomics Research

Sharon A. Doyle

Summary

Ligation-independent cloning (LIC) is a simple, rapid, and efficient method for high-throughput cloning. In this system, linear plasmid vector and insert DNA are treated to generate complementary single-stranded overhangs that anneal during a short incubation.

The LIC system is adaptable for use with any vector following an alteration of the vector sequence. This chapter describes the creation of an LIC-compatible vector, with tips on how to make any vector LIC-enabled. It also includes a protocol for generating high-quality linearized vector template for the LIC reaction. Lastly, a step-by-step protocol of the LIC reaction is outlined, with useful tips and tricks for optimization and screening.

Key Words: Ligation-independent cloning; T7 expression system; pET vector; hexa-histidine tag.

1. Introduction

Many proteomics initiatives require the production of large collections of expression clones. Although traditional methods of cloning, such as restriction enzyme-based cloning, have been well established on a small scale, they are not adaptable for a high-throughput environment. Methods that facilitate cloning in a high-throughput manner are vital to the success of these initiatives.

This chapter describes ligation-independent cloning (LIC) ***(1,2)***, an ideal cloning strategy for high-throughput proteomics. In this system, linear plasmid vector and insert DNA are treated to generate complementary single-stranded overhangs that anneal during a short incubation. In addition to the ease of LIC due to the speed and efficiency of the annealing reaction, other attributes make it an optimal system for cloning large numbers of cDNAs. Because restriction enzymes are not used, insert sequences do not need to be screened for internal sites or modified prior to cloning. Any vector can be made LIC-compatible,

From: *Methods in Molecular Biology, vol. 310: Chemical Genomics: Reviews and Protocols*
Edited by: E. D. Zanders © Humana Press Inc., Totowa, NJ

allowing for the production of proteins with different affinity tags, often using the same insert PCR product. The additional amino acids added to the encoded protein by the LIC sequences are minimal and can be modified if necessary, which is often important when N- or C-terminal extensions affect protein function or stability.

2. Materials

1. pET30a (Novagen, Madison, WI) or other expression vector system.
2. *Eschericia coli* (*E. coli*) strain NovaBlue competent cells (Novagen, Madison, WI).
3. Oligonucleotide primers.
4. BseR I restriction enzyme.
5. dNTPs, HotstarTaq polymerase (Qiagen, Valencia, CA), Platinum Pfx polymerase (Invitrogen, Carlsbad, CA).
6. Mung bean nuclease, T4DNA polymerase (Novagen, Madison, WI).
7. Agarose gel electrophoresis equipment.
8. Luria-Bertani (LB) media and LB agar plates, SOC media.
9. Kanamycin.
10. Sodium dodecyl sulfate (SDS).
11. Dithiothreitol (DTT).
12. Polymerase chain reaction (PCR) cleanup and agarose gel extraction systems (Qiagen, Valencia, CA).

3. Methods

First, a method to generate an LIC vector and prepare it for cloning will be described to illustrate how any vector system used for protein expression can be modified for LIC. Next, the generation of suitable insert sequences will be described, followed by the LIC reaction, and confirmation of positive clones.

3.1. Expression Plasmid

3.1.1. Construction of a Ligation-Independent Cloning-Compatible Plasmid

LIC can be used with any plasmid vector following an alteration of the sequence using standard molecular-biology techniques. The LIC region inserted into the plasmid must contain a unique restriction enzyme cleavage site, used to linearize the plasmid DNA (*see* **Note 1**), flanked by approx 12–15 base pairs (bp) on each side that act as the annealing sites for the LIC reaction. These flanking regions are used to generate long, single-stranded overhangs used in the LIC reaction by treatment with a DNA polymerase with 3' to 5' exonuclease activity in the presence of one of the dNTPs; the polymerase chews back the 3' ends of the DNA until it encounters the dNTP present in the reaction. Therefore, the

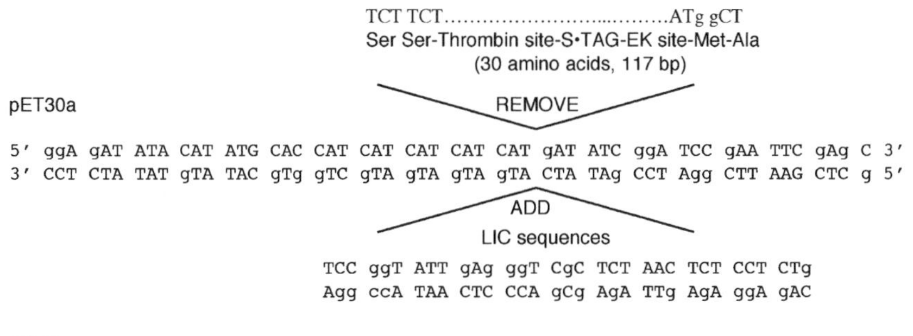

pNHis

Met His His His His His His Ser Gly Ile Glu Gly Arg

ATG CAC CAT CAT CAT CAT CAT TCC **ggT ATT gAg ggT CgC..cDNA_TAA..GGC** TCT AAC TCT CCT CTg

TAC gTg gTA gTA gTA gTA gTA Agg CCA TAA CTC CCA **gCg..cDNA_ATT..CCg AgA TTg AgA ggA gAC**

Fig 1. Construction of the ligand-independent cloning (LIC) vector containing the N-terminal hexahistidine affinity tag. The new vector, pNHis, encodes a protein with an N-terminal extension of six histidine residues followed by six additional amino acids that encode a Factor Xa cleavage site. A stop codon was added to the gene sequence so that the 3' LIC sequence did not add six extra amino acids to the C-terminus of the protein sequence.

sequence composition of the region flanking the restriction cleavage site must be devoid of one dNTP to allow the enzyme to chew back a minimum of 12 bp *(3)*. Several LIC sequences have been used that provide different affinity tags and protease cleavage sites *(4)*. However, the efficiency of the LIC reaction should be carefully assessed when new sequences are designed, because lower cloning efficiencies are problematic in a high-throughput environment.

Plasmids in the pET family of expression vectors, used widely for protein production, utilize the T7 promoter in conjunction with T7 RNA polymerase to drive high-level expression of heterologous proteins in *E. coli* *(5–7)*. To generate an LIC-compatible vector that incorporates an N-terminal hexahistidine tag into the protein product and minimizes extraneous amino acids, pET30a was modified to produce pNHis (**Fig. 1**). The sequence following the hexahistidine tag sequence was removed and replaced with 5' TCCGGTATTGAGGGTCGC TCTAACTCTCCTCTG 3' using standard molecular-biology techniques. This sequence contains a BseR I recognition sequence and cleavage site, which is absent from the pET30a sequence, and the flanking sequences that are used to create the overhanging ends for the annealing reaction. This sequence encodes the Factor Xa site (IEGR) that may be used to cleave the hexahistidine tag and LIC sequence encoding amino acids from the protein product. A diagram showing the details of the LIC reaction (outlined below) using the pNHis vector is

pNHIS Vector:

5′ ATGCACCATCATCATCATCAT**TCCGGTATTGAGGGTCGCTCTAACTCTCCTCTG**GATATCGGATCCGAATTCGAGC 3′
3′ TACGTGGTAGTAGTAGTAGTA**AGGCCATAACTCCCAGCGAGATTGAGAGGAGAC**CTATAGCCTAGGCTTAAGCTCG 5′

Step 1. Linearize (BseR I)

5′ ATGCACCATCATCATCATCAT**TCCGGTATTGAGGGTCG** **CTCTAACTCTCCTCTG**GATATCGGATCCGAATTCGAGC 3′
3′ TACGTGGTAGTAGTAGTAGTA**AGGCCATAACTCCCA** **GCGAGATTGAGAGGAGAC**CTATAGCCTAGGCTTAAGCTCG 5′

Step 2. Create blunt ends

5′ ATGCACCATCATCATCATCAT**TCCGGTATTGAGGGT** **CTCTAACTCTCCTCTG**GATATCGGATCCGAATTCGAGC 3′
3′ TACGTGGTAGTAGTAGTAGTA**AGGCCATAACTCCCA** **GAGATTGAGAGGAGAC**CTATAGCCTAGGCTTAAGCTCG 5′

Step 3. LIC exonuclease treatment (dCTP)

5′ ATGCACCATCATCATCATCAT**TCC** **CTCTAACTCTCCTCTG**GATATCGGATCCGAATTCGAGC 3′
3′ TACGTGGTAGTAGTAGTAGTA**AGGCCATAACTCCCA** **C**CTATAGCCTAGGCTTAAGCTCG 5′

cDNA Insert:

Step 1. PCR amplify with LIC primers

5′ *GGTATTGAGGGTCGC* *cDNA TAA* *GGCTCTAACTCTCCTCT* 3′
3′ *CCATAACTCCCAGCG* cDNA *ATT* *CCGAGATTGAGAGGAGA* 5′

Step 2. LIC exonuclease treatment (cGTP)

5′ *GGTATTGAGGGTCGC* *cDNA TAA* *GG* 3′
3′ *GCG* *cDNA ATT* *CCGAGATTGAGAGGAGA* 5′

Final Product:

5′ ATGCACCATCATCATCATCAT**TCC***GGTATTGAGGGTCGC* *cDNA TAA* *GG***CTCTAACTCTCCTCTG**GATATC 3′
3′ TACGTGGTAGTAGTAGTAGTA**AGGCCATAACTCCCA***GCG* *cDNA ATT* *CCGAGATTGAGAGGAGA***C**CTATAG 5′

Fig 2. Ligation-independent cloning (LIC) using the pNHis vector. The bold type denotes the LIC sequences of the vector, which contains the BseR I recognition sequence (underlined). The complementary LIC sequences of the insert DNA are shown in italics.

shown (**Fig. 2**). Several hundred micrograms of plasmid DNA were generated following confirmation of the sequence.

3.1.2. Preparation of Linear Plasmid DNA for LIC

1. Digest vector DNA (10 μg) with BseR I for 6 h at 37°C in a 200 μL reaction volume (*see* **Note 2**).
2. Add 20 U mung bean nuclease and incubate at 30°C for 3 h.
3. Add SDS to a final concentration of 0.01% (w/v) to inactivate the nuclease.
4. Isolate the linear vector from uncut and supercoiled vector on a 2% agarose gel at 100 V for approx 4 h, or until well separated. Use undigested vector as a negative control (*see* **Note 3**).

5. Extract the linear vector DNA from the agarose gel using standard molecular-biology techniques (QIAQuick gel extraction kit [Qiagen] or similar). Resuspend the DNA in H_2O, and keep the DNA concentration as high as possible.
6. Determine the absorbance at 260 nm, calculate the concentration, and store in aliquots at –20°C.
7. Test the vector DNA to ensure that all undigested DNA has been removed by performing the LIC reaction (*see* **Subheading 3.3.**) with LIC-treated linear DNA and no insert DNA. Following transformation into NovaBlue cells, there should be very few transformants (fewer than 10 colonies is optimal).

3.2. Preparation of cDNA Inserts for LIC

1. Amplify genes of interest using a high-fidelity polymerase (such as Platinum Pfx polymerase) with primers that contain 5' extensions for LIC. For the pNHis vector, the forward primer uses the extension 5' GGTATTGAGGGTCGC, followed by the cDNA sequence starting with the second codon. The reverse primer uses the extension 5' AGAGGAGAGTTAGAGCCTTA. Note that a stop codon is added to this sequence so that the product does not contain additional amino acids encoded by the LIC sequence (*see* **Note 4**).
2. Analyze the PCR products by agarose gel electrophoresis. If products are highly pure, concentrate the sample using a 96-well PCR cleanup protocol such as QIAquick (Qiagen) (*see* **Note 5**). Alternatively, products may be isolated by agarose gel electrophoresis and extracted and concentrated using an extraction protocol such as MinElute (Qiagen). Highly concentrated PCR products produce optimal results (0.1 pmol/µL), so resuspension volumes should be chosen accordingly (*see* **Note 6**).

3.3. Ligation-Independent Cloning

1. Separate reactions are performed for plasmid vector and insert DNA as follows. In one tube, 0.1 pmol linear vector DNA (400 ng for pNHis), 2 µL 10X T4 DNA polymerase buffer, 2 µL 25 m*M* dCTP (*see* **Note 7**), 1 µL 100 m*M* DTT, and 0.4 µL LIC qualified T4 DNA polymerase (1.25 U) are added to a final volume of 20 µL. In another tube (or reaction plate) 0.2 pmol linear insert DNA (*see* **Note 8**), 2 µL 10X T4 DNA polymerase buffer, 2 µL 25 m*M* dGTP (*see* **Note 7**), 1 µL 100 m*M* DTT, and 0.4 µL LIC qualified T4 DNA polymerase (1.25 U) are added to a final volume of 20 µL. Incubate the reactions at room temperature (22°C) for 40 min.
2. To stop the separate LIC reactions, heat inactivate at 75°C for 20 min.
3. To anneal the vector and insert, combine 2 µL (0.01 pmol) of LIC vector and 2 µL (approx 0.02 pmol) of LIC insert in a tube and incubate at room temperature for 10 min. A negative control sample of 2 µL vector and 2 µL H_20 should be included.
4. To stop the annealing reactions, add 1.3 µL of 25 m*M* ethylenediamine tetraacetic acid (EDTA) and incubate at room temperature for 10 min. Reactions may be stored at –20°C prior to transformation if desired.

5. Transform 1.3 µL of the annealing reaction (diluted 1:5 with H_2O) into NovaBlue competent cells and plate the entire reaction onto LB agar plates containing 50 µg/mL kanamycin. Incubate overnight at 37°C.

3.4. Identification of Positive Clones

It is beneficial to identify clones that contain the correct insert prior to plasmid purification and sequencing. The following protocol for colony PCR utilizes primers that anneal to the vector sequence, which is advantageous because one optimized set of PCR conditions is used. This allows for easy preparation, and avoids problems related to different primer conditions. In addition, a negative control of vector DNA can serve as a built-in control for the reaction. Primers should be chosen that produce a product of approx 100 bp when vector DNA without insert is amplified. To amplify the pNHis vector, the T7 promoter (5' TAATACGACTCACTATAGGG 3') and T7 terminator (5' GCTAGTTATT GCTCAGCGG 3') primers were used.

1. Make a master mix of PCR reagent. For each sample to be amplified, add: 5 µL 10X HotstarTaq buffer, 5 µL dNTP mix (2 m*M* each), 2.5 µL 5 µ*M* T7 promoter primer, 2.5 µL 5 µ*M* T7 terminator primer, 35 µL sterile H_2O, and 0.25 µL (1.25 U) HoststarTaq polymerase.
2. Aliquot 50 µL of the master mix into each well of the PCR plate. Pick large colonies with sterile pipet tips and add to the appropriate wells.
3. Cycle under the following conditions: 95°C, 15 min; 94°C, 1 min; 55°C, 1 min; 72°C, 1 min (repeat **steps 2–4** 35 times); 72°C, 7 min.
4. Run 5 µL of each reaction on a 1% agarose gel to check for the presence of insert DNA of the proper size.

4. Notes

1. Any restriction enzyme can be used at this step, depending on the sequences that one is using for LIC. The desired product is blunt-ended DNA, so choose a blunt-cutting enzyme if possible. Another useful trick is to choose an enzyme with a recognition sequence that occurs a distance from the cleavage site; this way, the recognition sequence does not need to fit in with the requirements of the flanking sequences.
2. If you are using the restriction enzyme BseR I or another enzyme that produces overhangs, do not alter the DNA concentration in this step, because activity of the mung bean nuclease that is added directly to this reaction is concentration dependent.
3. It is very important to separate the undigested or nicked plasmid DNA from the linear DNA at this step, since failure to do so produces background colonies when the DNA is used for LIC. Although this step is time consuming, 10 µg of purified vector DNA is enough to perform 250 reactions.
4. When an N-terminal affinity tag is used, it is beneficial to add the stop codon to the insert sequence so that no additional amino acids are added to the C-terminal end of the protein. If the same PCR product is going to be cloned into many vectors (with affinity tags at N- and C-terminal ends), the insert sequence cannot con-

tain the stop codon. In this case, the LIC sequence can be altered to add a stop codon in a position that minimizes extra tag sequences.

5. The PCR reactions must be buffer exchanged to remove unincorporated dNTPs, even if the PCR products are very pure.
6. We have found that cloning efficiency is reduced when increased volumes of purified PCR product are used, possibly as a result of carryover of contaminants from the cleanup reaction. Therefore, it is best to have the PCR product as concentrated as possible. The exact amount of DNA in this step can vary somewhat to allow for standardized pipeting volumes without compromising cloning efficiency.
7. Our LIC scheme uses dCTP in the vector sample and dGTP in the insert sample to create the single-stranded overhangs. This can vary, depending on the LIC sequence that is being used.
8. The exact amount of DNA in this step can vary somewhat to allow for standardized pipeting volumes without compromising cloning efficiency.

Acknowledgments

The author thanks Michael Murphy, Peter Beernink, and Paul Richardson for helpful suggestions and critical reading of the manuscript. This work was performed under the auspices of the US Department of Energy, Office of Biological and Environmental Research, by the University of California, Lawrence Livermore National Laboratory (contract No. W-7405-Eng-48), Lawrence Berkeley National Laboratory (contract No. DE-AC03-76SF00098), and Los Alamos National Laboratory (contract No. W-7405-ENG-36).

References

1. Aslanidis, C. and deJong, P. J. (1990) Ligation-independent cloning of PCR products (LIC-PCR). *Nucleic Acids Res.* **18,** 6069–6074.
2. Haun, R. S., Servanti, I. M., and Moss, J. (1992) Rapid, reliable ligation-independent cloning of PCR products using modified plasmid vectors. *Biotechniques* **13,** 515–518.
3. Aslanidis, C., de Jong, P. J., and Schmitz, G. (1994) Minimal length requirement of the single-stranded tails for ligation-independent cloning (LIC) of PCR products. *PCR Methods Appl.* **4(3),** 172–177.
4. Stols, L., Gu, M., Dieckman, L., Raffen, R., Collart, F., and Donnelly, M. (2002) A new vector for high-throughput, ligation-independent cloning encoding a tobacco etch virus protease cleavage site. *Protein Expression and Purification* **25,** 8–15.
5. Studier, F. W. and Moffatt, B. A. (1986) Use of bacteriophage T7 RNA polymerase to direct selective high-level expression of cloned genes. *J. Mol. Biol.* **189,** 113–130.
6. Rosenberg, A. H., Lade, B. N., Chui, D., Lin, S., Dunn, J. J., and Studier, F. W. (1987) Vectors for selective expression of cloned DNAs by T7 RNA polymerase. *Gene* **56,** 125–135.
7. Studier, F. W., Rosenberg, A. H., Dunn, J. J., and Dubendorff, J. W. (1990) Use of T7 RNA polymerase to direct expression of cloned genes. *Meth. Enzymol.* **185,** 60–89.

8

Screening for the Expression of Soluble Recombinant Protein in *Escherichia coli*

Sharon A. Doyle

Summary

Protein expression and purification have traditionally been time-consuming, case-specific endeavors, and are considered to be the greatest bottlenecks in most proteomics pipelines. *Escherichia coli* (*E. coli*) is the most convenient and cost-effective host, although optimal conditions for the expression of different proteins vary widely. Proteins vary in their structural stability, solubility, and toxicity in this environment, resulting in differing rates of protein degradation, formation into insoluble inclusion bodies, and cell death, thus affecting the amount of soluble protein that can be obtained from *E. coli* grown in culture. To take full advantage of a variety of strategies developed to improve the expression of soluble protein in *E. coli*, an easy, rapid means to test many growth parameters is necessary. This chapter describes a dot-blot expression screen to test the effects of growth and induction parameters on the yield of soluble protein. The expression screen is used to detect hexahistidine-tagged proteins expressed in *E. coli*; however, it is adaptable for the detection of other affinity tags or fusion partners that have suitable antibodies available. In this example, induction time and temperature are tested; however, it can be used to test additional parameters, such as affinity tag type and placement, *E. coli* host type, and growth medium formulations.

Key Words: Protein expression screen; hexahistidine tag; dot blot.

1. Introduction

Protein expression and purification have traditionally been time-consuming, case-specific endeavors, and are considered to be the greatest bottlenecks in most proteomics pipelines *(1)*. *Escherichia coli* (*E. coli*) is the most convenient and cost-effective host, although optimal conditions for the expression of different proteins vary widely. Proteins vary in their structural stability, solubility, and toxicity in this environment, resulting in differing rates of protein degradation,

From: *Methods in Molecular Biology, vol. 310: Chemical Genomics: Reviews and Protocols*
Edited by: E. D. Zanders © Humana Press Inc., Totowa, NJ

formation into insoluble inclusion bodies, and cell death, thus affecting the amount of soluble protein that can be obtained from *E. coli* grown in culture. A variety of *E. coli* strains designed to address many of these problems have been made, by allowing for disulfide bond formation in the *E. coli* cytoplasm, more efficient usage of codons that are rarely found in *E. coli* genes, and a reduction of specific *E. coli* proteases (Novagen, Stratagene). The T7 expression system ***(2–4)*** has also helped to alleviate many problems associated with recombinant expression by allowing for tightly regulated induction of expression. However, variability in the recombinant expression of proteins still exists. Induction strength and time, and growth temperature are just a few parameters that can have drastic effects on the amount of soluble protein that can be obtained from the *E. coli* host. Adding further complexity to this problem, affinity tags that are used as a means to standardize protein purification have varying effects on protein stability and solubility ***(5)***. The hexahistidine tag, which allows for protein to be adsorbed to Ni^{2+}-charged resin ***(6)***, is often used because it is small and often does not need to be removed by protease digestion prior to downstream experimentation. Another useful tag is the maltose binding protein (MBP) tag, which allows for the protein to be captured with amylose resin ***(7)***, and can stabilize and solubilize the fusion partner ***(8)***. A variety of affinity tags, in addition to their placement at the N-or C-terminal end of the coding sequence, must be tested for optimal protein expression.

To take full advantage of the variety of strategies developed to improve the expression of soluble protein in *E. coli*, an easy and rapid means to test many growth parameters is necessary. This chapter describes a dot-blot expression screen to test the effects of growth and induction parameters on the yield of soluble protein. The expression screen is used to detect hexahistidine-tagged proteins expressed in *E. coli*, but it is adaptable for the detection of other affinity tags or fusion partners that have suitable antibodies available. In this example, induction time and temperature are tested; however, it can be used to test additional parameters, such as affinity tag type and placement, *E. coli* host type, and growth medium formulations. Results of the screen may be used to segregate samples into groups for further parallel processing in an efficient high-throughput protein production pipeline.

2. Materials

1. Rosetta (DE3) (Novagen, Madison, WI) and BL21 Gold (DE3) (Stratagene, La Jolla, CA) *E. coli* competent cells or other expression host strain.
2. cDNA clones in T7 expression vector.
3. Luria-Bertani (LB) media and agar plates.

4. Chloramphenicol.
5. Kanamycin.
6. 24-well (7 mL) round bottom blocks.
7. Isopropyl-β-D-thiogalactopyranoside (IPTG).
8. Lysis buffer: 50 m*M* sodium phosphate buffer (pH 8.0), 300 m*M* NaCl, 10 m*M* imidazole. Adjust pH to 8.0 and store at 4°C.
9. Lysozyme.
10. Benzonse (Novagen, Madison, WI) or DNAse.
11. Protease inhibitor cocktail (for histidine-tagged proteins) (Sigma-Aldrich, St. Louis, MO).
12. Phenylmethylsulfonylfluoride (PMSF).
13. Protran nitrocellulose membrane (0.2-μ pore size) (Schleicher and Schuell, Keene, NH).
14. Tris-buffered saline (TBS): 6 m*M* Tris, 150 m*M* NaCl (pH 7.5).
15. TBS T/T: 20 m*M* Tris, 500 m*M* NaCl, 0.05% Tween-20, 0.2% Triton X-100 (pH 7.5).
16. Blocking buffer: TBS with 3% bovine serum albumin (BSA).
17. Penta His horseradish peroxidase (HRP) conjugate kit (Qiagen, Valencia, CA).
18. Metal-enhanced DAB substrate kit (Pierce, Rockford, IL).

3. Methods

3.1. Protein Expression and Cell Lysis

The following protocol describes the expression of proteins using the pNHis vector ***(11)*** in the *E. coli* strains Rosetta (DE3) pLysS and BL21 Gold (DE3), based on previous results (not shown). The strain Rosetta (DE3), used for the proteins of *Ciona intestinalis*, is optimized for the expression of proteins containing rare codons often found in genomes enriched with an abundance of either GC or AT base pairs. The strain BL21 Gold (DE3) is used for the expression of proteins from *Xylella fastidiosa*. Note that alternative expression plasmids and *E. coli* strains may be used (*see* **Note 1**), requiring adjustments to the general protocol such as the use of antibiotics.

1. Transform or electroporate Rosetta (DE3) pLysS or BL21 Gold (DE3) cells with pNHis expression plasmids following the manufacturer's protocol. Plate onto LB agar plates containing 50 μg/mL kanamycin and 50 μg/mL chloramphenicol (Rosetta cells only), and grow overnight at 37°C.
2. Pick a single colony of each sample and grow 5-mL cultures (in a 24-well block) of LB containing 50 μg/mL kanamycin and 50 μg/mL chloramphenicol (Rosetta cells only) under non-inducing conditions (no IPTG) by shaking at 180 rpm at 37°C overnight.
3. Identify the parameters that will be tested, such as inducer concentration, temperature, and time of culture growth following induction of protein expression (*see*

Note 2). Keep in mind that some sets of conditions will require a separate 24-well growth block. Aliquot 5 mL of LB containing 50 µg/mL kanamycin and 50 µg/mL chloramphenicol (Rosetta cells only) into each well of the blocks, and inoculate with 100 µL of the overnight starter cultures.

4. Grow cultures by shaking at 180 rpm at 37°C until an optical density (O.D.) at 600 nm wavelength of approx 0.6–0.8 is obtained. Add IPTG at the desired concentration and continue to shake at the desired temperature for the desired length of time. In this example, growth temperature (18, 25, 30, and 37°C) and time (4 h or overnight) following induction with 1 m*M* IPTG were tested.
5. Pellet the cells by centrifugation at 4°C and freeze at –70°C or in a dry ice/ethanol bath. Prepare fresh lysis buffer by adding 1 mg/mL lysozyme, 2.5 U/mL Benzonase® nuclease, 2 m*M* $MgCl_2$, 2 µL/mL protease inhibitor cocktail, and 1 m*M* PMSF. Resuspend the frozen pellets in 500 µL fresh lysis buffer and shake on a plate shaker at 4°C for 30 min to lyse the cells.
6. Remove 10 µL of crude lysate (spotted on the membrane as total protein) and remove the insoluble material by centrifugation. The supernatant (cleared lysate) and crude lysate are then used in the dot blot procedure described below to determine the amount of soluble and total recombinant protein, respectively.

3.2. Dot Blot Procedure

1. Spot 2 µL of crude and cleared lysate onto Protran® nitrocellulose membrane using a multichannel pipettor. Also spot a set of standards. This example spotted a sample protein, isocitrate dehydrogenase (IDH), in a range of 15–1500 ng. Allow the membrane to dry completely.
2. Incubate the membrane by gentle shaking at room temperature as follows: TBS, 5 min × 3; blocking buffer, 30 min; TBS T/T 5 min × 3; penta His HRP conjugate (1:1000 in blocking buffer) 30 min; TBS T/T, 5 min × 5.
3. Dilute the DAB concentrate (10X stock) with stable peroxide buffer to 1X following the manufacturer's protocol. Submerse the membrane in the DAB solution and develop for approx 3 min (**Fig. 1**). Rinse membrane with H_2O, incubate in a dark place until it is completely dry, and scan the image using a flatbed scanner.

4. Notes

1. A wide variety of *E coli* strains designed for protein expression are commercially available, derived from the *E. coli* strain BL21, that are tailored to facilitate disulfide bond formation, fine-tune protein expression levels, enhance the expression of proteins that contain rare codons, and other specific requirements. These strains are available as lambda DE3 lysogens, which carry a chromosomal copy of the T7 RNA polymerase gene under control of the lac promoter (inducible by IPTG). Once induced, the T7 RNA polymerase drives expression of genes that are under control of the T7 promoter, such as those cloned into the pET family of expres-

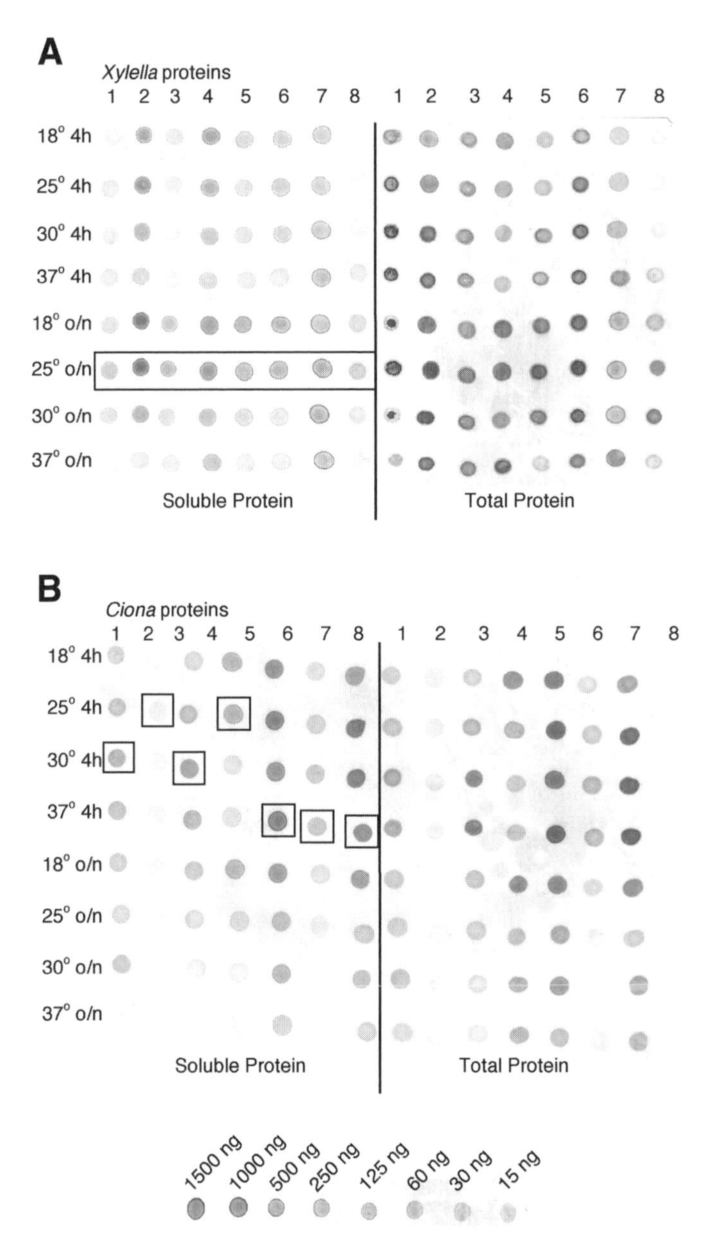

Fig. 1. Expression screen dot blots of **(A)** *Xylella* samples and **(B)** *Ciona* samples, grown for 4 h or overnight (o/n) at 18, 25, 30, and 37°C. The standard curve is shown below. The spots outlined with squares indicate samples that would be chosen for protein purification.

sion vectors (Novagen, Madison, WI). Dramatic differences in protein expression can be seen when different strains are used for the production of proteins in many cases; therefore, it is useful to test several strains when producing many proteins in a high-throughput format without knowledge of their specific characteristics.

2. Conditions most commonly tested are inducer (IPTG) concentration (0.1 m*M* to 1.0 m*M*), growth temperature after induction (18°C to 37°C), and duration of time following induction (4 h to overnight). Additional parameters to consider testing include *E. coli* strain, media formulation, and placement of affinity tag (N- or C-terminal).

Acknowledgments

The author thanks Jennifer Massi and Shirin Fuller for technical assistance, and Michael Murphy, Peter Beernink, and Paul Richardson for reading of the manuscript. This work was performed under the auspices of the US Department of Energy, Office of Biological and Environmental Research, by the University of California, Lawrence Livermore National Laboratory (contract No. W-7405-Eng-48), Lawrence Berkeley National Laboratory (contract No. DE-AC03-76SF00098), and Los Alamos National Laboratory (contract No. W-7405-ENG-36).

References

1. Pedelacq, J. D., Piltch, E., Liong, E. C., et al. (2002) Engineering soluble proteins for structural genomics. *Nat. Biotechnol.* **20,** 927–932.
2. Studier, F. W. and Moffatt, B. A. (1986) Use of bacteriophage T7 RNA polymerase to direct selective high-level expression of cloned genes. *J. Mol. Biol.* **189,** 113–130.
3. Rosenberg, A. H., Lade, B. N., Chui, D., Lin, S., Dunn, J. J., and Studier, F. W. (1987) Vectors for selective expression of cloned DNAs by T7 RNA polymerase. *Gene* **56,** 125–135.
4. Studier, F. W., Rosenberg, A. H., Dunn, J. J., and Dubendorff, J. W. (1990) Use of T7 RNA polymerase to direct expression of cloned genes. *Meth. Enzymol.* **185,** 60–89.
5. Braun, P. and LaBaer, J. (2003) High-throughput protein production for functional proteomics. *Trends in Biotechnology* **21(9),** 383–388.
6. Janknecht, R., de Martynoff, G., Lou, J., Hipskind, R., and Nordheim, A. (1991) Rapid and efficient production of native histidine-tagged protein expressed by recombinant vaccinia virus. *Proc. Natl. Acad. Sci. USA* **88,** 8972–8976.
7. Riggs, P. (1992) Expression and purification of maltose-binding protein fusions. In *Current Protocols in Molecular Biology* (Ausubel, F. A., Brent, R., Kingston, R. E., et al., eds.), Greene Publishing and Wiley-Interscience, New York, pp. 16.6.1–16.6.14.

8. Kapust, R. B. and Waugh, D. S. (1999) *Escherichia coli* maltose-binding protein is uncommonly effective at promoting the solubility of polypeptides to which it is fused. *Protein Sci.* **8,** 1668–1674.
9. Doyle, S. A., Murphy, M. B., Massi, J. M., and Richardson, P. M. (2002) High-throughput proteomics: a flexible and efficient pipeline for protein production. *J. Proteome Research* **1(6),** 531–536.

9

High-Throughput Purification of Hexahistidine-Tagged Proteins Expressed in *E. coli*

Michael B. Murphy and Sharon A. Doyle

Summary

This chapter describes a method for efficient high-throughput purification of hexahistidine-tagged proteins that are expressed in *Escherichia coli* (*E. coli*) using immobilized metal affinity chromatography (IMAC) *(2)* in a 96-well format. This approach is particularly suitable for proteomic applications that require modest amounts of highly purified proteins to be generated very efficiently. This approach is also very useful for identifying protein targets that are most amenable to scaled-up production for use in structural studies. The typical yield of proteins purified using this system is 50–150 μg, which is generally greater than that of many in vitro expression systems and much less costly. The method as described has been optimized for purifying approx 150 μg of hexahistidine-tagged protein, but the method is flexible, so that the amount of affinity matrix and culture volumes can be adjusted for optimal binding capacity and consequently highest purity. Although the method detailed here uses IMAC to purify hexahistidine-tagged proteins, this basic platform can be used with many other tags and affinity resins.

Key Words: IMAC; protein production; hexahistidine tag.

1. Introduction

Large collections of purified proteins have become essential to systems-biology programs for generating protein ligands and for validating protein interactions. These large protein collections will likewise benefit structural studies, and quests for protein-based therapeutics. Whether many protein targets need to be produced, such as a microbial proteome, or a few poorly expressing protein targets need to be expressed as soluble fragments (divide and conquer approach), efficient high-throughput processing can be a bottleneck *(1)*.

From: *Methods in Molecular Biology, vol. 310: Chemical Genomics: Reviews and Protocols*
Edited by: E. D. Zanders © Humana Press Inc., Totowa, NJ

This chapter describes a method for efficient high-throughput purification of hexahistidine-tagged proteins that are expressed in *Escherichia coli* (*E. coli*) using immobilized metal affinity chromatography (IMAC) *(2)* in a 96-well format. This approach is particularly suitable for proteomic applications that require modest amounts of highly purified proteins to be generated very efficiently. This approach is also very useful for identifying protein targets that are most amenable to scaled-up production for use in structural studies. The typical yield of proteins purified using this system is 50–150 μg, which is generally greater than that of many in vitro expression systems, and much less costly. The method as described has been optimized for purifying approx 150 μg of hexahistidine-tagged protein, but the method is flexible, so that the amount of affinity matrix and culture volumes can be adjusted for optimal binding capacity, and consequently highest purity. Although the method detailed here uses IMAC to purify hexahistidine-tagged proteins, this basic platform can be used with many other tags and affinity resins.

2. Materials

1. 24-well (7 mL) and/or 96-well (2 mL) round bottom blocks (Qiagen, Valencia, CA).
2. Genemate 96-well filter plate with GMF-5 filter (ISC Bioexpress, Kaysville, UT).
3. AirPore Tape (Qiagen, Valencia, CA).
4. 96-well plate shaker/vortex.
5. 96-well plate/block centrifuge.
6. 96-well filter-plate vacuum manifold and regulator (Qiagen, Valencia, CA).
7. Luria-Bertani (LB) media and LB agar plates.
8. Kanamycin.
9. Isopropyl-β-D-thiogalactopyranoside (IPTG).
10. Lysis buffer: 50 m*M* sodium phosphate buffer (NaH_2PO_4), pH 8.0; 300 m*M* NaCl; 20 m*M* imidazole; 0.05% polysorbate 20; 1 m*M* $MgCl_2$. Store at 4°C.
11. Wash buffer: 50 m*M* NaH_2PO_4, pH 8.0; 500 m*M* NaCl; 20 m*M* imidazole; 0.05% polysorbate 20. Store at 4°C.
12. Elution buffer: 50 m*M* NaH_2PO_4, pH 8.0; 150 m*M* NaCl; 500 m*M* imidazole; 10% glycerol. Store at 4°C.
13. Lysozyme (Sigma, St. Louis, MO).
14. Benzonase (Novagen, Madison, WI) or DNase.
15. Protease inhibitor cocktail (for histidine-tagged proteins) (Sigma-Aldrich, St. Louis, MO).
16. Phenylmethylsulfonylfluoride (PMSF).
17. Ni-NTA Superflow resin (Qiagen, Valencia, CA) or other IMAC resin.
18. BL21 Gold (DE3) competent cells (Stratagene, La Jolla, CA).
19. 3–15% polyacrylamide gels (BioRad, Hercules, CA).
20. GelCode Blue (Pierce, Rockford, IL).

21. Agilent 2100 Bioanalyzer and Protein 200 plus lab-chip kit (Agilent, Palo Alto, CA).

3. Methods

First, a general method of transformation of the expression plasmid into *E. coli* will be described, as well as appropriate conditions for growth and induction of the expression cultures. Then, the 96-well protein purification method is detailed. Last, analysis of the purified proteins is described using both lab-on-a-chip technology and traditional sodium dodecyl sulfide (SDS)-polyacrylamide gel electrophoresis (PAGE).

3.1. Transformation and Expression Induction

Expression plasmids, in our example pNHIS *(3)*, containing genes of interest fused to a hexahistidine coding tag are first purified from 5-mL cultures of a non-expression strain of *E. coli*, such as DH5α or Novablue (a K-12 strain from Novagen), grown in LB containing kanamycin for antibiotic selection. An appropriate *E. coli* strain for protein expression is then transformed with the purified expression plasmids, as described below. The *E. coli* cell strain that is most commonly used for protein expression driven by T7 RNA polymerase *(4)* is BL21 (DE3). This strain contains a chromosomal copy of the T7 RNA polymerase gene (indicated by the DE3), the expression of which is inducible by addition of IPTG. There are many derivative and mutant *E. coli* strains available that offer strategies for producing problematic proteins (*see* **Note 1**).

1. Add 1 μL of purified plasmid DNA to 25 μL of BL21 (DE3) competent cells in a 1.5-mL microfuge tube, gently mix, and place on ice for 10 min.
2. Place in a 42°C water bath for 30 s, then immediately back on ice for 2 min. Add 500 μL of SOC (or LB) and grow with vigorous shaking at 37°C for 30 min.
3. Plate 50 μL onto prewarmed LB agar plates containing 50 μg/mL kanamycin, and grow overnight at 37°C.
4. Next, grow starter cultures by picking a colony from the above plates into 2 mL LB plus 50 μg/mL kanamycin in sterile 24-well blocks. Cover the blocks with air-pore tape and grow overnight at 37°C, shaking at 180–200 rpm.
5. Aliquot 5 mL of LB (*see* **Note 2**) plus 50 μg/mL kanamycin into each of the wells of new 24-well blocks. Inoculate the 5 mL of LB with 100 μL of the starter cultures (*see* **Note 3**). Cover the blocks with air-pore tape and grow at 37°C at 180–200 rpm for approx 3 h until the optical density (O.D.) at 600 nm wavelength is 0.6–0.8. The OD_{600} should be checked every hour by removing 100 μL from several representative wells.
6. Add IPTG to induce protein expression, transfer the blocks to a shaker incubator that has been adjusted to the desired expression temperature, and grow for the desired amount of time. Each of these parameters should be optimized by the expression

screening technique described in the previous chapter (*see* **Note 4**). In the example in this chapter, protein expression was induced with 1 m*M* IPTG and cultures were grown for 6 h at 25°C.

7. Pellet the cells by centrifugation at 2500*g* for 10 min at 4°C and then carefully decant all media. Freeze the bacterial pellets at –80°C.

3.2. Protein Purification

Purification of the proteins from the cultures grown in 24-well blocks can be performed in 96-well filter plates using a vacuum manifold for column forming, washing, and elution steps, as outlined below. The hexahistidine-tagged proteins in cleared cell lysates are batch loaded onto Ni-NTA agarose in a 96-well filter plate. Columns are formed by applying low (200 mbar) vacuum pressure. These mini-columns are extensively washed, and subsequently the proteins are eluted into a microtiter plate.

1. Thaw bacterial pellets in 24-well blocks at room temperature. While waiting, add protease inhibitors, lysozyme, and Benzonase to the chilled lysis buffer: 1 mg/mL lysozyme, 10 units Benzonase/mL, 2 µL/mL protease inhibitor cocktail (no ethylenediamene tetraacetic acid [EDTA]), and 1 m*M* PMSF. Keep on ice.
2. Add 0.5 mL of chilled lysis buffer plus additives to each pellet and resuspend by pipetting, using a 12-channel pipet to mix a row of 6 wells with two tips in each of the 6 wells.
3. Vortex the 24-well blocks containing the resuspended pellets at 4°C on a plate shaker for 20 min. Performing the lysis at 4°C will help prevent protein degradation.
4. Centrifuge 24-well blocks at 2500*g* in a plate centrifuge for 20 min at 4°C. The pellets obtained can be saved to purify insoluble protein under denaturing conditions (not discussed here).
5. Equilibrate the Ni-NTA resin in lysis buffer and resuspend to obtain a 50% slurry. Add 50 µL of the 50% slurry of Ni-NTA resin, which is enough to bind approx 150 µg of protein, into the wells of a 96-well filter-plate (*see* **Note 5**). Liquid may begin to flow by gravity; in order to prevent loss of material, place the filter plate on a piece of parafilm.
6. Carefully transfer the cleared bacterial lysate supernatants from each of the 24 wells into the wells of a 96-well filter plate. Cover the filter-plate top with clear plate-seal tape, to prevent cross-contamination.
7. Mix the Ni-NTA resin with the cleared lysate in the 96-well filter plate on a piece of parafilm on a plate vortex for 30 min at 4°C.
8. Remove the 96-well filter plate from the plate vortex leaving the parfilm behind; place on the vacuum manifold (with a waste collection container below) and apply vacuum (approx 200 mbar) so that each well slowly drips over several minutes until all the lysate has filtered through, and release the vacuum after the material has flowed through.
9. Add 750 µL of wash buffer (*see* **Note 6**) into each well and apply the vacuum just as in **step 8**. Repeat the wash step two more times.

10. Fit a 96-well microtiter plate (200 μL) or 2-mL collection microtubes into the vacuum manifold, carefully aligning the drop-formers to the receiver plate wells. Add 100 μL elution buffer and incubate for 2 min. Then apply a vacuum of 200–400 mbar and release vacuum after material has flowed through. Store the plate of eluted proteins at 4°C for short term or until evaluated, and then at –80°C for long-term storage.

3.3. Determination of Protein Yield and Purity

Eluted material was assessed on an Agilent 2100 Bioanalyzer using a Protein 200 Plus LabChip (Caliper) to determine protein purity, concentration, and yield. However, traditional SDS-PAGE can also be used, if necessary. The LabChip system is a fluorescence-based detection system that offers the advantages of requiring only 1–4 μL of protein samples, as compared to 10 μL required for equivalent band identification on traditional SDS-PAGE gels, and results for a set of 10 samples are obtained in 30 min. The Biosizing software provides detailed tables of raw and analyzed data and gel-like images, and is convenient for data storage and retrieval. The Agilent 2100 Bioanalyzer system provides accurate protein concentrations, alleviating the need of running additional assays and using more of the samples. In addition, the concentration of each individual protein is determined from the fluorescent peak area, or band, on the gel-like image; thus, accurate concentrations of a protein of interest can be easily obtained in partially purified samples.

1. For SDS-PAGE, 10 μL of each protein sample were run on 3–15% gradient polyacrylamide gel. The gels were stained with GelCode Blue and scanned on Fluor-S MultiImager (**Fig. 1B**) (BioRad, Hercules, CA).
2. For the Bioanalyzer, the protein LabChip is prepared according to the manufacturer's instructions.
3. The samples are prepared by mixing 2 μL of eluted protein, 2 μL H_2O, and 2 μL of an SDS-based denaturing sample buffer containing β-mercaptoethanol as well as upper and lower protein mass standards. Samples are heated to 95°C for 2 min and spun briefly (*see* **Note 7**).
4. Samples and ladder are then diluted to 90 μL with H_2O, and then 6 μL of each diluted sample is loaded into a well of the LabChip according to the manufacturer's instructions. Dilution of the sample may seem counter-intuitive but is necessary to reduce the SDS concentration in the denaturation buffer.
5. The LabChip is then placed in an Agilent 2100 Bioanalyzer and the Protein 200 Plus program is run.
6. Agilent Biosizing software is used to determine the sizes of the proteins of interest by normalization against two internal standards of 6 and 210 kDa. The fluorescent peak identification settings were adjusted from the default settings to improve sensitivity—0.8 for the minimum peak height, 0.2 s for the minimum peak width, and 4 for the slope threshold (**Fig. 1A**).

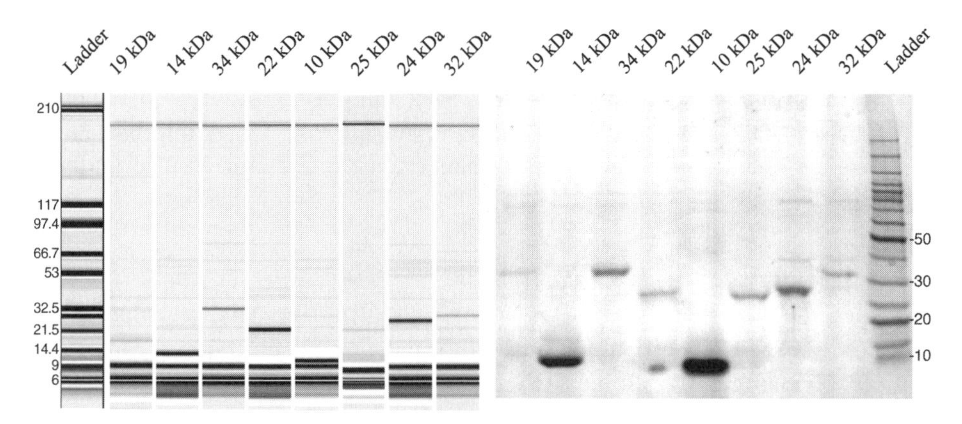

Fig. 1. Purified proteins imaged using **(A)** Agilent 2100 Bioanalyzer and **(B)** traditional sodium dodecyl sulfide-polyacrylamide gel electrophoresis. **(A)** Note that the lowest two bands and the uppermost band are internal standards for sizing and quantitation. **(B)** 3–15% polyacrylamide gel stained with GelCode Blue. In both panels, the predicted size of the protein is labeled at the top of each lane.

4. Notes

1. A variety of *E. coli* strains have been developed for protein expression that may improve the soluble yields of particular classes of proteins. We have found that many microbial proteins express well in the traditional BL21(DE3) strain. However, eukaryotic proteins often express better in strains that provide tRNAs for codons that *E. coli* rarely use, such as Rosetta (Novagen). Other examples include mutant strains that may improve the proper folding of some proteins by promoting disulfide bond formation, such as Origami and Origami B (Novagen), and mutant strains that facilitate IPTG concentration-dependent expression induction, such as Tuner (Novagen). Testing several expression strains from various vendors is often a good strategy to improve the odds of producing soluble protein, and further highlights the utility of high-throughput expression and purification techniques.
2. A variety of uniquely formulated media are available that may improve growth. For example, richer media may be used, such as Power, Turbo, or Superior broth (Athena ES, Baltimore, MD) to increase biomass and subsequent protein yield.
3. Replicates of each of the proteins can be set up in the 24-well blocks and then combined into the same well of a 96-well block to increase the yield of low-expressing samples.
4. The chapter on expression screening describes a simple method to quickly identify the optimal condition for expression induction. The amount of IPTG (0.1 to 1 m*M*), optimal temperature (18°C to 37°C), and length of the induction time (4 h to overnight) should be identified to improve the likelihood of producing soluble protein and improve the yields of soluble proteins. Additionally, purification from several protein expression conditions can be tested in parallel.

5. IMAC resins are available from many sources, and the binding capacity can vary. Additionally, use of metal chelation via tetradendate adsorbant (NTA) as opposed to tridendate (IDA), ensures that much less leaching of metal ions will occur. The purity of hexahistidine-tagged proteins is optimal from purification conditions in which the amount of the target protein is equal to or greater than the binding capacity of the column matrix. The amount of lysate and Ni-NTA Superflow agarose that we have suggested should be a guide. The expression screening method described in the previous chapter can be used to assess and then group samples with similar expression levels and then approximate the optimal amount of Ni-NTA Superflow agarose required.

   Additionally, we tested several 96-well filter plates and found that the Genemate plates have dropformers that perform better than others, with less cross-contamination and optimal recovery of eluent, and are less expensive than most other plates.
6. The compositions of the buffers can be varied. Adjusting the stringency of the wash—for example, by increasing the NaCl concentration to 500 m*M* or the imidazole concentration to 20 m*M*—can improve purity; however, lowering these concentrations may improve binding of some hexahistidine-tagged proteins. The QIAexpressionist handbook (Qiagen, Valencia, CA) provides some very useful suggestions for various buffer compositions.
7. The LabChip protocol recommends using 4 μL of protein sample; however, we find that using 1–2 μL of each sample plus H_2O works as well or better, and uses less material. This may be due to a dilution of components that affect the sensitivity of the assay, such as salt and imidazole. Many common reagents affect the sensitivity of the LabChip assay; therefore, care should be taken when the buffer composition is altered.

Acknowledgments

We thank Shirin Fuller for technical assistance, and Paul Richardson and Peter Beernink for helpful suggestions and critical reading of the manuscript. This work was performed under the auspices of the US Department of Energy, Office of Biological and Environmental Research, by the University of California, Lawrence Livermore National Laboratory under contract No. W-7405-ENG-48, Lawrence Berkeley National Laboratory under contract No. DE-AC03-76SF00098, and Los Alamos National Laboratory under contract No. W-7405-ENG-36.

References

1. Braun, P. and LaBaer, J. (2003) High throughput protein production for functional proteomics. *Trends Biotech.* **21(9),** 383–388.
2. Smith, M. C., Furman, T. C., Ingolia, T. D., and Pidgeon, C. (1988) Chelating peptide-immobilized metal ion affinity chromatography. A new concept in affinity chromatography for recombinant proteins. *J. Biol. Chem.* **263(15),** 7211–7215.

3. Doyle, S. A., Murphy, M. B., Massi, J. M., and Richardson, P. M. (2002) High-throughput proteomics: a flexible and efficient pipeline for protein production. *J. Proteome Res.* **1(6),** 531–536.
4. Studier, F. W. and Moffatt, B. A. (1986) Use of bacteriophage T7 RNA polymerase to direct selective high-level expression of cloned genes *J. Mol. Biol.* **189(1),** 113–130.

10

The Wheat Germ Cell-Free Expression System

Methods for High-Throughput Materialization of Genetic Information

Tatsuya Sawasaki, Mudeppa D. Gouda, Takayasu Kawasaki, Takafumi Tsuboi, Yuzuru Tozawa, Kazuyuki Takai, and Yaeta Endo

Summary

This chapter contains protocols for high-throughput protein production based on the cell-free system prepared from eukaryote wheat embryos.

Key Words: Cell-free protein synthesis; pure embryo isolation; wheat extract preparation; transcription and translation reactions.

1. Introduction

The need to translate genome information into proteins has driven the search for robust protein-expression technologies. Existing protein-expression technologies can be classified into chemical synthesis, in vivo expression, and cell-free protein synthesis. The chemical synthesis and in vivo expression methods have many limitations. Chemical synthesis is not practical for the synthesis of long peptides *(1)*, and in vivo expression can produce only those proteins which do not destroy host cell physiology *(2–4)*. Cell-free translation systems, in contrast, can synthesize proteins with high speed and accuracy, approaching in vivo rates *(5,6)*, and can express proteins that seriously interfere with cell physiology. However, they have been relatively inefficient because of their instability over time *(7)*. Our recent findings on endogenous inhibitors of translation in plants *(9,10)* and their efficient elimination from embryos prior to the preparation of the extract by extensive washing, has led to an extraordinarily stable and efficient translation system *(11)*. In addition to this, we have adapted the cell-

From: *Methods in Molecular Biology, vol. 310: Chemical Genomics: Reviews and Protocols*
Edited by: E. D. Zanders © Humana Press Inc., Totowa, NJ

free system to address the high-throughput needs of modern proteomics, by improving several critical parameters ***(12)***, such as: (a) optimization of the 5' and 3' UTRs of mRNA for the purpose of eliminating the 5'-^{7}mGpppG (cap) and poly(A)-tail (pA), thereby increasing translation initiation; and (b) design of split primers for polymerase chain reaction (PCR) to generate transcription templates directly from *Escherichia coli* cells carrying cDNA, thus bypassing the time-consuming cloning steps. As a result, our significantly improved wheat germ cell-free translation system, in combination with a well optimized expression vector, is helping to deliver high-throughput expression of many proteins in parallel and their characterization in terms of activity and solubility.

In this chapter, protocols for high-throughput protein production based on the cell-free system prepared from eukaryote wheat embryos will be discussed in detail. To begin, we give detailed protocols for the generation of desired open reading frame (ORF) with or without cloning. Then we describe the preparation of mRNA either by using PCR-generated DNA or by highly purified plasmid. Finally, detailed protocols will be presented for the cell-free translation of proteins by the following methods: batch, continuous-flow cell-free (CFCF) protein synthesis system, bilayer method, and an automated protein synthesizer, GenDecoder.

2. Materials

1. Wheat seeds.
2. Rotor speed mill (Pulverisette 14, Fritsh, Germany).
3. Sieve (710- to 850-μm mesh).
4. Cyclohexane and carbon tetrachloride.
5. Nonidet P-40.
6. Milli-Q water, freshly prepared.
7. Bronson model 2210 sonicator (Yamato, Japan).
8. Extraction buffer: 40 m*M* HEPES-KOH (pH 7.6), 100 m*M* potassium acetate, 5 m*M* magnesium acetate, 2 m*M* calcium chloride, 4 m*M* dithiothreitol (DTT), and 0.3 m*M* of each of the 20 amino acids.
9. Sephadex G-25 (fine) column and MicroSpin G-25 and Nick columns (Amersham Biosciences).
10. Plasmid of Ehime University (pEU) protein expression vector.
11. cDNA of protein to be synthesized.
12. Oligonucleotide primers.
13. PCR thermo-cycler MP (Takara, Otsu, Japan).
14. ExTaq DNA polymerase (Takara, Otsu, Japan).
15. 5X transcription buffer (TB): 400 m*M* HEPES-KOH (pH 7.8), 80 m*M* magnesium acetate, 10 m*M* spermidine, and 50 m*M* DTT.
16. Nucleotide tri-phosphates (NTPs) mix: a solution containing 25 m*M* each of ATP, GTP, CTP, and UTP.

17. SP6 RNA polymerase and RNasin (80 U/μL, Promega, Madison, WI).
18. Approx 99% ethanol and 7.5 *M* ammonium acetate.
19. 5X wheat expression buffer (WEB): 110 m*M* HEPES-KOH (pH 7.8), 400 m*M* potassium acetate, 8.5 m*M* magnesium acetate, 2 m*M* spermidine, 8.5 m*M* DTT, 1.2 m*M* amino acid mix, 6 m*M* ATP, 1.25 m*M* GTP, and 80 m*M* creatine phosphate.
20. 1 mg/mL creatine kinase.
21. Translational substrate buffer (TSB): 30 m*M* HEPES-KOH (pH 7.8), 100 m*M* potassium acetate, 2.7 m*M* magnesium acetate, 0.4 m*M* spermidine, 2.5 m*M* DTT, 0.3 m*M* amino acid mix, 1.2 m*M* ATP, 0.25 m*M* GTP, and 16 m*M* creatine phosphate.
22. 24-well microplate (Whatman, Inc., Clifton, NJ).

3. Methods

The following methods explain (a) cloning of the desired ORF into the pEU expression system and preparation of transcription template DNA by PCR for mRNA preparation, (b) generation of transcription template DNA directly from *E. coli* cells containing cDNA, (c) preparation of mRNA using transcription template DNA, and (d) translation of prepared mRNA either in batch mode, bilayer system, or by the CFCF method.

3.1. Purification of Embryos and Preparation of Extract

Isolation of embryos and preparation of extract were carried out as reported previously ***(11)***. The main aim is to remove contaminating endosperms from embryos by washing (**steps 5** and **6**).

3.1.1. Isolation of Wheat Embryo

1. Grind the wheat seeds in a mill (rotor speed mill).
2. Sieve through a 710- to 850-μm mesh.
3. Select embryo with the solvent flotation method by using a solvent containing cyclohexane and carbon tetrachloride (240:600 v/v).
4. Dry overnight in a fume hood.
5. Wash three times with 10 vol of sterile water under vigorous stirring.
6. Sonicate for 3 min in a 0.5% solution of Nonidet P-40 using a Bronson model 2210 sonicator.

3.1.2. Preparation of Wheat Embryo Extract

1. Grind 5 g of isolated wheat embryo to a fine powder in liquid nitrogen.
2. Add 5 mL of extraction buffer and vortex the mixture briefly.
3. Centrifuge the embryo lysate (30,000*g*, 30 min) and retain the supernatant.
4. Gel filter the supernatant using a sephadex G-25 (fine) column, equilibrated with two volumes of extraction buffer.
5. Collect the void fraction.

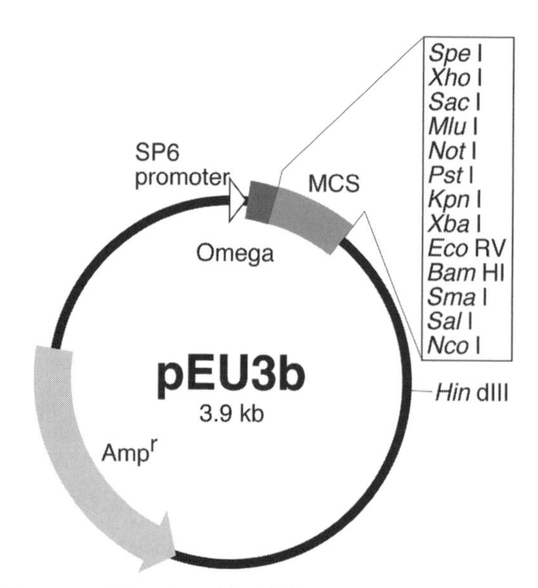

Fig. 1. Schematic diagram of the plasmid of Ehime University (pEU) expression vector.

6. Centrifuge the fraction (30,000*g*, 10 min) and retain the final supernatant.
7. Adjust to 200 A_{260}/mL with extraction buffer.
8. Divide into small aliquots, and store at –80°C until use.

3.2. DNA-Template Construction for Transcription

The 5' and 3' UTRs of eukaryotic mRNA are crucial for the regulation of gene expression by controlling mRNA translational efficiency, stability, and localization. The 5' cap and poly A (pA), found on almost all eukaryotic mRNAs, stimulate translation initiation and stabilize mRNA in synergy. However, for in vitro translation, the use of 5' cap and pA can be a major problem, as described Chapter 11 by Endo and Sawasaki. These problems have been effectively overcome by optimizing the 5' and 3' UTRs ***(12)***.

3.2.1. Cloning of Desired ORF Into the pEU3b Expression System

The optimized elements were assembled into a pSP65-derived vector and called pEU (**Fig. 1**). As shown in **Fig. 1**, the pEU3b has a multi-cloning site downstream of the SP6 promoter and omega sequences and an ampicillin resistance gene. By using this assembled vector, the desired genes can be expressed efficiently without cap and poly(dA/dT) ***(12)***. Cloning of the desired gene into the pEU3b expression system was achieved by using standard molecular-biology techniques. After verifying the DNA sequence, plasmid DNA containing the desired ORF was prepared by a CsCl method *(13)* or by a small-scale plasmid preparation kit. However, it should be noted that when plasmid is directly used for transcription, we preferred to use highly purified plasmid without the contaminating RNase used in commercial kits.

3.2.2. pEU-Based PCR

The plasmid prepared by commercial kits can be directly used for the preparation of mRNA. However, because of the low purity and concentration of plasmid, and contamination by RNase, it may not be suitable for transcription; in addition, the preparation of plasmid by density gradient centrifugation is tedious. To overcome these problems, we have optimized PCR-based linear template DNA generation by designing a set of universal primers for the pEU vector. The designed primers are SPu (5'-GCGTAGCATTTAGGTGACACT) and AODA2303 (5'-GTCAGACCCCGTAGAAAAGA). By using these primers, the desired DNA template can be easily generated and used directly for the preparation of mRNA without further purification, as follows:

1. Prepare 50 μL of PCR reaction mixture according to the manufacturer's instructions (Takara) by mixing 100 pg/μL of plasmid and 200 n*M* of each primer with 200 μ*M* of each dNTP and 1.25 U of ExTaq DNA polymerase.
2. Program the PCR thermo-cycler (Takara) for 1 min denaturation at 96°C followed by 25 cycles of amplification: 98°C for 10 s, 55°C for 30 s, and 72°C for 5 min, depending on the length of the ORF (1 kb per min).
3. The template DNA is then used for transcription after analyzing its integrity by agarose gel electrophoresis.

3.2.3. Direct PCR From cDNA or Colonies by the Split Primers PCR Technique

Under **Subheadings 3.2.1.** and **3.2.2.**, we discussed general cloning protocols and PCR-based template generation. Because cloning of cDNA into an expression vector is one of the most laborious and time-consuming steps of this process, direct synthesis from PCR-generated linear cDNAs is preferred *(14,15)*. These methods generally involve the incorporation of transcriptional and translational start and stop signals into primers for the generation of autonomous coding fragments by PCR. In the following subsections, therefore, we focus on the construction of template DNA for transcription directly from cDNA clones using the split-primers PCR technique *(12)*. The strategy for the construction of template DNA directly from the cDNA clones is depicted in **Fig. 2**.

3.2.3.1. Design of Primers

As shown in **Fig. 2**, the split-primers PCR technique involves four major steps:

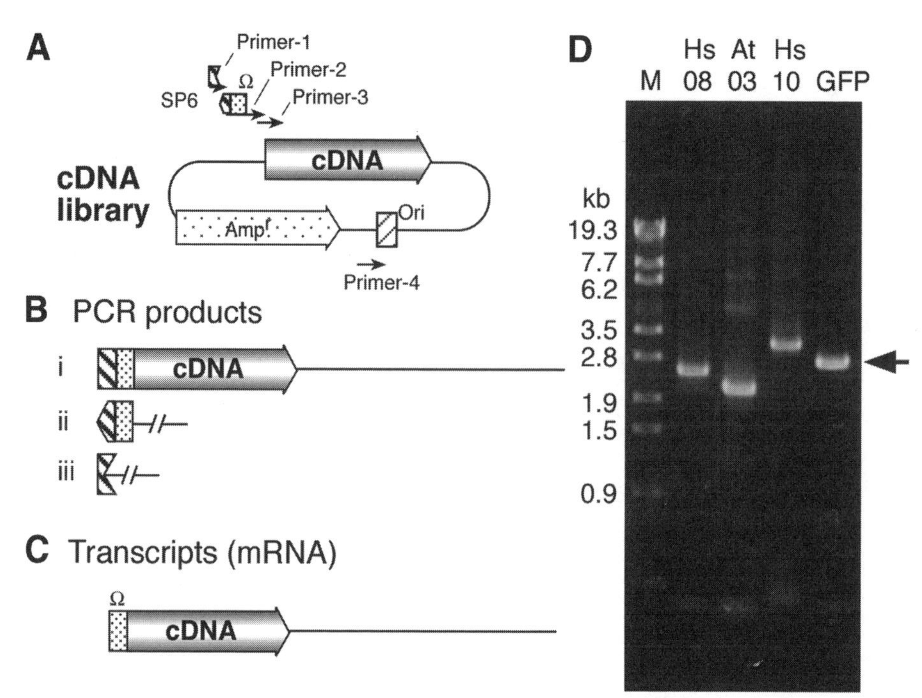

Fig. 2. Direct transcriptional template generation from cDNA library by using the split-primers polymerase chain reaction (PCR) technique. (*a–c*) A schematic representation of the split-primers design for equipping the cDNA sequences with required UTRs, where *b* and *c* are expected PCR-generated DNAs and mRNA, respectively. (*d*) Split-primer PCR-generated products.

1. Designing the target-specific primer (primer-3) in such a way that its 3' end has the target-specific sequence and the 5' end has a part of the omega sequence.
2. Primer-2 has the full-length omega sequence and a part of the SP6 promoter sequence at its 5' end.
3. Primer-1 has a part of the SP6 promoter and part of overlapping sequence at the 3' end.
4. Primer-4 is specific for the gene-carrying vector somewhere at the 3' UTR.

Therefore, for each clone of cDNA, primer-3 is the only specific primer, and the remaining primers are common for all cDNAs. Based on this strategy, the following four primers have been designed:

1. Primer-1: 5'-GCGTAGC<u>ATTTAGGTGACACT</u> (the underlined sequence is the 5' half of the promoter).
2. Primer-2: 5'-<u>*GGTGACACT*ATAGAA</u>GTATTTTTACAACAATTACCAACAACAACAACAAACAACAACAACATTACATTTTACATTCTACAACTA*CCACC*

CACCACCACCAATG (underlined sequence is the 3'-half of the promoter and the sequence in italic denotes the annealing region of primer-1 or primer-3).

3. Primer-3: 5'-CCACCCACCACCACCAatgnnnnnnnnnnnnnnn (the 5'-coding region of a target gene is in lowercase).
4. Primer-4: 5'-AGCGTCAGACCCCGTAGAAA based on reported PCR procedures ***(9–11)***.

3.2.3.2. First PCR

1. Prepare 60 µL of a PCR reaction by mixing 3 ng of plasmid or 3 µL of *E. coli* (overnight culture) as template, 200 µ*M* each of dNTP, 1.5 U of ExTaq DNA polymerase, 10 n*M* of primer-3 and primer-4, and the buffer supplied by the manufacturer.
2. Program the PCR thermocycler for 4 min denaturation at 94°C followed by 30 cycles of amplification: 98°C for 10 s, 55°C for 30 s, and 72°C for 5 min, depending on the length of gene (1 kb per min).

3.2.3.3 Second PCR

In order to introduce the omega and SP6 promoter sequence into the first PCR-amplified target gene, the 30 µL second PCR was carried out by mixing 3 µL of the first PCR product (without any purification), 100 n*M* primer-1 and -4, and 1 n*M* primer-2 with the same cycling conditions as the first.

3.3. Preparation of mRNA

The next step is the preparation of mRNA. Cell-free translation of proteins can be achieved mainly by three different modes—batch ***(11,12)***, bilayer ***(16)***, and continous-flow cell-free ***(18)***. Appropriate volumes of transcription products can be produced depending on the mode of translation.

1. Prepare transcription reactions as given below (**Table 1**).
2. Incubate the reaction mixture at 37°C for 3 h.
3. Centrifuge the mixture (6000*g*, 1 min) and retain the supernatant.
4. Add 3 vol of Milli-Q water, 1/7.5 vol of 7.5 *M* ammonium acetate, and 2.5 vol of approx 99% EtOH, and keep on ice for 10 min after mixing well.
5. Centrifuge the mixture (20,000*g*, 5 min) and discard the supernatant.
6. Wash the pellet with 500 µL of 70% ethanol and then spin for 5 min at 20,000*g*.
7. Dissolve the mRNA pellet in 28 µL of Milli-Q water for batch, bilayer, or dialysis mode of translation.

4. Translation Reactions

The final step is cell-free translation. Cell-free translation of proteins can be achieved using three different modes ***(11,12,15,16)***:

1. Batch mode translation, where, as the translation is carried out in a homogenous solution, protein synthesis proceeds until one of the substrates is used up or the

Table 1
Reaction Mixture for Transcription

Reagent	Mode of translation (total of 50 μL reaction)			
	Batch	Bilayer	CFCF	Final conc.
Template DNA	x μL	x μL	x μL	1/10 vol. (PCR)[a] 100 ng/μL (CP)[b]
5x TB	2	5	10	1x
25 m*M* NTP mix	1.2	3	6	3 m*M*
80 U/μL SP6 RNApoly	0.125	0.313	0.625	1 U/μL
80 U/μL RNain	0.125	0.313	0.625	1 U/μL
Milli-Q	up to 10	up to 25	up to 50 μL	

[a]PCR product (>10 ng/μL).
[b]High-quality circular plasmid (CP).
CFCF, continuous-flow cell-free protein synthesis system; PCR, polymerase chain reaction; TB, transcription buffer.

byproducts cause inhibition. Thus, the translation ceases in approx 3 to 4 h. This method is suitable for rapidly checking protein functions, such as enzymatic activity, or identifying binding partners such as DNAs or proteins.

2. Bilayer system, where the translation reaction is separated from translational substrate buffer (TSB) not by a semipermeable membrane but by carefully overlaying TSB ***(16)***.
3. The CFCF protein synthesis method, where the translation reaction is separated from TSB by a membrane ***(17)***.

Advantages of the last two methods are that there is constant removal of translational byproducts and replenishment of substrates over a period of time, leading to much higher protein yields. These two modes are preparative scale. However, the quality of the extract also plays the important role in allowing translation over a longer time ***(11)***. In the following subsections, we explain all three modes of protein translation in detail.

4.1. Batch Mode Cell-Free Translation

This protocol is for the 50 μL batch-type translation as reported elsewhere ***(11)***. Scale-up can be achieved if desired.

1. Prepare 50 μL of translational reaction mixture by combining 10 μL of wheat embryo extract (thus 20%), 2 μL of 1 mg/mL creatine kinase, 10 μL of 5X WEB, and 28 μL of dissolved mRNA from 10 μL transcription mixture. If necessary, add (^{14}C)leucine (final 2 μCi/mL).
2. Incubate at 26°C for 4 h (for results, *see* **Fig. 3** in Chapter 11).

4.2. Bilayer Method of Translation

The bilayer method uses no membrane *(16)*, but performs similarly to CFCF. In principle, exchange of substrates and dilution of byproducts take place between the translation mix and TSB by expansion of the translation mix, whereas as in the CFCF method, exchange takes place across the membrane by diffusion.

1. Prepare 50 μL of translational reaction mixture by combining 10 μL of wheat embryo extract (thus 20%), 2 μL of 1 mg/mL creatine kinase, 10 μL of 5X WEB, and 28 μL of dissolved mRNA from 25 μL of transcription mixture.
2. Add 250 μL of TSB in the U-shaped microtiter plate well.
3. Carefully place 50 μL of the reaction mixture at the bottom of microtiter plate well (as shown in **Fig. 3**).
4. Seal the plate and wrap with Saran Wrap to avoid evaporation.
5. Keep the plate in an incubator at 26°C for 16 to 18 h without shaking.

4.3. CFCF Method of Translation Using a Dialysis Cup

This protocol is for translating the proteins at preparative scale using 50-μL to 2-mL reaction volumes. The level of protein synthesis during a time course was analyzed by sodium dodecyl sulfate (SDS)-polyacrylamide gel electrophoresis (PAGE) and is shown in **Fig. 4**.

1. Prepare 50 μL of translational reaction mixture by combining 10 μL of wheat embryo extract (thus 20%), 2 μL of 1 mg/mL creatine kinase, 10 μL of 5X WEB, and 28 μL of dissolved mRNA from 50 μL transcription mixture.
2. Wash the dialysis cup with 200 μL of Milli-Q water three times.
3. Add 2 mL of TSB to a Whatman 24-well plate.
4. Add the reaction mixture into dialysis cup.
5. Place the cup into the Whatman plate.
6. Wrap with Saran Wrap to avoid evaporation.
7. Keep the 24-well plate in an incubator at 26°C for 24 h.

If the reaction is to proceed more than 48 h, add an additional quantity of mRNA in **step 1**.

5. Robotic Genome-Scale Protein Synthesizer, GenDecoder

We have recently established an automated protein synthesizer, GenDecoder (**Fig. 5**), based on our efficient and robust wheat germ cell-free translation system *(18)*. As shown in the flow chart (**Fig. 5A**), the system begins with transcription of DNA templates generated by the "split primer" PCR *(12)* in multi-well microtiter plates, and can cover all steps up to protein production, being equipped with all the essentials for the purification of mRNA, such as pipetting, mixing, incubating, and centrifuging, then subsequent translation using the bilayer method. Autoradiographic analysis has revealed that approx 95% of the tested

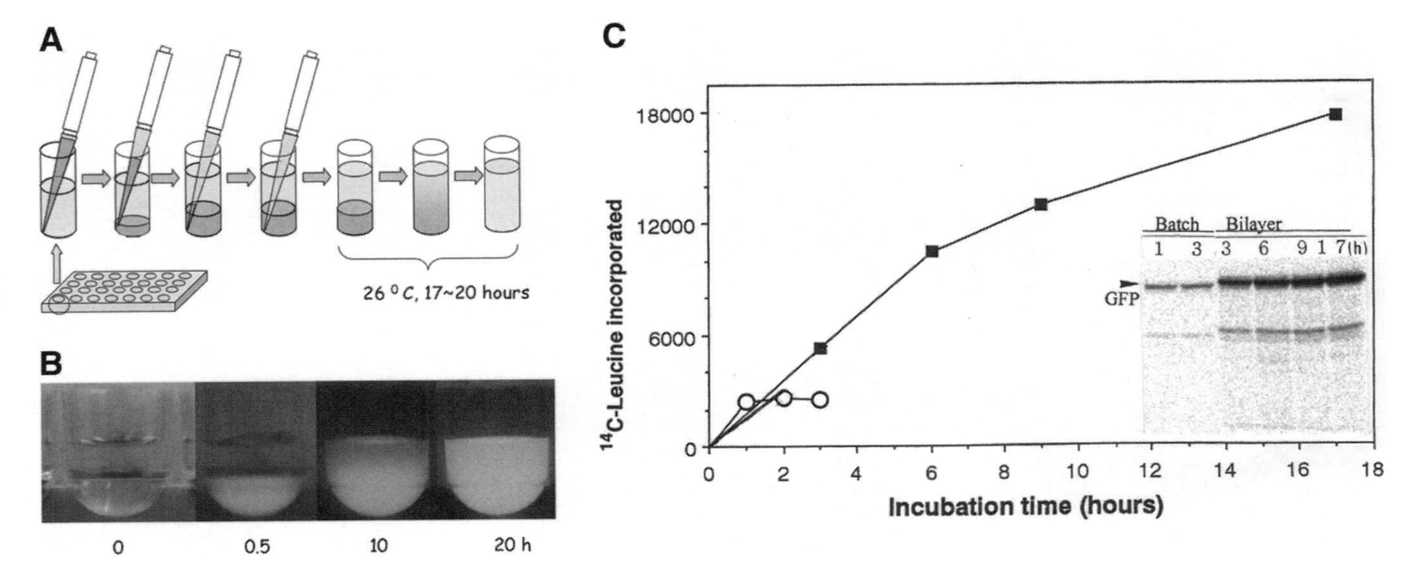
A
26 °C, 17~20 hours
B
0
0.5
10
20 h
C
18000
12000
6000
0
^{14}C-Leucine incorporated
0
2
4
6
8
10
12
14
16
18
Incubation time (hours)
Batch
Bilayer
1
3
3
6
9
1
7(h)
GFP

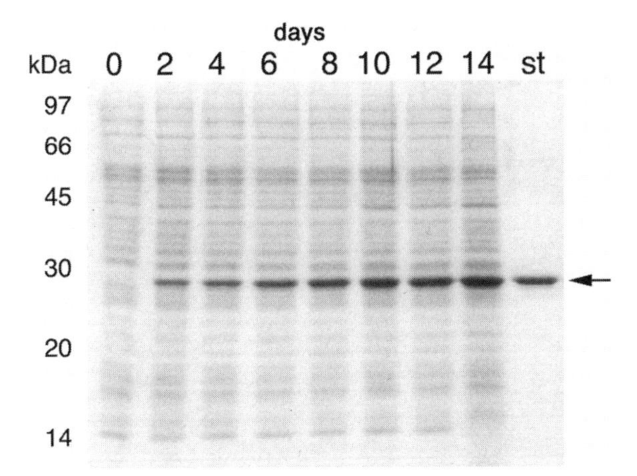

Fig. 4. Massive production of green fluorescent protein (GFP) by the continuous-flow cell-free method. Sodium dodecyl sulfide-polyacrylamide gel electrophoresis analysis of GFP produced during 14 d of reaction. mRNA produced by transcription of circular plasmid of Ehime University (pEU) was used for the translation reaction in the dialysis membrane system and was added every 48 h. A 0.1-μL aliquot of the mixture was run on the gel and protein bands were stained with Coomassie® Brilliant Blue. The arrow shows GFP and "st" designates an authentic GFP band (0.5 μg).

cDNA samples showed protein production. One-fourth of those samples were clearly visible with Coomassie® Brilliant Blue staining. The capability of the robotic protein expression was comparable with the manual operation, when *Arabidopsis* and human cDNAs were tested (**Fig. 5C**, arrowheads). Currently, the GenDecoder can handle four 96-well multi-titer plates (384 samples) in a fully automated manner in 24 h. Theoretically the power of the system can be further increased up to 3000 samples in 24 h without difficulty by improving the pipetting efficiency.

Fig. 3. *(Opposite page)* A novel bilayer cell-free protein synthesis. **(A)** Schematic illustration of the method. Wheat embryo cell-free system as described under **Subheading 4.2.** **(B)** Synthesis of green fluorescent protein (GFP) by bilayer mode. **(C)** The bilayer method (○), bilayer but mixed (■). For the measurement of (^{14}C)leucine incorporation, samples were vortexed and hot trichloroacetic acid-insoluble radioactivity in 5 μL in the batch reaction or 30 μL in the bilayer reaction, thus adjusted amount of extracts in each system. The inset shows autoradiograms, and arrowheads mark GFP. Reprinted from ***(16)*** by permission of Federation of the European Biochemical Societies.

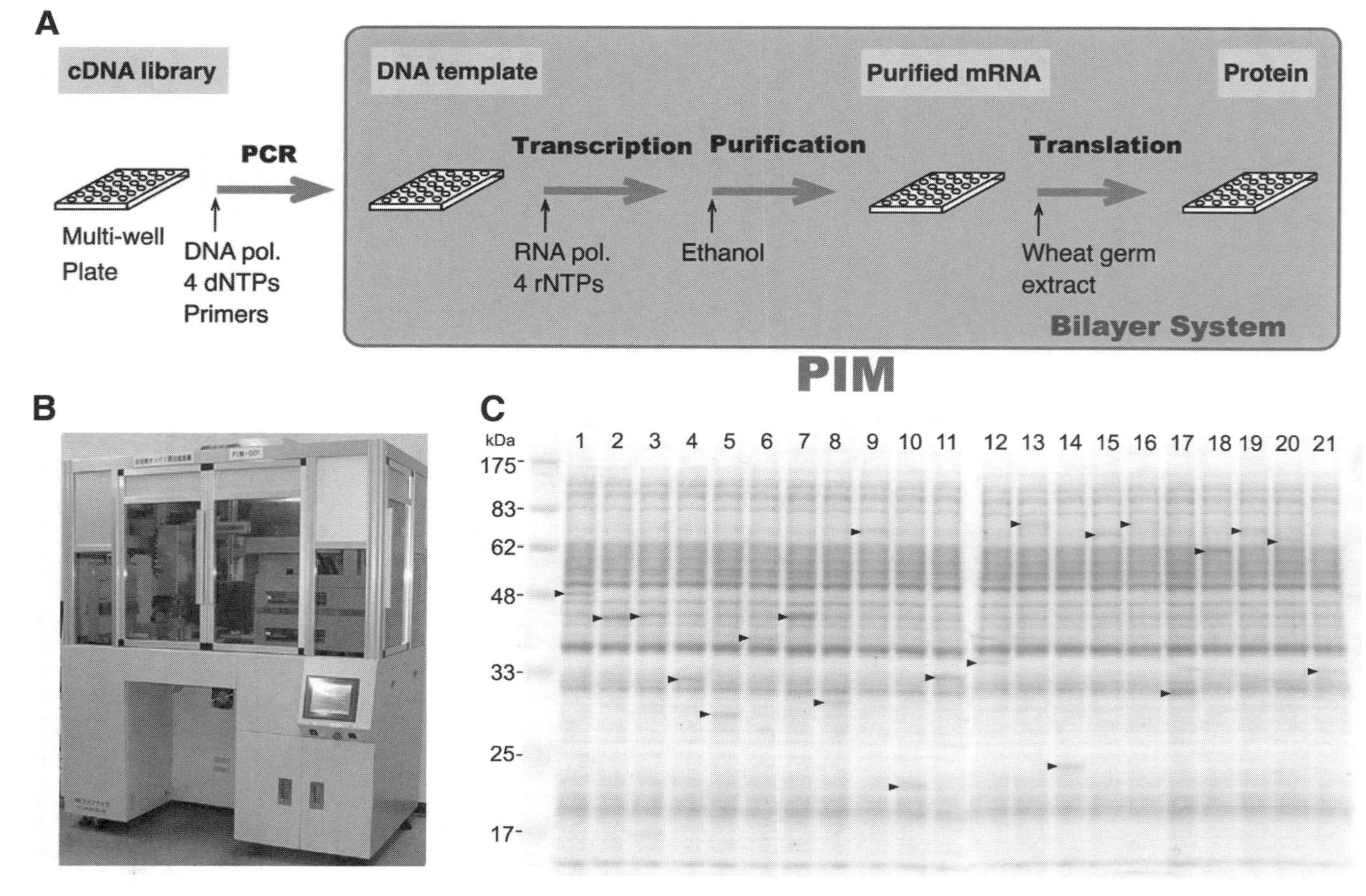
A
cDNA library
Multi-well
Plate
PCR
DNA pol.
4 dNTPs
Primers
DNA template
Transcription
RNA pol.
4 rNTPs
Purification
Ethanol
Purified mRNA
Translation
Wheat germ
extract
Protein
Bilayer System
PIM
B
C
kDa
175-
83-
62-
48-
33-
25-
17-
1 2 3 4 5 6 7 8 9 10 11 12 13 14 15 16 17 18 19 20 21

Because this protocol bypasses most of the time-consuming steps, GenDecoder may provide a powerful means of rapid and systematic screening in many applications *(5,6)*. These include high-throughput enzymatic testing of a large number of gene products for functional annotation, high-throughput crystallization of proteins for identification of their three-dimensional structure through nuclear magnetic resonance (NMR) or X-ray diffraction, rapid evolutionary design of proteins, construction of protein-protein interaction identification systems, and industrial-scale protein production systems. The information accumulated on gene product structure and function should revolutionize our understanding of biology and fundamentally alter the practice of medicine, possibly influencing other industries as well.

References

1. Blaschke, U. K., Silberstein, J., and Muir, T. W. (2000) Protein engineering by expressed protein ligation. *Methods Enzymol.* **328,** 478–496.
2. Henrich, B., Lubitz, W., and Plapp, R. (1982) Lysis of *Escherichia coli* by induction of cloned phi X174 genes. *Mol. Gen. Genet.* **185,** 493–497.
3. Golf, S. A. and Goldberg, A. L. (1987) An increased content of protease La, the lon gene product, increases protein degradation and blocks growth in *Escherichia coli. J. Biol. Chem.* **262,** 4508–4515.
4. Chrunyk, B. A., Evans, J., Lillquist, J., Young, P., and Wetzel, R. (1993) Inclusion body formation and protein stability in sequence variants of interleukin-1 beta. *J. Biol. Chem.* **268,** 18,053–18,061.
5. Kurland, C. G. (1982) Translational accuracy in vitro. *Cell* **28,** 201–202.
6. Pavlov, M. Y. and Ehrenberg, M. (1996) Mutants of EF-Tu defective in binding aminoacyl-tRNA. *Arch. Biochem. Biophys.* **328,** 9–16.
7. Roberts, B. E. and Paterson, B. M. (1973) Efficient translation of tobacco mosaic virus RNA and rabbit globin 9S RNA in a cell-free system from commercial wheat germ. *Proc. Natl. Acad. Sci. USA* **70,** 2330–2334.
8. Wool, I. G., Glück, A., and Endo, Y. (1992) Ribotoxin recognition of ribosomal RNA and a proposal for the mechanism of translocation. *Trends Biochem. Sci.* **17,** 266–269.

Fig. 5. *(Opposite page)* Pictorial representation of GenDecoder specialized for functional analysis of genetic information. **(A)** Flow chart of cell-free expression process. **(B)** A photograph of GenDecoder, dimensions 1.5 m (W), 0.85 m (D), and 1.8 m (H). The robot is equipped with three robotic arms for pipetting and plate transfer, one incubator for transcription and translation, maximum capacity four plates, and centrifuge for mRNA recovery after ethanol precipitation. Incubator capacity can be increased by handling plates manually. **(C)** Proteins in 1 μL from each translated mixture were analyzed as in **Fig. 4**, thus 0.2 μL of original translational mix. (Reprinted from **ref.** ***18*** with kind permission of Springer Science and Business Media.)

9. Barbieri, L., Battelli, M. G., and Stirpe, F. (1993) Ribosome-inactivating proteins from plants. *Biochim. Biophys. Acta* **1154,** 237–282.
10. Ogasawara, T., Sawasaki, T., Morishita, R., Ozawa, A., Madin, K., and Endo, Y. (1999) A new class of enzyme acting on damaged ribosomes: ribosomal RNA apurinic site specific lyase found in wheat germ. *EMBO J.* **18,** 6522–6531.
11. Madin, K., Sawasaki, T., Ogasawara, T., and Endo, Y. (2000) A highly efficient and robust cell-free protein synthesis system prepared from wheat embryos: plants apparently contain a suicide system directed at ribosomes. *Proc. Natl. Acad. Sci. USA* **97,** 559–564.
12. Sawasaki, T., Ogasawara, T., Morishita, R., and Endo, Y. (2002) A cell-free protein synthesis system for high-throughput proteomics. *Proc. Natl. Acad. Sci. USA* **99,** 14,652–14,657.
13. Sambrook, J. and Russell, D. W. (2001) *Molecular Cloning: A Laboratory Manual, 3rd ed.*, Cold Spring Harbor Laboratory Press, Cold Spring Harbor, NY.
14. Gurevich, V. V. (1996) Use of bacteriophage RNA polymerase in RNA synthesis. *Methods Enzymol.* **275,** 382–397.
15. Hanes, J. and Pluckthun, A. (1997) In vitro selection and evolution of functional proteins by using ribosome display. *Proc. Natl. Acad. Sci. USA* **94,** 4937–4942.
16. Sawasaki, T., Hasegawa, Y., Tsuchimochi, M., et al. (2002) A bilayer cell-free protein synthesis system for high-throughput screening of gene products. *FEBS Lett.* **514,** 102–105.
17. Spirin, A. S., Baranov, V. I., Ryabova, L. A., Ovodov, S. Y., and Alakhov, Y. B. (1988) A continuous cell-free translation system capable of producing polypeptides in high yield. *Science* **242,** 1162–1164.
18. Endo, Y. and Sawasaki, T. (2004) High-throughput, genome-scale protein production method based on the wheat germ cell-free expression system. *J. Struct. Funct. Genomics* **5(1–2),** 45–57.

11

Advances in Genome-Wide Protein Expression Using the Wheat Germ Cell-Free System

Yaeta Endo and Tatsuya Sawasaki

Summary

In the current post-genomic era, cell-free translation platforms are gaining importance in structural as well as functional genomics. They are based on extracts prepared from *Escherichia coli* cells, wheat germ, or rabbit reticulocytes, and when programmed with any mRNA in the presence of energy sources and amino acids, can synthesize the respective protein in vitro. Among the cell-free systems, the wheat germ-based translation system is of special interest due to its eukaryotic nature and robustness. This chapter outlines the existing protein production platforms and their limitations, and describes the basic concept of the wheat germ-based cell-free system. It also demonstrates how the conventional wheat germ system can be improved by eliminating endogenous inhibitors, by using an expression vector specially designed for this system and polymerase chain reaction-directed protein synthesis directly from cDNAs in a bi-layer translation system. Finally, a robotic procedure for translation based on the wheat germ extract and bi-layer cell-free translation is described.

Key Words: Cell-free protein synthesis; wheat embryo extract; 5' and 3' ORF design; pEU construction; bilayer reaction method; robotic automation.

1. Introduction

With the completion of genomic DNA sequencing for various species, attention has shifted towards the structural and functional characteristics of proteins. However, rapid progress in proteomics requires the following: (a) availability of many proteins, and (b) availability of sufficient amounts of proteins in naturally folded states. To meet these requirements, various existing protein production technologies have been adopted for genome-wide protein translation.

From: *Methods in Molecular Biology, vol. 310: Chemical Genomics: Reviews and Protocols*
Edited by: E. D. Zanders © Humana Press Inc., Totowa, NJ

1.1. Limitations of the Recombinant Protein Expression System

Currently, three strategies are being used for protein production: chemical synthesis, in vivo expression, and cell-free protein synthesis. Whereas chemical synthesis is not practical for the synthesis of longer peptides ***(1)***, in vivo expression systems have several inherent limitations. Although the recombinant *Escherichia coli* system seems to be the best option in terms of economy and ease of protein production, eukaryotic proteins often get precipitated and require time-consuming procedures to solubilize and refold those proteins ***(2)***. Mammalian cell expression systems can be effective tools, but are plagued by difficulties in purifying recombinant proteins, by limitations on the size of the recombinant protein expressed and expression time, and by unstable protein expression ***(3)***. Many of these obstacles have been overcome using expression systems from insect cells. The baculovirus/insect system can produce recombinant proteins in reasonable quantity and activity. However, the process is time-consuming owing to requirements for virus stock production and the need for sterile conditions to avoid contamination. In addition, because of its infectious nature, the virus destroys a significant percentage of the cell population and therefore produces a mixture of intracellular protein precursors (with uncleaved signal peptides), extracellular mature proteins, and proteolytic degradation products ***(4,5)***. This is obviously problematic for the preparation of active proteins. Furthermore, another common and even more serious disadvantage of in vivo expression systems is that they are limited to producing proteins that do not interfere with host cell physiology ***(6–8)***. Overall, protein production via in vivo systems has limitations in terms of high-throughput and functionally active protein production, low yield, economy, and/or ethical issues.

1.2. Cell-Free Translation Systems

Cell-free translation systems can produce proteins with high speed and accuracy, approaching in vivo rates ***(9,10)***, and they can express proteins that seriously interfere with cell physiology.

Although researchers realized in the early 1950s that protein biosynthesis can continue after cell disruption, the first cell-free translation systems were made from the cytoplasmic extract of rat liver cells ***(11)***. Later, cell-free translation systems from *E. coli* cells were prepared ***(12,13)***. However, these systems were based on the endogenous mRNA remaining attached to ribosomes while the extract was prepared. The landmark in the history of cell-free translation was the translation of an external source of mRNA in *E. coli* extracts ***(14)***. The authors demonstrated that the pre-incubation of extracts at physiological temperature could eliminate endogenous mRNA from the extract; ribosomes in the extract were found to accept either exogenous natural mRNA or synthetic polyribonucleotide as templates for protein synthesis. Then in the mid-1970s, cell-free systems based on wheat germ extracts ***(15)*** and rabbit reticulocyte lysates ***(16)*** were reported. Whereas the wheat germ extract could be used for the expression of external sources of mRNA without any pretreatment because of the low levels of endogenous mRNA, rabbit reticulocyte lysate required pretreatment with micrococcal nuclease. Even though attempts have been made to prepare the extracts from various prokaryotic and eukaryotic sources for cell-free translation, extracts from *E. coli*, wheat germ, and rabbit reticulocyte lysate remain the major platforms at the present time.

The productivity of the *E. coli* cell-free translation system has been widely acknowledged ***(17–20)***, but proper protein folding remains a substantial challenge, especially for the synthesis of multi-domain proteins from eukaryotic genomes, owing to the prokaryotic nature of the translation mechanism. Although attempts are being made to synthesize eukaryotic proteins in the *E. coli* cell-free system by adding external additives and molecular chaperones ***(21)***, these make multiple protein synthesis in parallel more complicated. On the other hand, the rabbit reticulocyte lysate cell-free expression system has not been adapted to provide preparative amounts of synthesized proteins because of its low efficiency, instability, high ribonuclease content, high endogenous globin content, and related ethical issues. The rabbit reticulocyte lysate is less available and more expensive than comparable amounts of wheat germ and *E. coli* extract. Therefore, the system is inefficient from the point of view of current demands for proteins. Many of these problems can be circumvented by using the wheat germ cell-free translation system. Because the system is derived from a eukaryotic plant source, it is especially powerful for synthesizing eukaryotic multi-domain proteins in folded state on a preparative scale. However, the conventional wheat germ system is plagued by a short half-life. We have been actively involved in improving it by understanding the system from a fundamental as well as technological view. Here we review how the system has been improved in terms of half-life, protein expression, and suitability of the system for automation.

1.3. Improvement of the Conventional Wheat Germ Cell-Free Translation System to Meet the Proteomics Demand

In order to improve the wheat germ cell-free translation to meet the high-throughput needs of proteomics, we addressed several critical parameters and solved them in a stepwise manner (**Fig. 1**) as follows.

1. Understanding the cause of the instability of wheat germ cell-free translation system over time. During our early investigations, we found that plants contain endogenous inhibitors of translation ***(22,23)*** and demonstrated that elimination of these inhibitors led to an extraordinarily stable and efficient translation system ***(24)***.

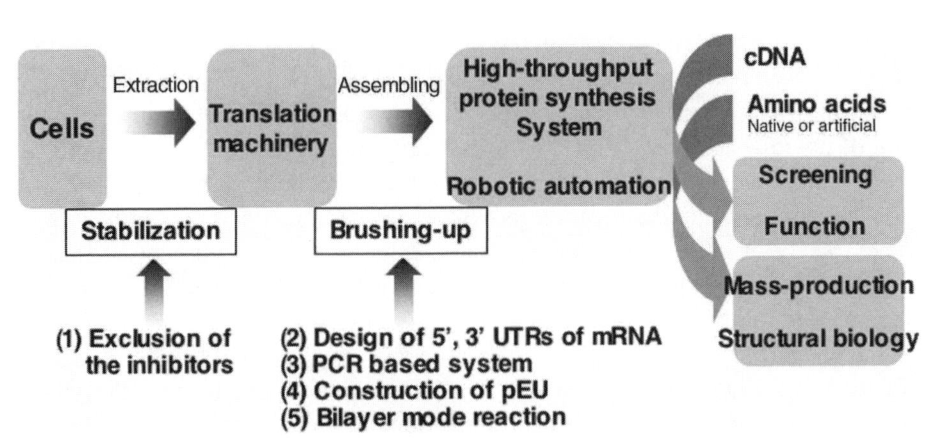

Fig. 1. Overview of the wheat germ cell-free translation system developed at our laboratory. The sequential establishment of the system started with stabilization, followed by integration with flexible plasmid of Ehime University (pEU) vector and finally high-throughput protein synthesis. (From **ref. *21a***, with permission from Elsevier.)

2. To improve the performance of the system further, several critical design parameters were considered ***(25)***: (a) elimination of the 5'-^{7}mGpppG (cap) and poly(A)-tail (pA) by optimizing the 5'- and 3'-UTRs of mRNA, thereby increasing translation initiation; (b) design of split primers for polymerase chain reaction (PCR) to generate transcription templates directly from *E. coli* carrying cDNAs, thus bypassing time-consuming cloning steps; (c) construction of an expression vector, specialized for the massive production of proteins.
3. We created an alternative to Spirin's continuous-flow cell-free (CFCF) protein synthesis method of cell-free translation by employing an innovative bi-layer ***(17)*** method without the use of any membrane
4. Automating the translation steps, starting with transcription in the wheat germ cell-free translation system (improved as above) using robotics ***(26)***.

The resulting system, with or without automation, was attractive for the systematic study of protein structure and function in the context of modern proteomics.

2. Preparation of a Highly Active and Robust Extract From Wheat Embryos

One of the most convenient and promising eukaryotic cell-free translation systems is based on wheat embryos, which store all the necessary components of translation. However, conventional wheat-germ extracts are plagued with a short half-life, and consequently, with low expression of proteins. In recent findings, we proposed that the possible cause for instability is the presence in the original endosperm extract of RNA *N*-glycosidase, tritin, and possibly other inhibitors of translation, such as thionin, ribonucleases, deoxyribonucleases, and proteases. To confirm the idea of tritin contamination of the extract, initially we measured the modification of ribosomes by tritin, a highly active and specific enzyme *(27)*. When germs were isolated from dry wheat seeds by conventional procedures *(28)*, microscopic examination revealed that the embryos contain some white material and a number of white and brownish granules (**Fig. 2A**). Analysis of ribosomal RNAs from protein synthesis reactions prepared from such samples showed the depurination of ribosomes (**Fig. 2B**). Based on the aniline-dependent appearance of specific RNA fragments (**Fig 2B**, arrow) and its intensity on the gel, it can be seen that 24% of the ribosome population had been depurinated at the end of the 4-h incubation. Furthermore, 7% of the population had already been depurinated at the start of the incubation. When RNA was extracted directly from embryos by guanidine isothiocyanate-phenol, little formation of the aniline-induced fragment was observed (**Fig 2B**, lanes 7 and 8). Thus, depurination must have started during the extract preparation, and continued during protein synthesis.

It is known that inactivation of any one ribosome among actively translating ribosomes results in cessation of translation through blockage of the respective polysomes *(27)*. Thus, the observed extent of depurination by tritin in those experiments can severely damage protein synthesis. After the unsuccessful attempt to neutralize the depurinating enzyme with synthetic RNA aptamers, expected to bind tightly to the ribosome-inactivating protein (RIP) *(29)*, we tried washing the embryos prior to preparation of extract. This was an improvement, because embryos that were extensively washed prior to extract preparation yielded better results *(24)* in terms of elimination of white patches on the embryos (**Fig. 2A**, left panel), and undetectable depurination of ribosome (**Fig. 2B**, lanes 10–12). Based on these results, it was concluded that endogenous RIP (tritin) and other inhibitors such as thionin, ribonucleases, deoxyribonucleases, and proteases were eliminated simultaneously from the embryo fraction by this simple washing procedure.

The cell-free system prepared from washed embryos has much higher translational activity than the conventional system (compare **Fig. 3A** and **B**). When 5'-capped dihydrofolate reductase (DHFR) mRNA containing 549 nt of 3' UTR with a pA tail was incubated with newly prepared as well as conventional extract, there was almost linear kinetics in DHFR synthesis over 4 h, compared with the regular system, which ceased to function after 1.5 h. Further, when washed extract in the reaction volume was increased to 48%, amino acid incorporation occurred initially at a rate twice that of 24% extract, and then stopped after 1 h. However, this pause was caused by a shortage of substrates rather than an irreversible inactivation of ribosomes or factors necessary for translation; addition of amino acids, ATP, and GTP after cessation of the reaction (arrow) restarted

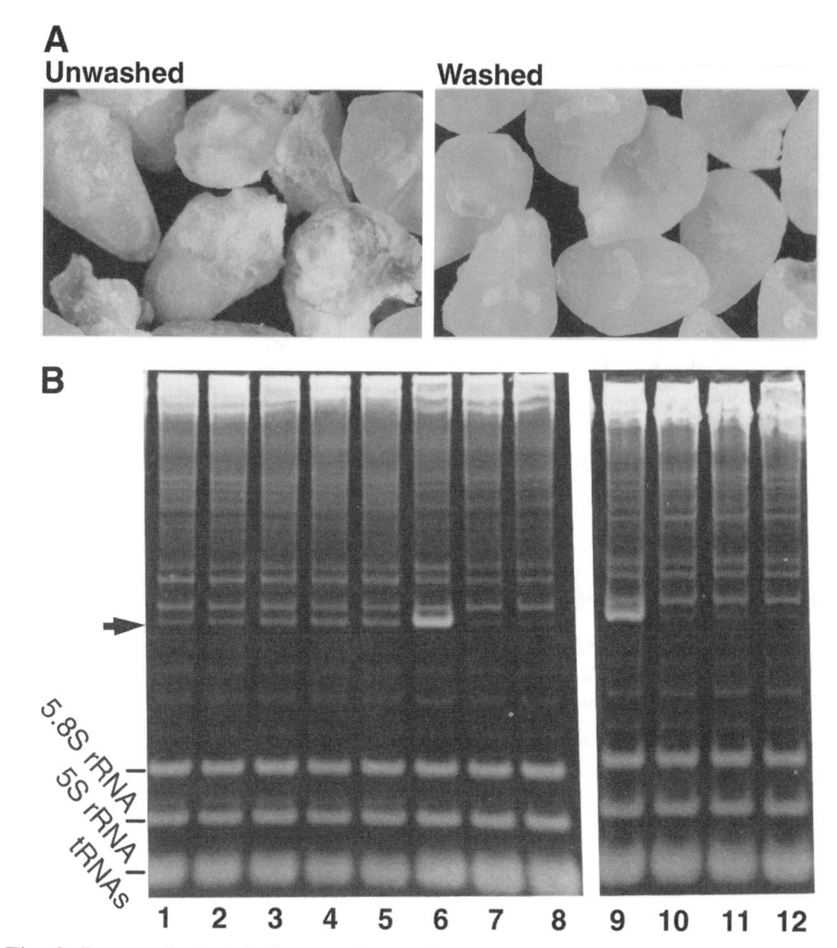

Fig. 2. Removal of tritin from embryos. Extracts were prepared from unwashed or washed embryos **(A)** and the depurination assay was performed **(B)**. Translation mixtures prepared with the extract from unwashed embryos were incubated for 0, 1, 2, 3, 4 h (lanes 1–5, respectively); mixtures with washed embryos were incubated for 0, 2, 4 h (lanes 10–12, respectively). Isolated RNA was treated with acid/aniline, and then separated on 4.5% polyacrylamide gels. Additionally, RNA was directly extracted from embryos with guanidine isothiocyanate-phenol and analyzed before (lane 7) and after (lane 8) treatment with acid/aniline. For the fragment marker, incubation was carried out in the presence of gypsophilin, a highly active ribosome-inactivating protein from *Gypsophila elegance*; the arrow indicates the aniline-induced fragment.

translation with kinetics similar to the initial rate. In contrast, when conventional extract was increased to 48%, protein synthesis decreased significantly compared with the 24% extract reaction. Also, the cessation of protein synthesis in the reaction with 24% extract could not be reversed by addition of more sub-

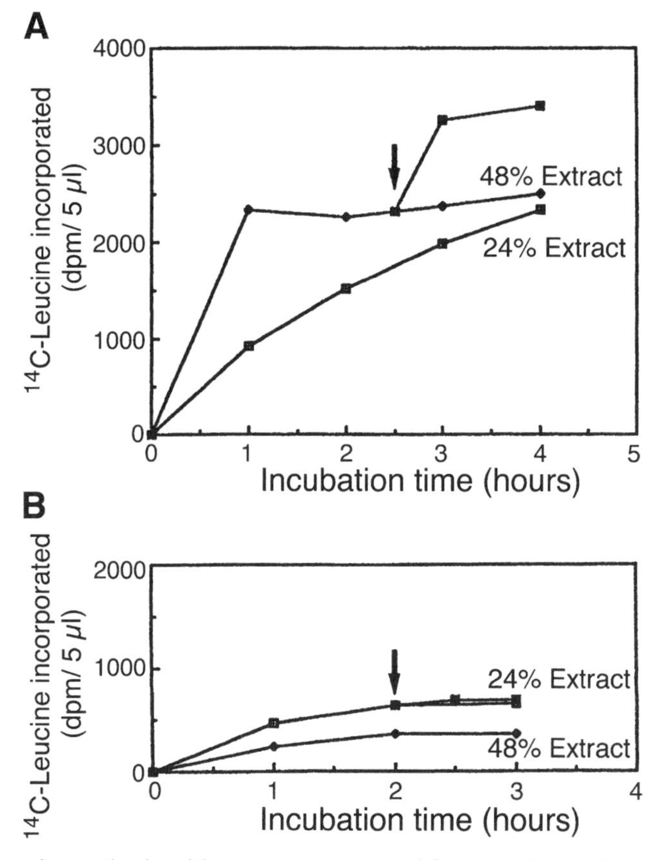

Fig. 3. Protein synthesis with an extract prepared from washed embryos. The batch system contains either 12 µL (24%) or 24 µL (48%) of extracts from washed **(A)** or unwashed wheat embryos **(B)**. Protein synthesis was measured as hot tricholoroacetic acid insoluble radioactivity. Arrows show addition of substrates.

strate, indicating irreversible damage of ribosomes by contaminants from endosperm (**Fig. 3B**). High protein synthesis activity with washed embryos was also confirmed by sucrose density gradient analysis, where significant formation of polysomes was observed after 1 h of incubation, and at 2 h a shift to heavier polysomes with a concomitant decrease of 80S monomers ***(24)***. Further, in order to check the long-term stability and productivity of this new cell-free translation, it was performed using Spirin's CFCF with continuous supply of substrates and removal of small byproducts ***(30)***; the reaction proceeded for more than 60 h, yielding 0.8–4 mg of functionally active proteins per mL of reaction mixture ***(24)***. The high efficiency of our system, therefore, can be attributed to at least two factors: first, high initiation, elongation, and termination rates (efficient usage and recycling of all translation factors, ribosomes, mRNA, tRNA, and so on); and second, low endogenous ribonuclease activity (retention of heavy polysomes for prolonged time). These results further support that the extensive washing of wheat embryos to eliminate endosperm contaminants resulted in extracts with high stability and activity.

3. Removal of cap and pA by Optimizing the mRNA 5' and 3' UTR

The 5' and 3' UTRs of eukaryotic mRNA play a crucial role in the regulation of gene expression by controlling mRNA translational efficiency, stability, and localization. However, for in vitro translation, the use of cap and pA has several limitations. The competition between cap molecule and capped mRNA to bind to the initiation factor eIF4E can lead to the inhibition of translation initiation and thus a narrow range of optimal concentration of mRNA. Also, accurate determination of mRNA prior to the start of each reaction for efficient translation becomes tedious when synthesizing many proteins at high-throughput from large numbers of genes, and producing large amounts of protein in CFCF, over a long period. In addition to these problems, long plasmid-encoded poly(dT/dA) is usually unstable during replication in host cells, and the cost of cap analog for mRNA synthesis is high.

To overcome these limitations, we endeavored to find 5'- and 3'-UTRs that enhance mRNA translation in the absence of cap and pA, by using pSP65 as a starting vector. We chose genomes of plant RNA viruses containing a positive-sense RNA lacking both cap and pA for this study ***(31)***. Initially focusing on the 5'-UTR by screening 5'-enhancer elements with less than 100 nts, we chose 71 nt UTR omega sequence (Ω71) of TMV as the most promising, owing to its known function as a general translation enhancer both in vivo and in vitro ***(31–33)***. A template coding for luciferase with the Ω71 5' UTR was synthesized by in vitro transcription whereas the 549 nts 3' UTR and a 100 nts pA were left intact. Activity assays in the batch system showed that this mRNA exhibited 34% higher activity than without cap and pA, observations consistent with an earlier report ***(32)***. Further, 77% improvement in the template activity was observed by inserting GAA at the 5'-end of Ω71 (GAA-Ω71) compared with control level. Next we examined the influence of the 3' UTR on mRNA activity. A systematic examination of the activity on luciferase mRNAs having different length 3'-UTRs and sequences with GAA-Ω71 as 5'-UTR revealed that the length of the 3'-UTR is more important than its sequence for efficient translation in the wheat cell-free system ***(25)***. These results are also consistent with studies reporting the presence of an exosome that degrades mRNA from the 3' end ***(34,35)***. As shown in **Fig. 4**, the optimal concentration for each mRNA template having GAA-Ω71 as 5'-enhancer and the 1626 nts (sequence-5) 3'-UTR

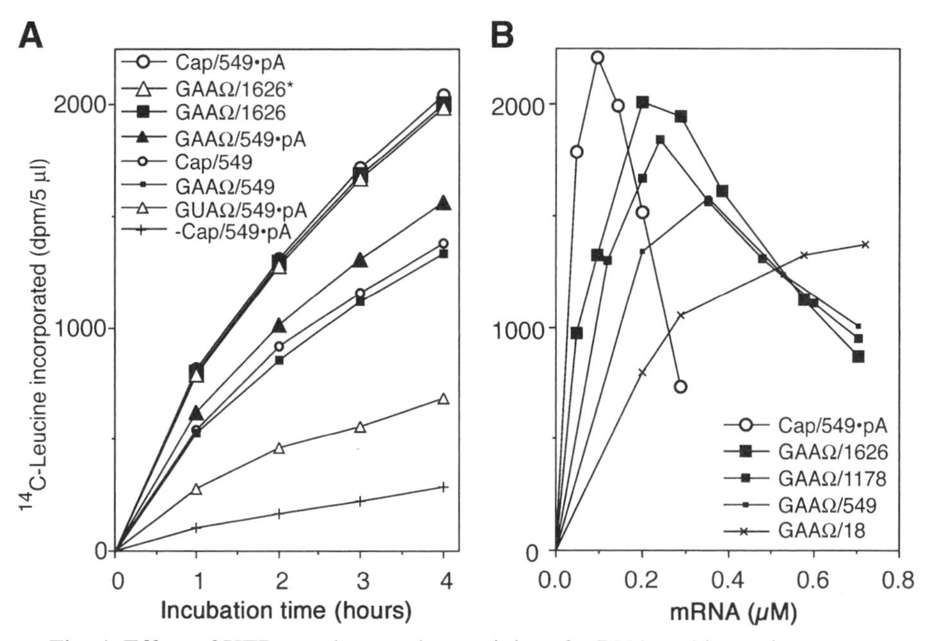

Fig. 4. Effect of UTRs on the template activity of mRNA and its optimum concentration. Protein synthesis activity was measured in the presence of [^{14}C]leucine (50 nCi/ 25 µL) using one of the following luciferase mRNAs: with 5'-cap and a 549 nts 3' UTR plus 100 nts of pA at the 3'-terminal (cap/549pA); with 5'-cap and 3'-549 but lacking pA (cap/549); with 5'-GUAΩ71 and 3'-549 plus pA (GUAΩ/549pA); with 5'-GAAΩ and 3'-18 but without pA (GAAΩ/18); with 5'-GAA-Ω and 3'-549 plus pA (GAAΩ/ 549 pA); with 5'-GAA-Ω and 3'-549 but without pA (GAAΩ/549); with 5'-GAA-Ω and 1178 3' nts (GAAΩ/1178); with 5'-GAA-Ω and 1626 3' nts (GAAΩ/1626); and with 1626 3' nts of a sequence from the negative strand of the plasmid DNA (GAAΩ/ N-1626). In **A**, reactions were carried out in 25 µL containing 2.5 pmol mRNAs (cap/ 549pA, cap/549), or 5 pmol of the other mRNAs. In **B**, 25-µL reaction mixtures containing 2.5 pmol mRNA (cap/549pA), or 5 pmol of other mRNAs. Incubations were done at 26°C for the indicated periods of time **(A)**, or 4 h in **(B)**, and hot-trichloroacetic acid-insoluble radioactivity was measured.

showed the highest template activity, similar to that of original mRNA with a 5' cap and a 549 nts 3' UTR with a pA of 100 nt. These optimized 5'- and 3'-UTRs were assembled in pSP65-derived vector and renamed plasmid of Ehime University (pEU). The designed templates with 5'- and 3'-UTR work well over a wide range of mRNA concentrations, allowing us to overcome the limitations associated with optimization of mRNA concentration, a highly desirable feature especially when aiming for high-throughput protein synthesis.

4. Split-Primer Design for the Construction of PCR-Generated Linear Template DNAs for Transcription and Translation

Cloning of cDNA into an expression vector is one of the most laborious and time-consuming steps of recombinant protein synthesis. Attempts have been made to direct synthesis from PCR-generated linear cDNAs *(18,36)*. These methods generally involve the incorporation of transcriptional and translational start and stop signals into primers for the generation of autonomous coding fragments by PCR. In order to establish a general and practical method for the cell-free expression of proteins, we focused our efforts on minimizing the primer-dimer effect during PCR by designing split primers *(25)*. **Figure 5** shows the results of protein synthesis in the wheat germ cell-free system in which mRNA from four different genes was transcribed following a typical PCR strategy *(25)* —four specific primers (**Fig. 5A-a**) with a unique primer for each gene (legend) and the remaining three in common for all genes. Each PCR product was used for transcription reactions, and synthesized RNAs were separated on agarose gels (**Fig. 5A-e**). Although **Fig. 5A-d** shows the PCR-generated DNA fragments as single major band in each lane, apparently with the correct size (**Fig. 5A-d**, arrow), the majority of the transcripts was shorter than 500 nts and migrated to the bottom of the gel (**Fig. 5A-e**, arrow). The upper bands of each lane (marked with asterisks) are mostly the DNA templates as confirmed by DNase-l treatment (results not shown). When the batch mode translated products are separated by sodium dodecyl sulfide (SDS)-polyacrylamide gel electrophoresis (PAGE) and visualized by autoradiography (**Fig. 5A-f**), the results show that for each mRNA, the main products are mostly short peptides of less than 10 kDa (arrowhead) rather than full-length proteins. Expected protein bands were still faint even after long exposure (asterisks in **Fig. 5A-f**). These unwanted bands on SDS-PAGE can be the cause of nonspecific amplification along with the full-length DNA template *(38)*. Possible artifacts include either DNAs with the 5'-UTRs with short parts of the ORF (**Fig. 5A-b-2**), or DNAs having the N-terminal portion but lacking the 5' enhancer (**Fig. 5A-b-3**). Although it is possible to partially overcome the above-mentioned problems with modified amplification conditions, such as the primer molar ratios to the cDNA and the annealing

---

Fig. 5. *(Opposite page)* Polymerase chain reaction (PCR)-based expression of cDNA information using conventional method **(A)** and a new strategy using split-type primers **(B)**. **a**. Design of the split-type primers for the introduction of the required UTRs into cDNA sequences. **b** and **c**. Expected PCR-generated DNAs and mRNA, respectively. **d**. PCR-generated DNA. **e**. A 10-μL aliquot from a PCR sample was used for the 100-μL transcription reaction, and transcripts were analyzed. **f**. All of the transcript was used for batch-mode translation (50 μL, 4 h), and products were analyzed by autoradiography. (From **ref.** ***36a***.)

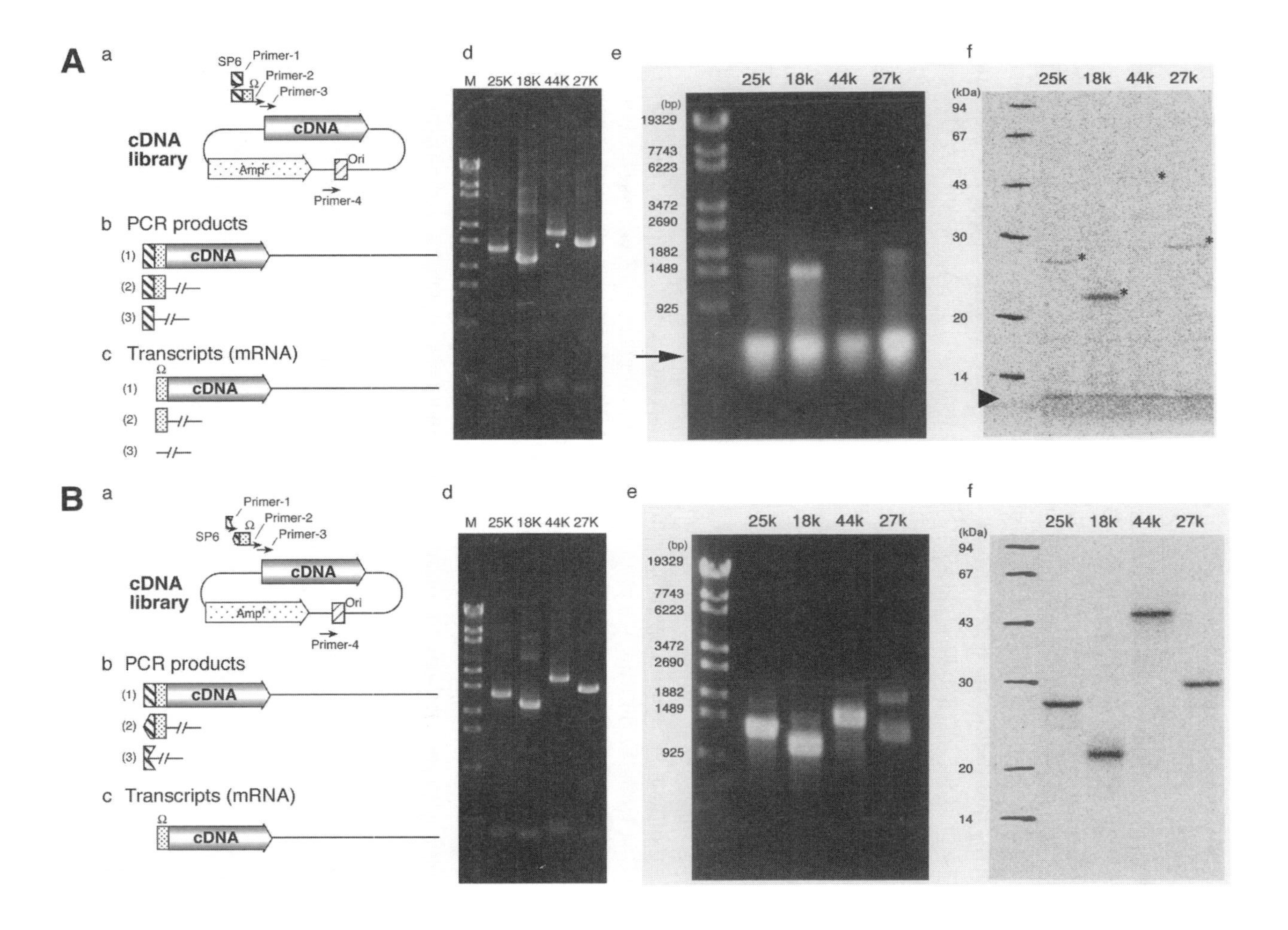
A
a
SP6
Primer-1
Ω
Primer-2
Primer-3
cDNA
cDNA library
Amp^r
Ori
Primer-4
b PCR products
(1)
cDNA
(2)
(3)
c Transcripts (mRNA)
Ω
(1)
cDNA
(2)
(3)
d
M 25K 18K 44K 27K
e
25k 18k 44k 27k
(bp)
19329
7743
6223
3472
2690
1882
1489
925
f
25k 18k 44k 27k
(kDa)
94
67
43
30
20
14
B
a
Primer-1
SP6
Ω
Primer-2
Primer-3
cDNA
cDNA library
Amp^r
Ori
Primer-4
b PCR products
(1)
cDNA
(2)
(3)
c Transcripts (mRNA)
Ω
cDNA
d
M 25K 18K 44K 27K
e
25k 18k 44k 27k
(bp)
19329
7743
6223
3472
2690
1882
1489
925
f
25k 18k 44k 27k
(kDa)
94
67
43
30
20
14

temperatures *(37)*, this strategy is impractical for high-throughput synthesis. Thus we explored a different approach to minimize nonspecific products.

The new strategy for PCR amplification introduces a split primer as illustrated in **Fig. 5B** *(25)*. The single primer for the promoter sequence was separated into two parts (**Fig. 5B-a**)—hence split-primer PCR. With this strategy, when the unpurified PCR products were used as transcription templates, RNA products migrated as single bands of the expected size (**Fig. 5B-e**) similar to DNA fragment patterns on the agarose gel (**Fig. 5B-d**). **Figure 5B-f** demonstrates efficient production of the same four proteins, each present as a single major band in the autoradiogram together with very few small peptides *(25)*. For a detailed protocol for the generation of PCR-based template DNA for cell-free translation, refer to Chapter 10. It is important to note here that reaction conditions for PCR and transcription were identical for all four genes and were not optimized for each gene.

To further confirm the suitability of our approach for high-throughput protein synthesis, we carried out cell-free reactions with *E. coli* cells carrying cDNAs *(25)*. All pipetting steps were performed automatically using a Qiagen BioRobot-8000 with standard 96-vessel titer plates for transcription, and 24 dialysis cups as translation chambers. For demonstration purposes, we chose 27 *Arabidopsis* and human genes with corresponding protein molecular weights from 10 to 85 kDa (**Table 1**). The unique primer (primer-3) for each of 13 DNAs (5 from *Arabidopsis* and 8 from human) was designed directly from data in GenBank and in AtMIPS. The other 14 cDNAs were randomly cloned from commercially available phage cDNA libraries of *Arabidopsis* plants and human T-cells. After they were sequenced, each primer-3 was designed using the sequencing data.

Two types of PCR-generated DNA constructs were prepared—one for the synthesis of an authentic protein, and the other for the synthesis of a fusion protein with glutathione *S*-transferase (GST) as a tag for affinity purification. PCR was carried out by employing the split-primer method, but the reaction was started with *E. coli* cells (*see* the legend to **Fig. 6**). After transcription (which was carried out in the wells of a microtiter plate), samples were precipitated and washed with ethanol to remove dNTPs, NTPs, and magnesium ions. The mRNAs were dissolved in water and transferred directly into dialysis cups, which contained the translation mixture. The cups were then dipped into translation buffer (**Fig. 6**). After 48 h of protein synthesis in the CFCF mode, 1-μL aliquots of three times diluted reaction mixture were separated by SDS-PAGE, and protein bands stained with Coomassie® brilliant blue (CBB) (**Fig. 6**). The results indicate that in 50 out of 54 cases (authentic and fusion proteins), the system worked very well, producing a single major protein band in each lane (asterisk), whereas in 4 cases (3 authentic proteins and 1 fusion), the amounts of protein were below the level of detection with CBB. The products were the correct size,

Table 1
Examples of Proteins Synthesized in the Polymerase Chain Reaction (PCR)-Directed Wheat Germ Cell-Free System

			Authentic		Fusion		
Proteins encoded by cDNAs	Annotation No.	M.W.	Total (mg)	Sup (%)	Total (mg)	Sup (%)	Clone Name
Arabidopsis							
(From GenBank and MIPS)							
chlorophyll a/b-binding protein	X56062	25,995	0.2	30	0.5	90	At01
Agamous-like gene 9 (AGL9)	AF015552	29,065	0.7	30	0.8	90	At02
Flowering locus T (FT)	AF152096	19,808	0.3	100	0.8	100	At03
HY5	AB005456	18,462	0.4	90	1.5	90	At04
Flowering locus F (FLF)	AF116527	21,864	0.2	100	0.4	100	At05
hypothetical protein	At1g69630[a]	11,311	0.1	40	0.1	100	At06
(From a commercial cDNA library[b])							
putative heat shock protein 40	AL021749	38,189	1.8	30	1.0	80	At07
heat shock protein 70-3	Y17053	71,144	0.9	100	–[d]	–[d]	At08
putative *s*-adenosylmethionine synthetase	AY037214	42,793	1.5	100	0.6	100	At09
NADPH thioredoxin reductase	Z23108	40,635	0.1	10	0.5	20	At10
putative ACC oxidase	AF370155	36,677	1.0	10	1.2	100	At11
putative fructokinase	AF387001	35,276	1.0	10	0.6	100	At12
rubisco activase	X14212	51,981	0.4	20	0.6	80	At13
glutaredoxin	At4g15660[a]	11,311	–[d]	–[d]	0.4	80	At14
chlorophyllase 2	AF134302	34,902	0.2	10	0.4	70	At15

(continued)

Table 1 (Continued)

Proteins encoded by cDNAs	Annotation No.	M.W.	Authentic Total (mg)	Authentic Sup (%)	Fusion Total (mg)	Fusion Sup (%)	Clone Name
Human							
(From GenBank)							
neuron-specific γ-2 enolase[e]	M22349	47,266	1.0	100	0.5	100	Hs01
ζ-crystallin/quinone reductase	L13278	35,205	2.3	80	1.3	100	Hs02
X11-like protein	AB014719	82,480	0.5	100	0.2	80	Hs03
importin α 1	NM_002266	57,859	–[d]	–[d]	0.2	30	Hs04
glyceraldehyde-3-phosphate dehydrogenase	M17851	36,051	0.4	70	0.9	100	Hs05
enolase 3[e]	NM_001976	46,956	1.7	80	0.9	100	Hs06
APBA3[e,f]	NM_004886	61,451	0.9	100	0.2	100	Hs07
JAK binding protein[e]	NM_003745	23,550	0.4	30	0.2	100	Hs08
(From a commercial cDNA library[c])							
phosphoglycerate kinase 1	XM_010102	43,965	1.0	100	0.7	100	Hs09
β-actin	X00351	41,735	1.3	100	0.3	100	Hs10
hypothetical protein FLJ10652	XM_006938	41,539	–[d]	–[d]	0.2	10	Hs11
hypothetical protein FLJ10559	XM_001479	35,237	0.7	50	1.0	70	Hs12

[a]MIPS *Arabidopsis thaliana* database MAtDB (http://www.mips.biochem.mpg. de/proj/thal/).
[b,c]Lambda ZAP II Library, products of Stratagene, Cat. no. 937010 and 936204, respectively.
[d]Below the detectable level.
[e]These genes were cloned from tissue (heart, brain, kidney, liver, placenta) cDNAs (BioChain Institute, Inc., Cat no. 0516001).
[f]Amyloid beta (A4) precursor protein-binding, family A, member 3.

as estimated by their mobility, and the yield was from 0.1 to 2.3 mg per mL of reaction volume, as estimated from densitometric scanning of the bands (**Table 1**). Maximum yield occurred in the supernatant, but some protein was in the insoluble phase after centrifugation of the sample at 30,000*g* for 15 min. We did not observe any significant differences in the productivity and solubility of *Arabidopsis* compared with human proteins. Translation of the fusion mRNAs encoding GST at the amino terminus led to products whose mobility was shifted by about 30 kDa (**Fig. 6**). Fusion proteins stayed soluble even for proteins that were insoluble when synthesized without GST.

5. Preparative Scale Protein Production and Quality of Synthesized Proteins

One of the challenging steps in the synthesis of proteins on a preparative scale is the stability of the translation system and the integrity of mRNA during the course of translation. As mentioned above, the efficacy of the wheat germ cell-free translation system has been significantly improved by eliminating the endogenous inhibitors, allowing the system to be used for the production of large amounts of proteins. In order to prove this, we used GFP as a model protein in the CFCF system. When the reaction was carried out with GFP mRNA transcribed from pEU circular DNA, large amounts of protein were obtained (*see* Chapter 10 by Sawasaki et al.). Interestingly, the reaction was continued up to 14 d with mRNA added every 2 d. Densitometric quantification of the CBB-stained band indicated that the active GFP yield after 14 d of translation was 9.7 mg per mL of reaction volume. The total amount of protein produced was more than that of the endogenous protein in the reaction mixture (7.5 mg/mL). This indicates that a twofold purification of the product is sufficient to purify it to homogeneity.

In order to verify the quality of proteins and general applicability of the cell-free system, we tested a subset of proteins for their functional integrity as enzymes *(25)*. As shown in **Fig. 7A**, each of five *Arabidopsis* protein kinases were synthesized effectively at the expected size (asterisk). Four out of five of these proteins displayed autophosphorylation activity after incubation with [γ-^{32}P]-ATP and magnesium (**Fig. 7B**). These results suggested that the *Arabidopsis* kinase domains of these products are folded into active forms in the cell-free system.

We next examined the folding of proteins produced in the cell-free system *(25)*. The system produced all of the gene products (numbered 03, 07, 09, 02, 10, and 12) in amounts of 0.6, 0.9, 0.3, 4.1, 2.6, and 0.8 mg, respectively, per 1-mL reaction in 48 h (data not shown here). Among the gene products expressed is flowering locus T protein (FT protein, 19.8 kDa), originally named from the phenotype of late-flowering mutants *(38)*. Because this protein has not been characterized biochemically, we chose this protein for nuclear magnetic resonance

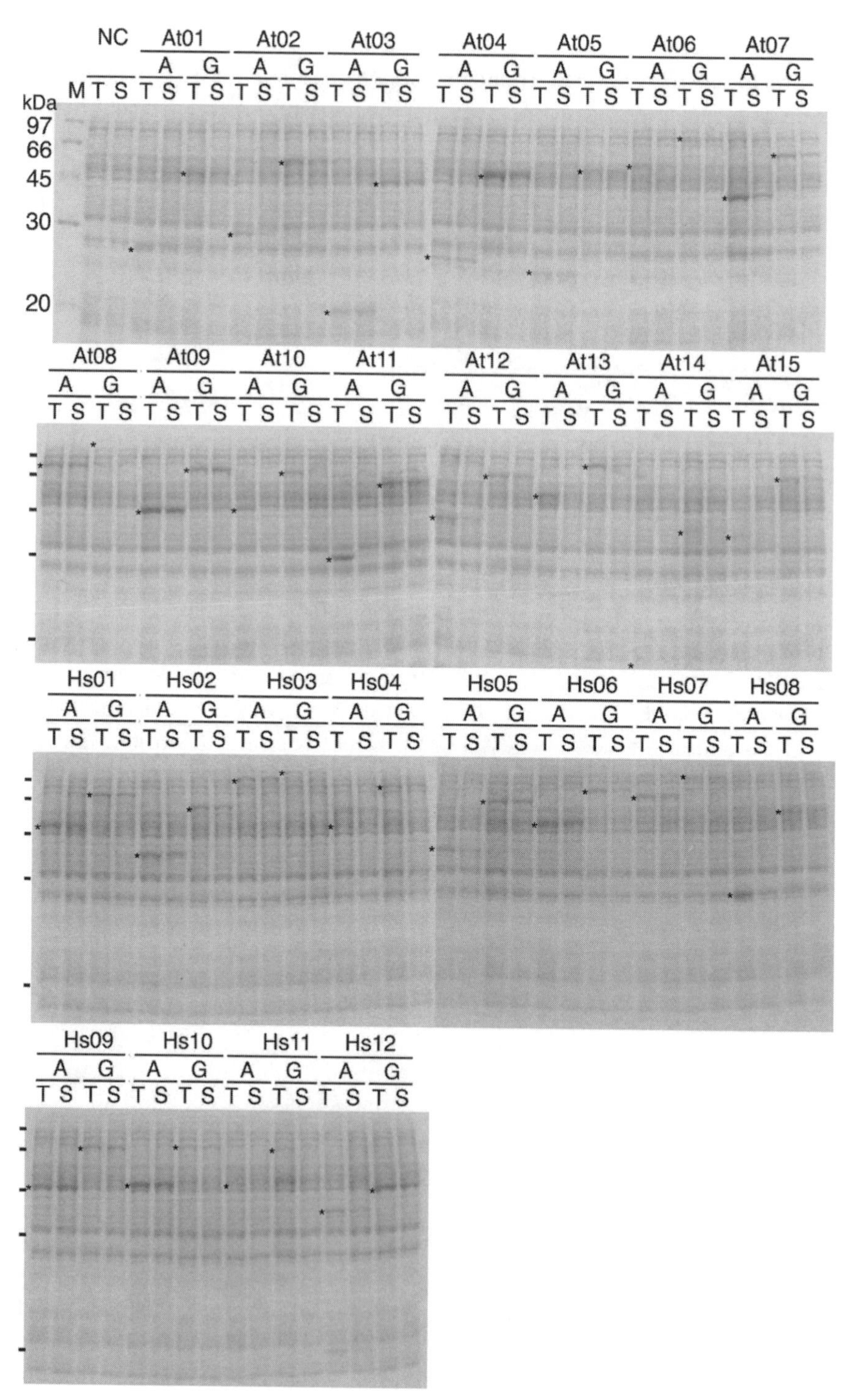
NC
At01
At02
At03
At04
At05
At06
At07
A
G
M T S T S T S
kDa
97
66
45
30
20
At08
At09
At10
At11
At12
At13
At14
At15
Hs01
Hs02
Hs03
Hs04
Hs05
Hs06
Hs07
Hs08
Hs09
Hs10
Hs11
Hs12

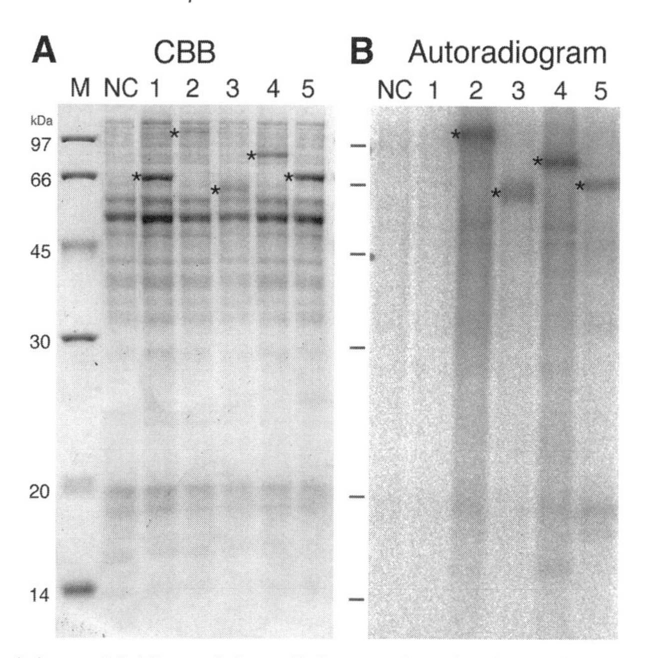

Fig. 7. Activity and folding of the cell-free produced polypeptides. Autophosphorylation activity of five *Arabidopsis* protein kinases. Sodium dodecyl sulfide-polyacrylamide gel electrophoresis and Coomassie® Brilliant Blue-stained gel of the partially purified products marked with asterisks **(A)**, and the autoradiogram **(B)**. Lanes 1–6 represent At1g07150, At5g49760, At2g02800, At5g62710, and At4g35500, respectively. NC denotes samples from the reaction mixture incubated in the absence of mRNA. M: protein size marker. Note that product of At1g07150 (lane 1) did not show activity.

(NMR) studies. The protein was mostly recovered in soluble form (*see* **Fig. 6**, lanes T, S of At03). To probe its folding state, heteronuclear single-quantum coherence (HSQC) with ^{15}N-labeled FT protein (four amino acids—Gly, Ala, Leu, and Gln, replaced with ^{15}N-labeled versions) was measured by NMR. The distribution of resonances in the 2D ^{15}N-^{1}H correlation spectrum shows a reasonable number of signals and indicates that the protein is folded in solution

---

Fig. 6. *(Opposite page)* A high-throughput production method for screening proteins from cDNA libraries. Authentic **(A)** and glutathione *S*-transferase (GST)-fused (G) proteins in the reaction mixtures after a semi-automated polymerase chain reaction/transcription and translation from 54 different cDNAs separated by sodium dodecyl sulfide-polyacrylamide gel electrophoresis and stained with Coomassie® Brilliant Blue. T and S, respectively, mark total translation product and the supernatant fraction after centrifugation at 30,000*g* for 15 min.

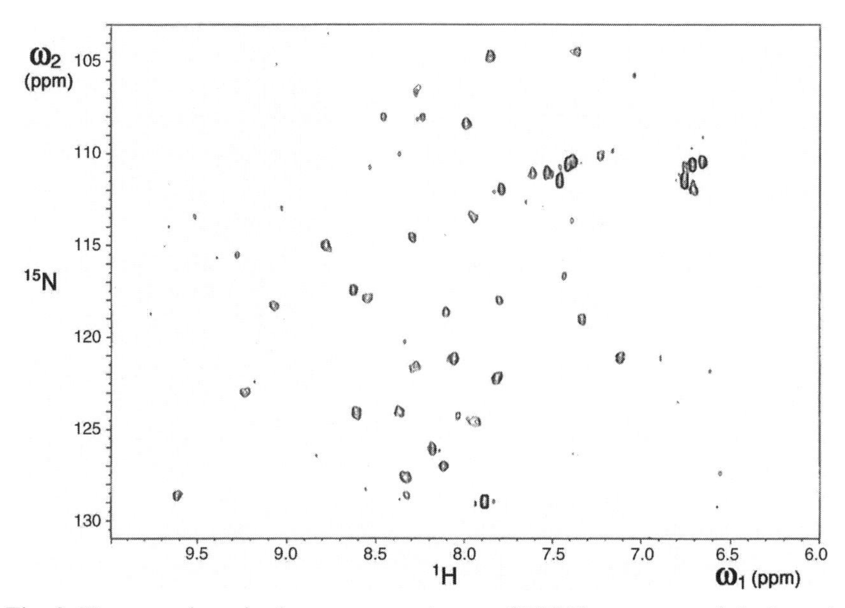

Fig. 8. Heteronuclear single-quantum coherenc (HSQC) spectrum of the hypothetical protein of the flowering locus T protein produced in the cell-free system. The FT protein was synthesized in the same way as in **Fig. 6** except that Ala, Leu, Gly, and Gln in both translation and substrate mixture were replaced with their ^{15}N-labeled forms (Isotec, Inc.). After incubation for 48 h, the reaction mixture (1 mL) was dialyzed against 10 m*M* phosphate buffer (pH 6.5) overnight, and then centrifuged at 30,000*g* for 10 min. The supernatant containing 30 μ*M* of the protein was directly subjected to nuclear magnetic resonance spectroscopy. The spectrum was recorded on a Bruker DMX-500 spectrometer at 25°C, and 2048 scans were averaged for the final ^{1}H-^{15}N HSQC spectrum.

(**Fig. 8**). These results indicate that the solubility of protein in the reaction mixture may be an indicator of "foldedness."

6. The Bilayer Reaction and the Robotic Automation of Protein Synthesis

The CFCF described by Spirin *(30)* is efficient for protein production using the efficient wheat germ cell-free system, as shown in the previous section. However, this system may not be suitable for massive screening of cDNA libraries for structural genomics and production of gene products. Because the CFCF apparatus is equipped with a semipermeable membrane in the reaction chamber, the system is quite complicated to manipulate. The delicate nature of the membrane can pose problems for automating the translation process. Also,

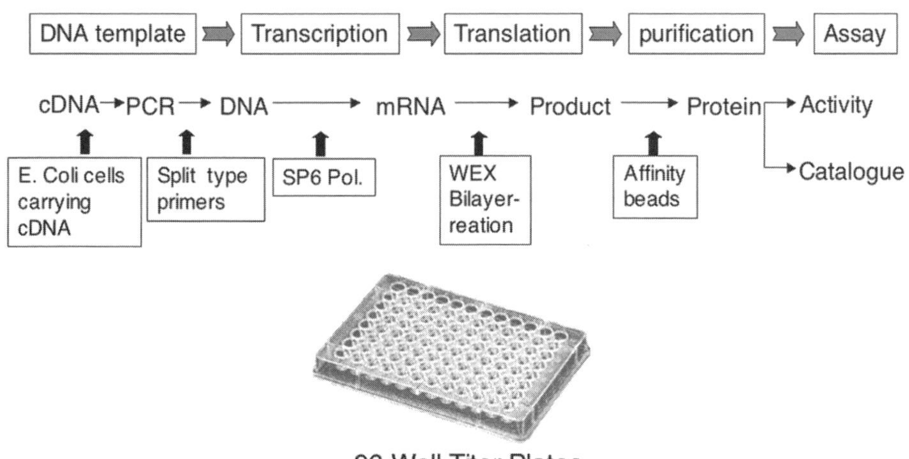

Fig. 9. Procedure for protein synthesis using the wheat embryo cell-free system. WEX is an abbreviation for wheat expression system.

for functional analysis of enzymes, tens of micrograms of proteins are required. However, the regular batch-mode reaction works for only a few hours, and the amount of synthesized protein is insufficient for biochemical analysis. Therefore, there is a need to invent a reaction system that is simpler and more cost-effective than CFCF, yet capable of producing proteins in greater quantities than batch mode, and approaching the yields of the CFCF method.

The bi-layer-translation method can synthesize protein without any membrane (*see* Chapter 10 *[17]*). To initiate the translation, the substrate mixture is simply overlaid onto the translation mixture. The system works for more than 10 times longer than batch-mode reactions, yielding sub-milligram quantities of proteins sufficient for functional analysis and determining solubility of gene products from cDNA libraries. Work to further improve the bi-layer system is in progress in our laboratory, where yields similar to those of the CFCF method have been obtained.

Attempts were then made to automate the whole translation based on this bi-layer principle. The flow chart (**Fig. 9**) summarizes our current protein synthesis processes, in which reactions start with transcription of DNA templates generated by split-primer PCR in multi-well microtiter plates. Protein synthesis in the robotic system begins with transcription of DNA templates generated by PCR. The synthesizer, named GenDecoder (*see* Chapter 10), can cover all steps preceding protein purification; it is equipped with all the essential operations (pipetting, mixing, incubating, centrifuging) for mRNA purification and finally translation. Autoradiographic analysis revealed that approx 95% of the tested cDNA samples showed protein production. One-fourth of those samples were clearly visible with CBB staining. The performance of the robotic protein expression system is comparable with manual operations using *Arabidopsis* and human cDNAs. Currently, GenDecoder can handle four 96-well microtiter plates (384 samples) in a fully automated manner in 24 h. The system can theoretically be further increased to 3000 samples without difficulty in 24 h by improving the pipetting efficiency.

In summary, the improved wheat germ cell-free translation system developed at our laboratory should be able to synthesize proteins in a folded state from genomes of various species. The system can be used for efficient high-throughput production of proteins directly from cDNAs (and genes) and to implement the proteins in chemical genomics applications such as compound screening. Although the system has not yet been optimized for posttranslational modifications such as disulfide bond formation and glycosylation, our preliminary results ***(39)*** show that it is possible to introduce disulfides into single-chain antibodies. Work at our laboratory is in progress to improve this. With the current version of GenDecoder, most of the transcription and translation processes can be performed efficiently by an automated robotic protein synthesizer. We anticipate that during the next decade, protein synthesis technologies will experience hundred-fold increases in throughput and cost reductions. The information on structure and function of gene products accumulated should revolutionize our understanding of biology and fundamentally alter the practice of medicine, and influence other industries as well. Finally, we would like to mention here that the system could possibly be applied toward creating an infinite number of novel sequences that were not selected during billions of years of evolution.

References

1. Blaschke, U. K., Silberstein, J., and Muir, T. W. (2000) Protein engineering by expressed protein ligation. *Methods Enzymol.* **328,** 478–496.
2. Baneyx, F. (1999) Recombinant protein expression in *Escherichia coli. Curr. Opin. Biotechnol.* **10,** 411–421.
3. Barnes, M. D., Bentley, C. M., and Dickson, A. J. (2003) Stability of protein production from recombinant mammalian cells. *Biotechnol. Bioeng.* **81,** 631–639.
4. Hsu, T. A. and Betenbaugh, M. J. (1997) Coexpression of molecular chaperone BiP improves immunoglobulin solubility and IgG secretion from *Trichoplusia ni* insect cells. *Biotechnol. Prog.* **13,** 96–104.
5. Ailor, E. and Betenbaugh, M. J. (1999) Modifying secretion and post-translational processing in insect cells. *Curr. Opin. Biotechnol.* **10,** 142–145.
6. Henrich, B., Lubitz, W., and Plapp, R. (1982) Lysis of *Escherichia coli* by induction of cloned phi X174 genes. *Mol. Gen. Genet.* **185,** 493–497.

7. Goff, S. A. and Goldberg, A. L. (1987) An increased content of protease La, the lon gene product, increases protein degradation and blocks growth in *Escherichia coli. J. Biol. Chem.* **262,** 4508–4515.
8. Chrunyk, B. A., Evans, J., Lillquist, J., Young, P., and Wetzel, R. (1993) Inclusion body formation and protein stability in sequence variants of interleukin-1 beta. *J. Biol. Chem.* **268,** 18,053–18,061.
9. Kurland, C. G. (1982) Translational accuracy in vitro. *Cell* **28,** 201–202.
10. Pavlov, M. Y. and Ehrenberg, M. (1996) Mutants of EF-Tu defective in binding aminoacyl-tRNA. *Arch. Biochem. Biophys.* **328,** 9–16.
11. Littlefield, J. W., Keller, E. B., Gross, J., and Zamecnik, P. C. (1955) Studies on cytoplasmic ribonucleoprotein particles from the liver of the rat. *J. Biol. Chem.* **217,** 111–123.
12. Schachtschabel, D. and Zillig, W. (1959) Investigations on the biosynthesis of proteins. I. Synthesis of radiocarbon labeled amino acids in proteins of cell-free nucleoprotein-enzyme-system of *Escherichia coli. Hoppe-Seyler's Z. Physiol. Chem.* **314,** 262–275.
13. Lamborg, M. R. and Zamecnik, P. C. (1960) Amino acid incorporation into the protein by extracts of *E. coli. Biochim. Biophys. Acta* **42,** 206–211.
14. Nirenerg, M. W. and Matthaei, J. H. (1961) The dependence of cell-free protein synthesis in *E. coli* upon naturally occurring or synthetic polyribonucleotides. *Proc. Natl. Acad. Sci. USA* **44,** 1588–1602.
15. Roberts, B. E. and Paterson, B. M. (1973) Efficient translation of tobacco mosaic virus RNA and rabbit globin 9S RNA in a cell-free system from commercial wheat germ. *Proc. Natl. Acad. Sci. USA* **70,** 2330–2334.
16. Pelham, H. R. and Jackson, R. J. (1976) An efficient mRNA-dependent translation system from reticulocyte lysates. *Eur. J. Biochem.* **67,** 247–256.
17. Sawasaki, T., Hasegawa, Y., Tsuchimochi, M., Kamura, N., Ogasawara, T., and Endo, Y. (2002) A bilayer cell-free protein synthesis system for high-throughput screening of gene products. *FEBS Lett.* **514,** 102–105.
18. Hanes, J. and Plukthun, A. (1997) In vitro selection and evolution of functional proteins by using ribosome display. *Proc. Natl. Acad. Sci. USA* **94,** 4937–4942.
19. Noren, C. J., Anthony-Cahill, S. J., Griffith, M. C., and Schultz, P. G. (1989) A general method for site-specific incorporation of unnatural amino acids into proteins. *Science* **244,** 182–188.
20. Wilson, D. S., Keefe, A. D., and Szostak, J. W. (2001) The use of mRNA display to select high-affinity protein-binding peptides. *Proc. Natl. Acad. Sci. USA* **98,** 3750–3755.
21. Ryabova, L. A., Desplancq, D., Spirin, A. S., and Plukthun, A. (1997) Functional antibody production using cell-free translation: effects of protein disulfide isomerase and chaperones. *Nat. Biotechnol.* **15,** 79–84.

21a. Sawasaki, T., Hasegawa, Y., Morishita, R., Seki, M., Shinozaki, K., and Endo, Y. (2004) Genome-scale, biochemical annotation method based on the wheat germ cell-free protein synthesis system. *Phytochemistry* **65,** 1549–1555.

22. Barbieri, L., Battelli, M. G., and Stirpe, F. (1993) Ribosome-inactivating proteins from plants. *Biochim. Biophys. Acta* **1154,** 237–282.
23. Ogasawara, T., Sawasaki, T., Morishita, R., Ozawa, A., Madin, K., and Endo, Y. (1999) A new class of enzyme acting on damaged ribosomes: ribosomal RNA apurinic site specific lyase found in wheat germ. *EMBO J.* **18,** 6522–6531.
24. Madin, K., Sawasaki, T., Ogasawara, T., and Endo, Y. (2000) A highly efficient and robust cell-free protein synthesis system prepared from wheat embryos: plants apparently contain a suicide system directed at ribosomes. *Proc. Natl. Acad. Sci. USA* **97,** 559–564.
25. Sawasaki, T., Ogasawara, T., Morishita, R., and Endo, Y. (2002) A cell-free protein synthesis system for high-throughput proteomics. *Proc. Natl. Acad. Sci. USA* **99,** 14,652–14,657.
26. Endo, Y. and Sawasaki, T. (2004) High-throughput, genome-scale protein production method based on the wheat germ cell-free expression system. *J. Struct. Funct. Genomics* **5(1–2),** 45–57.
27. Wool, I. G., Glück, A., and Endo, Y. (1992) Ribotoxin recognition of ribosomal RNA and a proposal for the mechanism of translocation. *Trends Biochem. Sci.* **17,** 266–269.
28. Johnston, F. B. and Stern, H. (1957) Mass isolation of viable wheat embryos. *Nature* **179,** 160–161.
29. Hirao, I., Madin, K., Endo, Y., Yokoyama, S., and Ellington, A. D. (2000) RNA aptamers that bind to and inhibit the ribosome-inactivating protein, pepocin. *J. Biol. Chem.* **275,** 4943–4948.
30. Spirin, A. S., Baranov, V. I., Ryabova, L. A., Ovodov, S. Y., and Alakhov, Y. B. (1988) A continuous cell-free translation system capable of producing polypeptides in high yield. *Science* **242,** 1162–1164.
31. Zaccomer, B., Haenni, A. L., and Macaya, G. (1995) The remarkable variety of plant RNA virus genomes. *J. Gen. Virol.* **76,** 231–247.
32. Gallie, D. R. and Walbot, V. (1992) Identification of the motifs within the tobacco mosaic virus 5'-leader responsible for enhancing translation. *Nucleic Acids Res.* **20,** 4631–4638.
33. Gallie, D. R. (1996) Translational control of cellular and viral mRNAs. *Plant Mol. Biol.* **32,** 145–158.
34. Beelman, C. A. and Parker, R. (1995) Degradation of mRNA in eukaryotes. *Cell* **81,** 179–183.
35. Jacobs, J. S., Anderson, A. R., and Parker, R. P. (1998) The 3' to 5' degradation of yeast mRNAs is a general mechanism for mRNA turnover that requires the SKI2 DEVH box protein and 3' to 5' exonucleases of the exosome complex. *EMBO J.* **17,** 1497–1506.
36. Gurevich, V. V. (1996) Use of bacteriophage RNA polymerase in RNA synthesis. *Methods Enzymol.* **275,** 382–397.
36a. Endo, Y. and Sawasaki, T. (2003) High-throughput, genome-scale protein production method based on the wheat germ cell-free expression system. *Biotechnol. Adv.* **21,** 695–713.

37. Erlich, H. A., Gelfand, D., and Sninsky, J. (1991) Recent advances in the polymerase chain reaction. *Science* **252,** 1643–1651.
38. Kobayashi, Y., Kaya, H., Goto, K., Iwabuchi, M., and Araki, T. (1999) A pair of related genes with antagonistic roles in mediating flowering signals. *Science* **286,** 1960–1962.
39. Kawasaki, T., Gouda, M. D., Sawasaki, T., Takai, K., and Endo, Y. (2003) Efficient synthesis of a disulfide-containing protein through a batch cell-free system from wheat germ. *Eur. J. Biochem.* **270,** 4780–4786.

12

Production of Proteins for NMR Studies Using the Wheat Germ Cell-Free System

Toshiyuki Kohno

Summary

This chapter describes protocols for preparing ^{15}N-labeled proteins (ubiquitin is used as an example) using *Escherichia coli* cells (with purification) and the wheat germ cell-free system (without purification). A comparison of ^{1}H-^{15}N heteronuclear single-quantum coherence (HSQC) spectra of yeast ubiquitin prepared using each method indicates that this wheat germ cell-free system may be used for rapid nuclear magnetic resonance analyses of proteins without purification.

Key Words: Cell-free protein synthesis; stable extract from wheat embryos; NMR; ubiquitin.

1. Introduction

Nuclear magnetic resonance (NMR) methods have been developed to determine the three-dimensional structures of proteins *(1–3)*, to estimate the protein folding *(1)*, and to discover high-affinity ligands for proteins *(4)*. However, one of the problems of applying such NMR methods to proteins is the requirement for milligram quantities of purified ^{15}N- and/or ^{13}C-labeled proteins of interest. In order to prepare stable isotope-labeled proteins, the *Escherichia coli* overexpression system has been widely used because it offers the advantages of large quantities of the protein and low-cost labeling. However, the *E. coli* overexpression system has some drawbacks: (1) this system can produce only those proteins that do not affect the physiology of the host cell, (2) the protein of interest should be extensively purified before NMR measurements because all of the proteins in the host cells are labeled with stable isotope, and (3) eukaryotic proteins often do not assume correct folding when expressed in such a prokaryotic system. Recently, a novel wheat germ cell-free protein synthesis system

From: *Methods in Molecular Biology, vol. 310: Chemical Genomics: Reviews and Protocols*
Edited by: E. D. Zanders © Humana Press Inc., Totowa, NJ

was developed *(5,6)*. The advantage of our system over others is its ability to produce eukaryotic proteins in their native conformation. In addition to this important point, our system also exhibits several attractive features for preparing samples for NMR: (1) it is stable over long periods and contains a very low amount of degradative enzymes such as proteases or ribonucleases *(6)*; (2) it can produce proteins in mg per mL of reaction volume *(6,7)*; (3) when it is used for stable isotope labeling, only the protein of interest can be labeled, and therefore, the synthesized protein may be analyzed by hetero-nuclear NMR measurements without purification *(7)*.

2. Materials

1. Yeast ubiquitin cDNA.
2. *E. coli* strains JM109 and BL21(DE3).
3. Oligonucleotide primers.
4. Restriction enzymes, T4 DNA polymerase, and T4 DNA ligase.
5. Agarose and DNA sequencing gel equipment.
6. Luria-Bertani (LB) media.
7. Ampicillin and kanamycin
8. Isopropyl-β-D-thiogalactopyranoside (IPTG).
9. Hen egg lysozyme.
10. Sonicator.
11. Purification buffer A: 50 m*M* sodium acetate (pH 5.0), 2 m*M* dithiothreitol.
12. Purification buffer B: 50 m*M* Tris-HCl (pH 8.0), 2 m*M* dithiothreitol.
13. SP Sepharose FF (Amersham Biosciences, Piscataway, NJ).
14. HiLoad 26/60 Superdex 75 pg (Amersham Biosciences).
15. Chromatography equipment.
16. pEU3b vector *(6)*.
17. 1 *M* HEPES-KOH (pH 7.8).
18. Magnesium acetate.
19. Spermidine trihydrochloride.
20. Dithiothreitol (DTT).
21. ATP, UTP, CTP, and GTP.
22. SP6 RNA polymerase and RNase inhibitor (Takara, Kyoto, Japan; or Promega, Madison, WI).
23. RNase-free water.
24. RiboGreen RNA Quantitation Kit (Molecular Probes, Eugene, OR).
25. Fluorometer TD-360 (Turner Designs, Sunnyvale, CA).
26. Potassium acetate.
27. ^{15}N-labeled amino acids (Cambridge Isotope Laboratories, Andover, MA)
28. Creatine phosphate.
29. Wheat germ extract (Cell Free Science, Yokohama, Japan).
30. Creatine kinase (Roche, Indianapolis, IN).

31. Wheat germ transfer RNA (Sigma-Aldrich, St. Louis, MO).
32. DispoDialyzer (1 mL, MWCO 8,000) Dialysis Tube (Spectrum Laboratories, Rancho Dominguez, CA).
33. Magnetic stirrer and stirrer bar.
34. Sodium dodecyl sulfate(SDS)-polyacrylamide gel electrophoresis (PAGE) equipment.
35. Centricon YM-3 centrifugal filter devices (Millipore, Bedford, MA).
36. NMR buffer: 50 m*M* sodium phosphate (pH 6.5), 0.1 *M* NaCl at 25°C.
37. MicroSpin G-25 columns (Amersham Bioscienes).
38. NMR test tubes (Shigemi, Tokyo, Japan).
39. 4, 4-Dimethyl-4-silapentane-1-sulfonic acid (DSS).

3. Methods

The methods described below outline (1) the preparation of ^{15}N-labeled protein with *E. coli* cells, (2) the construction of the expression plasmid for the wheat germ cell-free system, (3) the preparation of mRNA for the wheat germ system, (4) protein synthesis by the wheat germ system, (5) sample preparation for NMR, and (6) NMR analysis of the proteins.

3.1. Preparation of ^{15}N-Labeled Yeast Ubiquitin with the E. coli *In Vivo Protein Expression System*

For comparison, the preparation of ^{15}N-labeled yeast ubiquitin with the *E. coli* in vivo expression system is briefly described in this section.

3.1.1. Construction of Expression Plasmid

Yeast ubiquitin cDNA was cloned from the total DNA of *Saccharomyces cerevisiae* using polymerase chain reaction (PCR) with primers designed according to the reported DNA sequence of the protein ***(8)***. The PCR primers, 5'-GGG CATATGCAAATTTTTGTCAAGAC-3' and 5'-GCGGTCGACTTATCCACC TCTCAATCTCAACA-3' were designed to introduce an *Nde*I and a *Sal*I site to the N-terminus and the C-terminus, respectively. The PCR product containing the cDNA for yeast ubiquitin was ligated into the pET-24a vector (Novagen; San Diego, CA) to produce pET-24a/ubiquitin ***(9,10)***.

3.1.2. Protein Expression in E. coli *Cells*

1. *E. coli* BL21(DE3) cells harboring the pET-24a/ubiquitin plasmid were cultured at 37°C overnight in 5 mL of LB medium containing 50 µg/mL kanamycin, and then were transferred to 2 L of M9 minimal medium supplemented with trace elements and $^{15}NH_4Cl$ (1 g/L) ***(9,11,12)***.
2. When the OD_{600} reached 1.0, IPTG (final 0.1 m*M*) was added, and protein expression was induced for 12 h at 37°C.
3. The cells were harvested by centrifugation (20,000*g*, 30 min) and stored at –80°C.

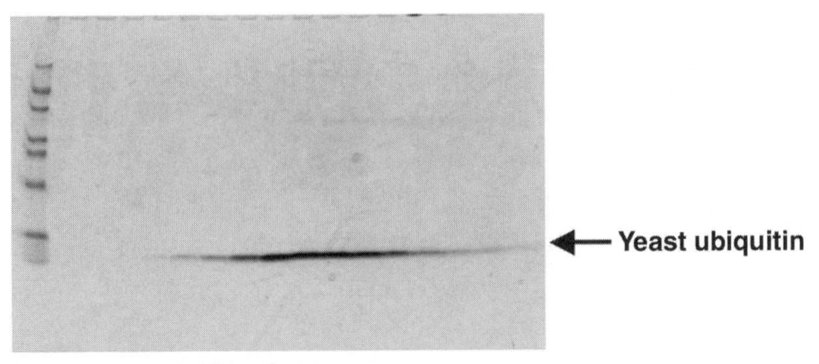

Fig. 1. SP Sepharose FF cation exchange chromatography. Fractions collected from a SP Sepharose FF column were resolved in a 15–25% gradient sodium dodecyl sulfate (SDS)-polyacrylamide gel electrophoresis (PAGE) gel and stained with Coomassie blue.

3.1.3. Protein Purification

1. The harvested cells were dispersed in 100 mL of purification buffer A containing 50 m*M* sodium acetate (pH 5.0), 2 m*M* DTT, and then disrupted using 10 mg of hen egg lysozyme and 15 min of sonication.
2. After centrifugation, the supernatant was incubated at 85°C for 5 min and chilled on ice.
3. After a further centrifugation step, the supernatant was loaded onto a 25-mL column of SP Sepharose FF (Amersham Biosciences) equilibrated with purification buffer A.
4. Following a wash step with 100 mL of buffer B, the ubiquitin was eluted by a sodium chloride gradient from 0 to 500 m*M* in purification buffer A (200 mL) (**Fig. 1**). The ubiquitin was eluted at about 150 m*M* of sodium chloride.
5. The fractions containing ubiquitin were lyophilized and redissolved in 10 mL of purification buffer B containing 50 m*M* Tris-HCl (pH 8.0), 2 m*M* DTT.
6. After centrifugation, the supernatant was loaded onto a HiLoad 26/60 Superdex 75 pg column (Amersham Biosciences) equilibrated with purification buffer B (**Fig. 2**).
7. The fractions containing ubiquitin were dialyzed twice against 2 L of deionized water and lyophilized.

3.2. Plasmid Template for Wheat Germ Cell-Free System

The construction of the plasmid template for yeast ubiquitin by wheat germ system is described under **Subheadings 3.2.1.–3.2.4.**, including (a) the description of the vector, (b) the description of the yeast ubiquitin DNA, (c) cloning, and (d) the preparation of plasmid.

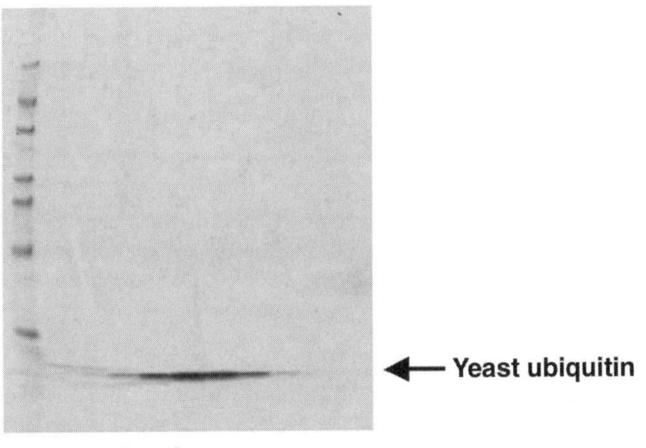

Fig. 2. Gel filtration chromatography. Fractions collected from a HiLoad 26/60 Superdex 75 pg column were resolved in a 15–25% gradient sodium dodecyl sulfate (SDS)-polyacrylamide gel electrophoresis (PAGE) gel and stained with Coomassie blue.

3.2.1. pEU3b Cell-Free Expression Vector

The plasmid of Ehime University (pEU3b) (*see* **Fig. 3**) cell-free expression vector is based on the SP6 transcription system developed by Sawasaki et al. ***(6)***. In general, the 5'-cap structure of mRNA plays a crucial role in eukaryotic translation initiation; therefore, it is necessary to introduce the cap structure into the mRNA by RNA polymerase if one wants to use eukaryotic protein synthesis systems in vitro. It is, however, inconvenient to introduce the cap structure by RNA polymerase in vitro. The pEU3b vector includes a GAA sequence flanking a 71-base, naturally occurring translational enhancer ***(13,14)***—the omega (Ω) sequence from tobacco mosaic virus—as a replacement of the 5'-cap ***(6)***, just after the SP6 phage promoter sequence in the 5'-untranslated region. This construct shows sufficient translation activity without the 5'-cap structure ***(6)***. This vector also contains an extremely long 3' untranslated region in order to stabilize mRNA against 3' exonuclease that may be included in wheat germ extract ***(6,15)***. Multiple cloning sites between 5'- and 3'- untranslated regions are introduced for cloning convenience (*see* **Fig. 3** and **Note 1**).

3.2.2. cDNA

Yeast ubiquitin cDNA was obtained from pET-24a/ubiquitin as described under **Subheading 3.1.1.**

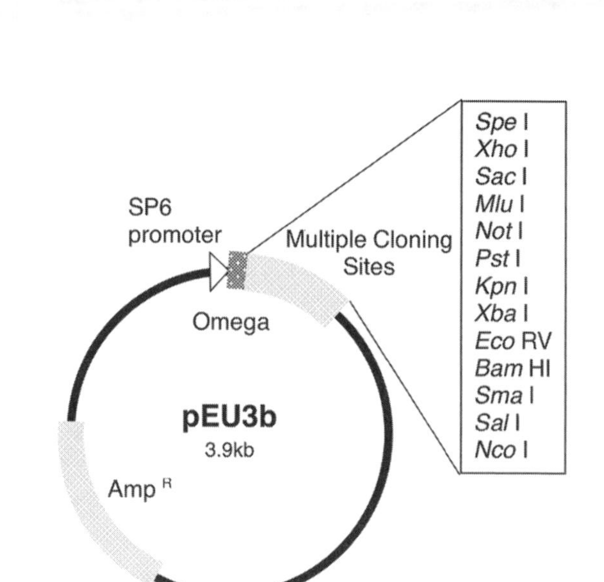

Fig. 3. Schematic drawing of the pEU3b cell-free expression plasmid.

3.2.3. Cloning

DNA manipulations were performed by standard recombinant DNA methods ***(16)*** to construct the expression plasmid, and are not described here in detail owing to space limitations.

1. The pEU3b plasmid was digested with *Spe* I and blunted with T4 polynucleotide kinase. The plasmid was then digested with *Sal* I. The coding region for yeast ubiquitin was prepared as follows.
2. The pET-24a/ubiquitin was digested by *Nde* I, blunted with T4 polynucleotide kinase, and then digested with *Sal* I.
3. The digested pEU3b plasmid and the coding ubiquitin region were ligated with T4 DNA ligase to produce pEU3b/ubiquitin.
4. The DNA was transformed into *E. coli* JM109 cells by standard methods ***(16)***. The *E. coli* JM109 cells were plated on LB plates containing ampicillin (50–100 μg/mL) and incubated overnight at 37°C.
5. Single colonies were selected and grown overnight in LB with ampicillin.
6. The plasmid DNA was then isolated and checked for the presence of the insert and for the correct orientation using restriction enzyme digestions and DNA sequencing.

3.2.4. Plasmid Preparation

1. The JM109 cells harboring the plasmid pEU3b/ubiquitin were cultured at 37°C overnight in 100 mL of LB medium containing 50 μg/mL ampicillin.

Table 1
5X Transcription Buffer (TB)

Stock	Reagent	Final concentration	
1.0 *M*	HEPES-KOH (pH 7.8)	400 m*M*	400 μL
1.0 *M*	Magnesium acetate	80 m*M*	80 μL
100 m*M*	Spermidine trihydrochloride	10 m*M*	100 μL
1.0 *M*	Dithiothreitol	50 m*M*	50 μL
	RNase-free water		370 μL
		Total	1.0 mL

Table 2
Transcription Solution

Stock	Reagent	Final concentration	
5 X	5X TB	1X	80 μL
25 m*M* each	ATP, CTP, GTP, UTP	5.0 m*M*	80 μL
70 U/μL	RNase inhibitor	1.0 U/μL	6 μL
50 U/μL	SP6 RNA Polymerase	1.5 U/μL	12 μL
1.0 μg/μL	pEU3b/ubiquitin	0.1 μg/μL	40 μL
	RNase-free water		182 μL
		Total 400 μL	

2. The cells were harvested and the plasmid was extracted according to the standard alkaline lysis with SDS method ***(16)***.
3. The extracted plasmid was further purified according to equilibrium centrifugation using CsCl-ethidium bromide gradients ***(16)***. Alternatively, the plasmid pEU3b/ubiquitin may be purified with commercially available ion-exchange resin (Wizard Plus SV, Promega) ***(16)*** (*see* **Note 2**).
4. Finally, the plasmid was dissolved in RNase-free water. The final concentration of the plasmid was adjusted to 1.0 μg/μL.

3.3. mRNA Transcription for the Wheat Germ Cell-Free System

The preparation of mRNA of yeast ubiquitin is described under **Subheadings 3.3.1.–3.3.2.**, including (a) the transcription of yeast ubiquitin gene with the plasmid pEU3b/ubiquitin by SP6 bacteriophage RNA polymerase, and (b) the quantitation of transcribed mRNA by a fluorometer.

3.3.1. mRNA Transcription

1. Prepare 400 μL of transcription solution for the yeast ubiquitin mRNA by mixing the ingredients listed in **Tables 1** and **2**.

2. Incubate the transcription solution for 5 h at 40°C. White precipitant is due to magnesium pyrophosphate after the reaction (*see* **Note 3**).
3. Centrifuge the transcription solution (20,000*g*, 5 min) at 4°C and remove the white pellet.
4. Put the supernatant on ice, and add 110 µL of 7.5 *M* ammonium acetate and 1.0 mL of 100% ethanol (*see* **Note 4**). Put the tube on ice for 10 min. Centrifuge the solution (20,000*g*, 20 min) at 4°C and discard the supernatant. Rinse the pellet with approx 300 µL of ice-cold 70% ethanol, then centrifuge (20,000*g*, 1 min) again. Remove the supernatant and dry the pellet. Re-suspend the pellet with 200 µL of RNase-free water.
5. The synthesized mRNA can be stored at –80°C for several weeks.

3.3.2. Quantitation of Synthesized mRNA

1. Dilute the RiboGreen Reagent (Molecular Probes) 200-fold. For example, to prepare enough working solution to assay five samples in 200-µL volumes, add 2.5 µL RiboGreen RNA quantitation reagent to 497.5 µL TE.
2. Prepare a 2-µg/mL solution of ribosomal RNA standard provided in TE. Prepare a series of standard RNA concentrations from 0 to 2 µg/mL. Mix equal volumes (100 µL) of diluted RiboGreen Reagent and standard RNAs. Measure the sample fluorescence using a TD-700 fluorometer (excitation approx 480 nm, emission approx 520 nm) and prepare the standard curve.
3. Take 5 µL of the transcribed ubiquitin mRNA and dilute 200-fold with TE. Mix the diluted mRNA with 100 µL of RiboGreen Reagent, and measure fluorescence. Determine the concentration of the mRNA by comparing the intensity of fluorescence with the standard curve determined above.
4. If you want to verify that the mRNA is not degraded by trace contamination by RNases, analyze the mRNA by agarose gel electrophoresis using standard protocols ***(16)***. Usually, ladder bands or smear bands around 1kb–3kb are visible when good mRNA is obtained (*see* **Fig. 4**). If RNases are present, some bands smaller than 1 kb are visible.
5. Adjust the concentration of mRNA to approx 0.5 to 1.0 µg/µL. The mRNA can be stored for several weeks at –80°C.

3.4. Protein Synthesis by the Wheat Germ Cell-Free System

Protein synthesis procedures using the wheat germ cell-free system are described as follows, including (a) the description of the wheat germ cell-free protein synthesis system developed by Madin et al., (b) the preparation of solutions for protein synthesis, and (c) the analysis of synthesized proteins.

3.4.1. The Wheat Germ Cell-Free Protein Synthesis System

The wheat germ cell-free protein synthesis system has been widely used to express various kinds of proteins. The activity of this system, however, was very low. Madin et al. found that this low activity was due to contamination by

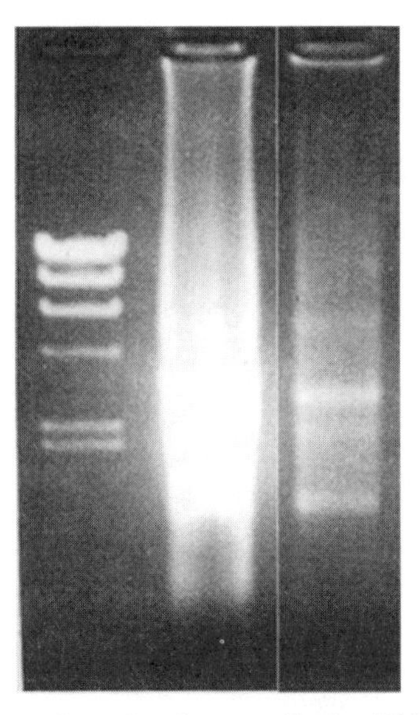

Fig. 4. Agarose gel electrophoresis of transcribed mRNA in vitro. Left lane: DNA marker λ/*Hind* III digest. Middle and right lanes: mRNA of yeast ubiquitin mRNA at various concentrations transcribed from pEU3b/ubiquitin.

ribosome-inactivating proteins from endosperm ***(5,17–20)***. They improved this system by extensively washing the embryos, thereby eliminating such ribosome-inactivating proteins, and showing that the system was far more active ***(5)***. This improved cell-free system can synthesize large proteins with a speed and accuracy approaching that of in vivo translation ***(6,21,22)*** and can synthesize proteins that would otherwise interfere with host cell physiology. With this system, the translation reaction proceeds over a week, when it is performed in a dialysis bag with a continuous supply of substrates and the removal of small byproducts ***(23,24)***, yielding mg quantities of proteins per mL. Therefore, we can now use this wheat germ system to prepare large amounts of protein for structural analyses (*see* **Note 5**).

3.4.2. Preparation of Solutions for Protein Synthesis

1. Prepare 5X translation solution described in **Table 3**. 5X translation solution can be stored at –20°C.
2. Mix 200 µL of 5X translation solution, 325 µL of wheat germ extract (OD = 150), 10 µL of wheat germ tRNA (20 µg/µL), 10 µL of creatine kinase (40 µg/µL), 14 µL of RNase inhibitor (40 U/µL), and 165 µg of the transcribed yeast ubiquitin mRNA. Adjust the volume to 1.0 mL with RNase-free water.

Table 3
5X Translation Solution

Stock	Reagent	Final concentration	
1.0 *M*	HEPES-KOH (pH 7.8)	100.0 m*M*	100 µL
4.0 *M*	Potassium acetate	384.0 m*M*	96 µL
1.0 *M*	Magnesium acetate	8.0 m*M*	8 µL
100 m*M*	Spermidine trihydrochloride	2.0 m*M*	20 µL
1.0 *M*	Dithiothreitol	10.0 m*M*	10 µL
2.5 m*M* each	^{15}N-labeled amino acids	1.2 m*M*	480 µL
100 m*M*	ATP	60 m*M*	60 µL
20 *M*	GTP	1.3 m*M*	66 µL
500 m*M*	Creatine phosphate	80.0 m*M*	160 µL
		Total	1.0 mL

Table 4
Dialysis Buffer

Stock	Reagent	Final concentration	
1.0 *M*	HEPES-KOH (pH 7.8)	30 m*M*	1.5 mL
4.0 *M*	Potassium acetate	100 m*M*	1.25 mL
1.0 *M*	Magnesium acetate	2.7 m*M*	135 µL
100 m*M*	Spermidine trihydrochloride	0.4 m*M*	200 µL
1.0 *M*	Dithiothreitol	2.5 m*M*	125 µL
2.5 m*M* each	^{15}N-labeled amino acids	0.3 m*M*	6.0 mL
100 m*M*	ATP	1.2 m*M*	600 µL
20 m*M*	GTP	0.25 m*M*	625 µL
500 m*M*	Creatine phosphate	16 m*M*	1.6 mL
0.5%	Sodium azide	0.005%	500 µL
	RNase-free water	37.5 mL	
		Total	50.0 mL

3. Load the 1-mL DispoDialyzer (Spectrum Laboratories) tube with the translation solution. Dialyze the solution against approx 15 mL of the dialysis buffer (*see* **Table 4**) in a sterile tube. Incubate the translation solution at 26°C for 2 d in the dialysis buffer with stirring.
4. After 2 d, add concentrated yeast ubiquitin mRNA to the translation solution and change the dialysis buffer. Incubate at 26°C for 2 more days as before (*see* **Note 6**).

3.4.3. Analysis of Synthesized Proteins

1. Recover the translation solution with a long, narrow plastic pipet. Take 5 µL of the protein synthesis solution. Centrifuge (20,000*g*, 20 min) and retain both supernatant and pellet.

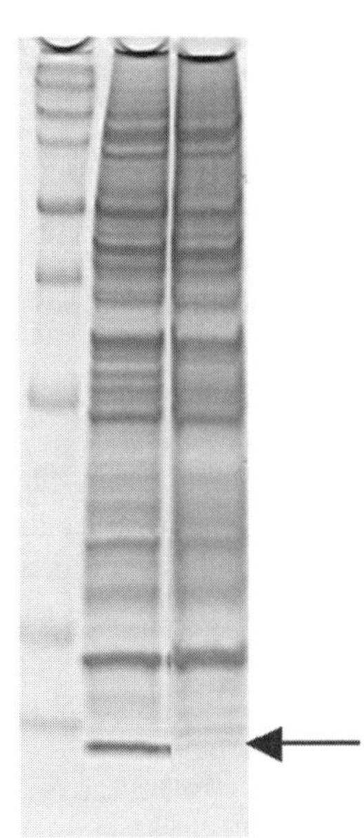

Fig. 5. Sodium dodecyl sulfate (SDS)-polyacrylamide gel electrophoresis (PAGE) analysis of ^{15}N-labeled yeast ubiquitin synthesized by wheat germ cell-free protein synthesis system. Left lane: molecular weight markers. Middle lane: yeast ubiquitin translation solution. Right lane: wheat germ alone (control). Yeast ubiquitin is indicated by an arrow.

2. Dissolve the pellet with 5 μL of water. Identify both the supernatant and pellet fractions or both the total and the supernatant fractions by Coomassie blue staining of a 15–25% gradient SDS-PAGE gel (*see* **Fig. 5**). Verify that yeast ubiquitin is synthesized and is in the supernatant fraction.
3. Estimate the concentration of the synthesized yeast ubiquitin by comparing the band in SDS-PAGE gel with those of molecular-weight markers.
4. The synthesized solution may be stored for a week at 4°C or for several months at –20°C.

3.5. Sample Preparation for NMR

The procedures for sample preparation of yeast ubiquitin for NMR analysis are described as follows.

3.5.1. NMR Sample Preparation of Yeast Ubiquitin Purified From E. coli

1. Measure approx 2.2 mg of the lyophilized ^{15}N-labeled yeast ubiquitin purified from *E. coli*.
2. Dissolve the ubiquitin with 225 μL of NMR sample buffer. Add 25 μL of 10 m*M* DSS (chemical shift internal standard) dissolved in D_2O.

3.5.2. NMR Sample Preparation of Yeast Ubiquitin Synthesized by Wheat Germ Cell-Free System

1. Put 1 mL of protein synthesis solution into Centricon YM-3 and centrifuge (5000*g*) at 4°C.
2. Centrifuge until the solution is concentrated to the volume of 200 μL (a few hours).
3. Equilibrate two MicroSpin G-25 Columns with NMR sample buffer. At first, re-suspend the resin in the column by vortexing gently.
4. Snap off the bottom closure and place the column in 1.5-mL microcentrifuge tubes. Then, prespin the column for 1 min at 735*g*.
5. Discard the solution in the microcentrifuge tubes and apply 250 μL of NMR sample buffer to each column. Spin the columns for 1 min at 735*g* and discard the buffer in the microcentrifuge tubes. Repeat this procedure eight times. Alternatively, if available, a MicroPlex 24 Vacuum (Amersham Biosciences) system may be useful to equilibrate MicroSpin G-25 Columns.
6. Place the columns in new 1.5-mL microcentrifuge tubes. Apply 100 μL of the synthesized protein to each tube.
7. Spin the columns for 2 min at 735*g*. Collect the NMR sample from the bottom of the microcentrifuge tubes (*see* **Note 7**). Adjust the sample volume to 225 μL with NMR sample buffer. Add 25 μL of 10 m*M* DSS (chemical shift internal standard) dissolved in D_2O.
8. Alternatively, buffer exchange can be done by repeated dilution with NMR sample buffer and concentration using Centricon YM-3. For example, four repeats of five-fold dilution and concentration with NMR sample are sufficient for buffer exchange.

3.6. NMR Analyses

The synthesized yeast ubiquitin labeled with ^{15}N with both the *E. coli* and wheat germ cell-free systems can now be analyzed by NMR. Here, the outline of the procedure for NMR ^{1}H-^{15}N heteronuclear single-quantum coherence (HSQC) measurements and the comparison of both spectra are described.

3.6.1. NMR Measurements

All NMR measurements were performed on a Bruker Avance500 spectrometer at 30°C. A two-dimensional ^{1}H-^{15}N HSQC spectrum ***(25)*** was acquired with 64 (t_1) × 1024 (t_2) complex points.

The data were processed using NMRPipe ***(26)*** on a Linux workstation. The ^{1}H shifts were referenced to the methyl resonance of DSS used as an internal standard. The ^{15}N chemical shifts were indirectly referenced using the ratio (0.101329118) of the zero-point frequencies to ^{1}H ***(27)***.

3.6.2. Comparison of ^{1}H-^{15}N HSQC Spectra of Yeast Ubiquitin

Figure 6 shows the ^{1}H-^{15}N HSQC spectrum of purified yeast ubiqutin prepared by *E. coli* in vivo system. The signals are assigned according to the described

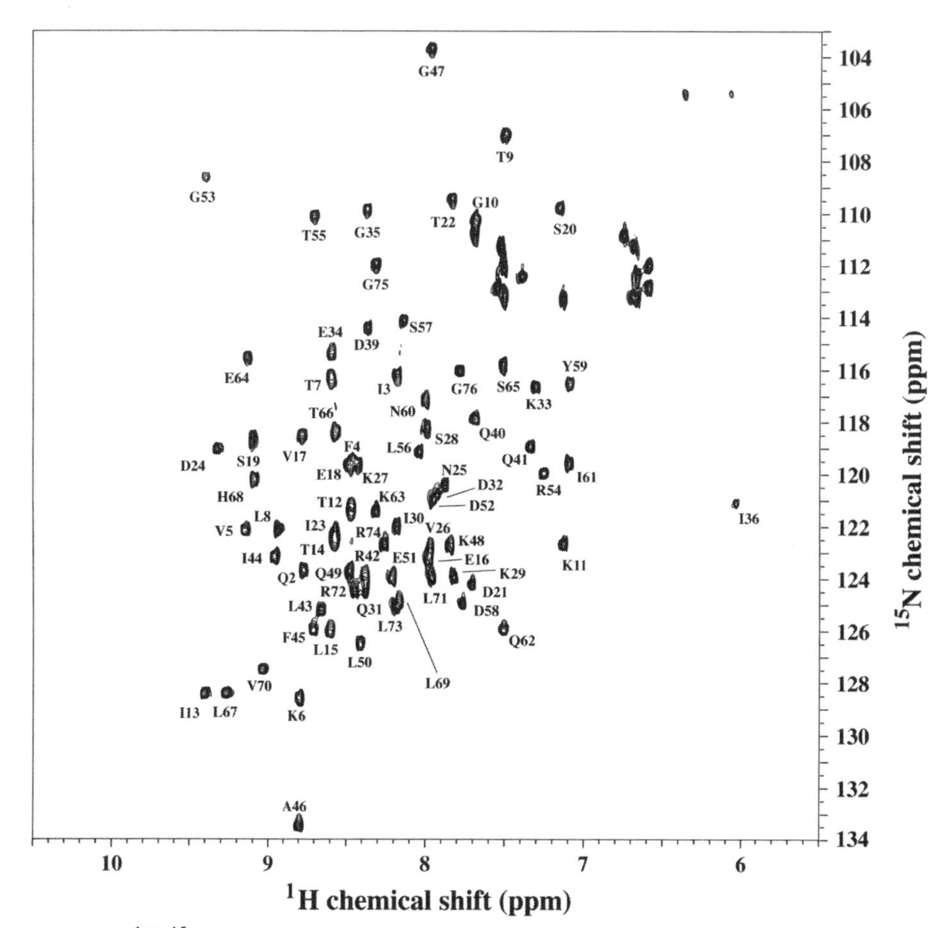

Fig. 6. ^{1}H-^{15}N heteronuclear single-quantum coherence (HSQC) spectrum of yeast ubiquitin overexpressed in *Escherichia coli* cells and purified (1.0 m*M*, 128 [t1] × 1024 [t2] complex points, 64 scans), obtained at the ^{1}H resonance frequency of 500 MHz. Spectral widths are 1600 and 6250 Hz in F_1 and F_2, respectively.

method *(9)*. **Figure 7** shows the ^{1}H-^{15}N HSQC spectrum of nonpurified yeast ubiqutin prepared by wheat germ in vitro system. All the signals visible in **Fig. 7** match those visible in **Fig. 6**, and no other signals derived from some contaminations of other proteins are visible. This indicates that yeast ubiquitin prepared by the wheat germ system assumes correct folding, and that no purification is necessary to measure the ^{1}H-^{15}N HSQC spectrum of proteins prepared by the wheat germ system. This feature may be useful for screening the "foldedness" *(1)* of proteins of interest very rapidly by NMR.

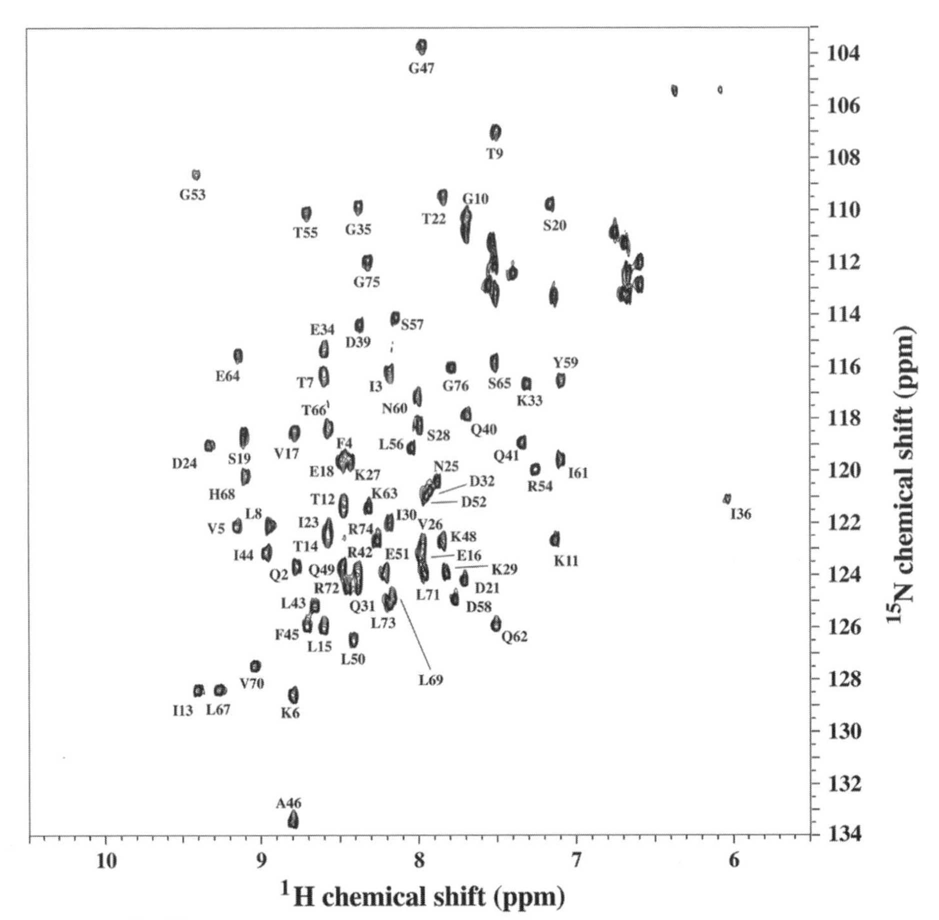

Fig. 7. ^{1}H-^{15}N heteronuclear single-quantum coherence (HSQC) spectrum of yeast ubiquitin synthesized by wheat germ cell-free protein synthesis system, not purified (0.1 m*M*, 128 [t1] × 512 [t2] complex points, 512 scans), obtained at the ^{1}H resonance frequency of 500 MHz. Spectral widths are 1600 and 6250 Hz in F_1 and F_2, respectively.

4. Notes

1. The pEU3b cell-free expression vector is very effective for producing large amounts of proteins in the wheat germ system *(6)*. Another cell-free expression vector, pEU3-NII, is also commercially available (included in the cell-free kit) from Toyobo (Osaka, Japan). This pEU3-NII vector is based on the T7 transcription system, which differs from pEU3b.
2. In order to obtain large amount of proteins with this wheat germ cell-free protein synthesis system, even trace contaminations of ribonucleases should be avoided throughout the transcription and translation steps. Therefore, great care must be

taken if plasmid purification kits are used for template DNA preparations, because these kits always use a solution including RNase A in the first step of plasmid purification.

3. SP6 RNA polymerase is usually used at 37°C. However, this enzyme is 30% more active at 40°C than at 37°C. Therefore, 40°C is used in this protocol.
4. In the ethanol precipitation step, the use of sodium acetate instead of ammonium acetate should be avoided, because trace sodium salt may interfere with the translation reaction.
5. Wheat germ extract is commercially available from Cell Free Science (Yokohama, Japan) or from Toyobo (Proteios; Osaka, Japan). Wheat germ extract may be prepared according to the protocol described in **ref. 5**.
6. With this wheat germ cell-free system, the translation reaction proceeds over a week, when mRNA is supplied and the dialysis buffer is changed every 2 d. If much more protein is needed, continue the translation reaction. Some antibiotics may be added to the dialysis solution in order to avoid bacterial contamination.
7. White precipitation of proteins from wheat germ extract may be seen when the buffer of the translation solution is exchanged for that with lower pH. Remove the precipitation by centrifugation (20,000*g*, 15 min).

Acknowledgments

The author thanks Drs. Y. Endo and T. Sawasaki for providing the pEU3b plasmid and useful discussions. I also thank Mr. T. Ogasawara and Dr. H. Morita for useful discussions, Dr. T. Sakamoto for preparation of ^{15}N-labeled ubiquitin from *E. coli*cells, Mss. M. Sugai and C. Komatsu for preparing plasmids pET-24a/ubiquitin and pEU3b/ubiquitin, and Ms. R. Tanaka for the preparation of ^{15}N-labeled ubiquitin with the wheat germ cell-free system. This work was supported in part by a grant from National Project on Protein Structural and Functional Analyses.

References

1. Montelione, G. T., Zheng, D., Huang, Y. J., Gunsalus, K. C., and Szyperski, T. (2000) Protein NMR spectroscopy in structural genomics. *Nat. Struct. Biol.* **7 Suppl,** 982–985.
2. Yokoyama, S., Hirota, H., Kigawa, T., et al. (2000) Structural genomics projects in Japan. *Nat. Struct. Biol.* **7 Suppl,** 943–945.
3. Wuthrich, K. (2000) Protein recognition by NMR. *Nat. Struct. Biol.* **7,** 188–189.
4. Shuker, S. B., Hajduk, P. J., Meadows, R. P., and Fesik, S. W. (1996) Discovering high-affinity ligands for proteins: SAR by NMR. *Science* **274,** 1531–1534.
5. Madin, K., Sawasaki, T., Ogasawara, T., and Endo, Y. (2000) A highly efficient and robust cell-free protein synthesis system prepared from wheat embryos: plants apparently contain a suicide system directed at ribosomes. *Proc. Natl. Acad. Sci. USA* **97,** 559–564.

6. Sawasaki, T., Ogasawara, T., Morishita, R., and Endo, Y. (2002) A cell-free protein synthesis system for high-throughput proteomics. *Proc. Natl. Acad. Sci. USA* **99,** 14,652–14,657.
7. Morita, E. H., Sawasaki, T., Tanaka, R., Endo, Y., and Kohno, T. (2003) A wheat germ cell-free system is a novel way to screen protein folding and function. *Protein Sci.* **12,** 1216–1221.
8. Ozkaynak, E., Finley, D., and Varshavsky, A. (1984) The yeast ubiquitin gene: head-to-tail repeats encoding a polyubiquitin precursor protein. *Nature* **312,** 663–666.
9. Sakamoto, T., Tanaka, T., Ito, Y., et al. (1999) An NMR analysis of ubiquitin recognition by yeast ubiquitin hydrolase: evidence for novel substrate recognition by a cysteine protease. *Biochemistry* **38,** 11,634–11,642.
10. Rajesh, S., Sakamoto, T., Iwamoto-Sugai, M., Shibata, T., Kohno, T., and Ito, Y. (1999) Ubiquitin binding interface mapping on yeast ubiquitin hydrolase by NMR chemical shift perturbation. *Biochemistry* **38,** 9242–9253.
11. Kobayashi, T., Yano, T., Mori, H., and Shimizu, S. (1979) Cultivation of microorganisms with a DO-stat and a silicone tubing sensor. *Biotechnol. Bioeng. Symp.* 73–83.
12. Kohno, T., Kusunoki, H., Sato, K., and Wakamatsu, K. (1998) A new general method for the biosynthesis of stable isotope-enriched peptides using a decahistidine-tagged ubiquitin fusion system: an application to the production of mastoparan-X uniformly enriched with ^{15}N and $^{15}N/^{13}C$. *J. Biomol. NMR* **12,** 109–121.
13. Gallie, D. R. and Walbot, V. (1992) Identification of the motifs within the tobacco mosaic virus 5'-leader responsible for enhancing translation. *Nucleic Acids Res.* **20,** 4631–4638.
14. Gallie, D. R. (1996) Translational control of cellular and viral mRNAs. *Plant. Mol. Biol.* **32,** 145–158.
15. Jacobs, J. S., Anderson, A. R., and Parker, R. P. (1998) The 3' to 5' degradation of yeast mRNAs is a general mechanism for mRNA turnover that requires the SKI2 DEVH box protein and 3' to 5' exonucleases of the exosome complex. *EMBO J.* **17,** 1497–1506.
16. Sambrook, J. and Russell, D. W. (2001) *Molecular Cloning, A Laboratory Manual, 3rd Ed.*, Cold Spring Harbor Laboratory Press, Cold Spring Harbor, NY.
17. Endo, Y. and Tsurugi, K. (1987) RNA N-glycosidase activity of ricin A-chain. Mechanism of action of the toxic lectin ricin on eukaryotic ribosomes. *J. Biol. Chem.* **262,** 8128–8130.
18. Endo, Y., Mitsui, K., Motizuki, M., and Tsurugi, K. (1987) The mechanism of action of ricin and related toxic lectins on eukaryotic ribosomes. The site and the characteristics of the modification in 28S ribosomal RNA caused by the toxins. *J. Biol. Chem.* **262,** 5908–5912.
19. Wool, I. G., Gluck, A., and Endo, Y. (1992) Ribotoxin recognition of ribosomal RNA and a proposal for the mechanism of translocation. *Trends Biochem. Sci.* **17,** 266–269.

20. Barbieri, L., Battelli, M. G., and Stirpe, F. (1993) Ribosome-inactivating proteins from plants. *Biochim. Biophys. Acta* **1154,** 237–282.
21. Kurland, C. G. (1982) Translational accuracy in vitro. *Cell* **28,** 201–202.
22. Pavlov, M. Y. and Ehrenberg, M. (1996) Rate of translation of natural mRNAs in an optimized in vitro system. *Arch. Biochem. Biophys.* **328,** 9–16.
23. Spirin, A. S., Baranov, V. I., Ryabova, L. A., Ovodov, S. Y., and Alakhov, Y. B. (1988) A continuous cell-free translation system capable of producing polypeptides in high yield. *Science* **242,** 1162–1164.
24. Kigawa, T., Yabuki, T., Yoshida, Y., et al. (1999) Cell-free production and stable-isotope labeling of milligram quantities of proteins. *FEBS Lett.* **442,** 15–19.
25. Schleucher, J., Schwendinger, M., Sattler, M., et al. (1994) A general enhancement scheme in heteronuclear multidimensional NMR employing pulsed field gradients. *J. Biomol. NMR* **4,** 301–306.
26. Delaglio, F., Grzesiek, S., Vuister, G. W., Zhu, G., Pfeifer, J., and Bax, A. (1995) NMR pipe: a multidimensional spectral processing system based on UNIX pipes. *J. Biomol. NMR* **6,** 277–293.
27. Wishart, D. S., Bigam, C. G., Yao, J., et al. (1995) ^{1}H, ^{13}C, ^{15}N chemical shift referencing in biomolecular NMR. *J. Biomol. NMR* **6,** 135–140.

13

Adenoviral Expression of Reporter Proteins for High-Throughput Cell-Based Screening

Jon Hoyt and Randall W. King

Summary

Recombinant adenoviruses are versatile tools for gene delivery into mammalian tissue-culture cells. Adenoviruses can infect both dividing and nondividing cells, with efficiencies approaching 100%. Here we describe the use of adenoviruses to express reporter proteins in high-throughput cell-based assays.

Key Words: Adenovirus; reporter protein; luciferase; green fluorescent protein; high-throughput; cell-based screening.

1. Introduction

High-throughput cell-based screening requires an efficient and reliable method for measuring the activity of a specific biological pathway. Reporter protein screens were initially developed to monitor pathways that control transcription in cells. Typically, a promoter or enhancer sequence is cloned upstream of a reporter protein such as β-galactosidase, luciferase, or green fluorescent protein (GFP) *(1)*. Stimulation of the pathway then leads to increased or decreased expression of the reporter protein, which can be measured using a plate reader. Reporter protein screens can also be used to monitor pathways that control the translation or stability of proteins of interest. The stability characteristics of a protein can often be conferred upon a reporter protein by producing an in-frame fusion. For example, fusing a mitotic cyclin to GFP results in cell-cycle-regulated proteolysis of the reporter protein *(2)*. Protein localization can also be used to monitor the activity of biological pathways, as illustrated by the transcription factor NF-AT, which localizes to the nucleus following activation of the protein phosphatase calcineurin. By fusing NF-AT to GFP, it is possible to use imaging approaches to measure the activation or inhibition of

From: *Methods in Molecular Biology, vol. 310: Chemical Genomics: Reviews and Protocols*
Edited by: E. D. Zanders © Humana Press Inc., Totowa, NJ

this pathway *(3)*. The development of automated, high-throughput microscopes has facilitated cellular imaging of reporter protein localization as a primary screening method *(4,5)*.

Each of these screening approaches requires a method for introducing the reporter construct into the cell. Available techniques include transient transfection, stable cell line construction *(6)*, or adenoviral delivery. Transient transfection requires large amounts of DNA, and can be inefficient and lead to variable results. Stable cell line construction can be problematic if the reporter construct inhibits cell proliferation, and can be time-consuming if a variety of different cell lines are to be screened. Adenoviral expression provides efficient expression in a wide variety of mammalian cell lines, at efficiencies that approach 100%. Unlike retroviruses, adenoviruses can infect both dividing and nondividing cells. Furthermore, recombinant adenoviruses are relatively stable, and virus can be generated easily in quantities sufficient for screening.

Here we describe the construction and use of adenoviruses to express luciferase- or GFP-fusion proteins for high-throughput cell-based screens. We constructed adenoviruses that express a cyclin B1–luciferase reporter protein or unmodified luciferase. This enabled us to screen for small molecules that specifically up-regulated levels of the cyclin B1–luciferase fusion protein without affecting luciferase. A similar approach was used to construct adenoviruses that express cyclin A–GFP or cyclin B1–GFP fusion proteins. These viruses were used in an imaging-based screen to identify small molecules that specifically upregulated these proteins or altered their subcellular localization.

2. Materials

1. Cell lines: 293 (ATCC CRL-1573), HeLa (ATCC CCL-2), BS-C-1 (ATCC CCL26).
2. Dulbecco's modified Eagle's medium (DMEM) with L-glutamine and 4.5 gm/L glucose. Add penicillin-streptomycin (Gibco BRL, Gaithersburg, MD) from 100X stock. Add fetal bovine serum (FBS) to a final concentration of 10%.
3. Phosphate-buffered saline (PBS), calcium and magnesium free.
4. Trypsin-ethylenediamine tetraacetic acid (EDTA) in Hank's balanced salt solution, without calcium or magnesium (Invitrogen, Carlsbad, CA).
5. Adenovirus precipitation solution: 20% polyethylene glycol (PEG 8000) in 2.5 *M* sodium chloride.
6. Luciferin dilution buffer (500 mL): combine 118 mg coenzyme A, 10 mL of 1 *M* tricine, 10 mL of 267 m*M* magnesium sulfate, 1 mL of 0.05 *M* EDTA, 2.55g dithiothreitol (DTT), and 155 mg of ATP. Bring final volume to 500 mL with distilled water. Adjust pH to 7.8 with potassium hydroxide pellets.
7. Luciferin stock solution (10X): add 75 mg of luciferin (Molecular Probes, Eugene, OR) to 50 mL of luciferin dilution buffer to make a 10X stock. Store at –20°C (*see* **Note 1**).

8. Phenylmethylsulfonyl fluoride (PMSF) stock (100X): add 174 mg PMSF to 10 mL 100% ethanol. Store at 4°C.
9. Lysis buffer: combine 7.5 mL of 15 m*M* magnesium sulfate, 25 mL of 25 m*M* diglycine (pH 7.8), and 0.8 mL of 4 m*M* ethyleneglycol tetraacetic acid (EGTA). Add distilled water to a final volume of 500 mL. Adjust pH to 7.8 with potassium hydroxide. Store at 4°C. On day of use, add Triton X-100 to 1%, PMSF to 1X from 100X stock, and PEG 8000 to a final concentration of 20%.
10. Cell staining and fixation solution: 8% formaldehyde in PBS. Add Hoecsht 33342 (Molecular Probes, Eugene, OR) to a final concentration of 2 μg/mL (*see* **Note 2**).
11. For luciferase assays, use white 384-well tissue culture treated assay plates (Costar, Acton, MA). For imaging assays, use black plates with clear bottoms. Either plastic-bottomed (cat. no. 3712, Costar) or glass-bottomed (MGB 101-12, Matrical, Spokane, WA) plates can be used.
12. High-throughput microscope: multiple vendors provide automated microscopes suitable for high-throughput screening. We used a Nikon TE300 inverted microscope outfitted with an automated stage, piezoelectric z-motor, and scientific-grade camera (Universal Imaging, Downingtown, PA). Instrument control and data analysis were managed using MetaMorph screening software (Universal Imaging).
13. Plate reader: Luminescence in 384-well plates was measured on an Analyst GT (Molecular Devices, Sunnyvale, CA).
14. Multichannel pipettor or automated liquid dispenser for filling 384-well plates. Examples of relatively inexpensive dispensers that work well for filling plates include the Multidrop dispenser (Titertek, Huntsville, AL) or the Precision 2000 (Bio-Tek, Winooski, VT).

3. Methods

3.1. Cell Culture

Culture cells in DMEM with 10% FBS, penicillin, and streptomycin. Grow cells at 37°C in a humidified incubator containing 5% CO_2. Cells should be split at a 1:5 ratio every 3–5 d to prevent cells from becoming too confluent (*see* **Note 3**).

3.2. Adenovirus Construction

Adenoviruses were constructed using the AdEasy system, provided as a generous gift by Bert Volgelstein *(7)*. The AdEasy system is now commercially available from ATCC (Manassas, VA), Stratagene (La Jolla, CA), and Qbiogene (Carlsbad, CA). Reporter proteins were constructed by fusing human cyclin A or cyclin B1 to the N-terminus of EGFP or luciferase using traditional cloning methods. Luciferase was cloned into the pShuttle CMV vector from psP-luc-NF (Promega Madison, WI). EGFP was cloned into the pShuttle-CMV from pEGFP-N1 (Clontech, Palo Alto, CA) (*see* **Note 4**).

3.3. Adenovirus Amplification and Purification

The procedure described as follows should provide 8 mL of virus at a titer between 1 × 10^7 and 1 × 10^{10} viruses per mL. This viral stock should be sufficient to screen several hundred 384-well plates.

1. Infect five 150-mm dishes of 293 cells that are 70–80% confluent with 100 μL viral supernatant per dish. Incubate 2–4 d until cells have rounded up and are mostly detached as a result of infection (*see* **Note 5**).
2. Use a rubber policeman to scrape off remaining cells. Do not use trypsin, because it will inactivate virus. Collect suspension and pellet cells by centrifugation at 1500*g* for 5 min. Save supernatant. Lyse remaining cells by freezing pellet in liquid nitrogen, followed by thawing in 37°C water bath. Repeat two additional times. Combine with previous supernatant and spin as before to pellet cell debris.
3. Add one-half volume adenovirus precipitation buffer to the supernatant. Incubate at 4°C overnight on an orbital shaker.
4. Pellet virus at 20,000*g* for 10 min and discard supernatant into bleach. A white pellet should be visible on the side or bottom of the tube.
5. Resuspend pellet in 8 mL PBS and centrifuge at 4000*g* for 10 min. Save supernatant.
6. Pass supernatant through a 0.2-μm syringe filter. Store aliquots at –80°C.

3.4. Optimization of Expression Level for Assays

Before screening, it is necessary to determine the optimal amount of adenovirus to use for each assay and for each cell line. If the goal is to identify small molecules that increase expression of the reporter protein, it may be optimal to adjust adenovirus concentrations such that the reporter protein is expressed at low levels in the absence of screening compounds. On the other hand, if the goal is to identify compounds that suppress expression of the reporter protein, it may be preferable to use higher amounts of virus to give a higher signal in the absence of compounds. For cellular imaging screens that use GFP reporters, we titrate the virus to the lowest concentration required for efficient expression in most cells in the population, as judged by imaging. The following protocols can be used to optimize expression levels for assays.

1. Grow cells in a T-75 flask until they are 70–80% confluent. Aspirate media and wash cells once with 15 mL PBS. Add 2 mL trypsin-EDTA and incubate plates at 37°C for 2–5 min. Resuspend cells in DMEM to a final concentration of 3.3 × 10^5 cells per mL (HeLa cells) or 2.64 × 10^5 per mL (BS-C-1 cells). Dispense 30 μL of cell suspension into each well (providing 10,000 cells per well for HeLa cells, or 8000 cells per well for BS-C-1). For luciferase assays, dispense cells into 384-well white plates; for imaging assays, use 384-well black plates with clear bottoms.
2. In a separate 96- or 384-well plate, prepare serial dilutions of the virus in tissue-culture medium. We have typically used 10-fold serial dilutions, although threefold dilutions may enable a more precise determination of optimal virus concentration.

3. Transfer 30 μL of diluted virus to the wells of the plate containing cells, for a total volume of 60 μL per well.
4. Incubate cells for 4 h at 37°C in a tissue-culture incubator.
5. Aspirate medium and replace it with 30 μL of tissue-culture medium.
6. Incubate cells for 18 h, or a period of time appropriate for your assay.
7. Follow the appropriate protocols below to determine expression or activity of the reporter protein.

3.5. High-Throughput Screening

3.5.1. Luciferase Assay

1. Grow HeLa cells in 150-mm tissue-culture dishes so that they are 70–80% confluent the day before screening. Approximately 2×10^9 cells (10–12 150-mm plates) are needed for a 50-plate assay using 384-well plates. Aspirate media and wash cells once with 15 mL PBS. Add 2 mL trypsin-EDTA per plate and incubate plates at 37°C for 2–5 min. Resuspend cells in DMEM to a final concentration of 2×10^5 cells per mL. Add appropriate volume of adenovirus to cell suspension.
2. Dispense 50 μL of cell suspension into each well of a white 384-well plate using a multichannel pipettor or an automated liquid dispenser. This will result in a final concentration of 10,000 cells per well.
3. Incubate plates for 4 h in a tissue-culture incubator (37°C).
4. Aspirate media to remove virus (*see* **Note 6**). Be careful not to disrupt cell monolayer.
5. Dispense 30 μL of DMEM into each well using a liquid dispenser.
6. Transfer test compounds into each well (*see* **Note 7**).
7. Incubate plates for 18 h, or for a period of time appropriate for your assay.
8. Aspirate media and wash wells twice with 90 μL of PBS (*see* **Note 8**).
9. Warm lysis buffer to room temperature. Add 20 μL lysis buffer per well and incubate plates at room temperature for 5 min.
10. Dilute 10X luciferin stock to 1X concentration in luciferin dilution buffer. Add 20 μL of this solution to each well and incubate plates at room temperature for 2 min (*see* **Note 9**).
11. Read luminescence with plate reader. *See* **Fig. 1** for results from a typical assay.

3.5.2. Imaging Assay

1. Grow B-SC-1 cells in 150-mm plates to 70–80% confluence the day before screening. Aspirate media and wash cells once with 15 mL PBS. Add 2 mL trypsin-EDTA per plate and incubate plates at 37°C for 2–5 min. Resuspend cells in DMEM to a final concentration of 2.6×10^5 cells per mL.
2. Add appropriate amount of adenovirus to cell suspension.
3. Dispense 30 μL of cell suspension into each well of a black, clear-bottomed 384-well plate using a multichannel pipettor or an automated plate dispenser. This will result in a final concentration of 8000 cells per well.

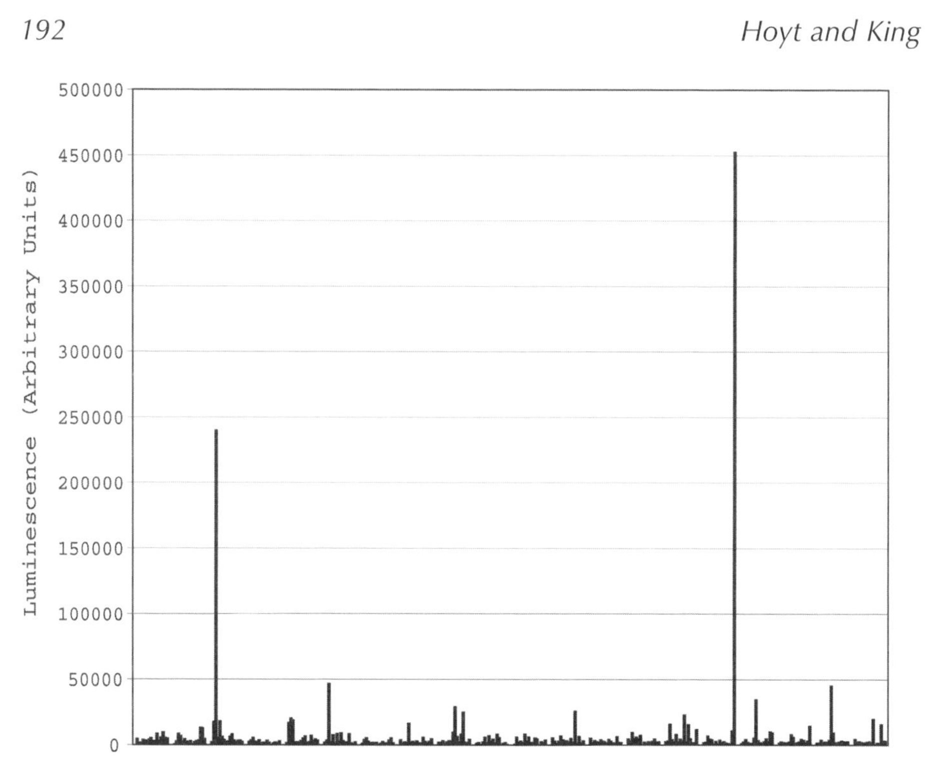

Fig. 1. A sample plate from a screen to identify small molecules that upregulate the expression of a cyclin-luciferase reporter protein. HeLa cells were infected with an adenovirus expressing cyclin-luciferase, as described in the methods section. The graph shows the activity of all wells from a single 384-well plate. Several compounds that upregulate cyclin-luciferase expression were identified.

4. Transfer test compounds into each well (*see* **Note 7**).
5. Incubate plates for 18 h.
6. Add 30 μL of fixation solution to each well and incubate plates at room temperature for 30 min.
7. Aspirate fix solution and wash once with 50 μL PBS.
8. Add 50 μL of PBS to each well. Seal plates and store at 4°C (*see* **Note 10**).
9. Read plates on automated microscope. *See* **Fig. 2** for an illustration of typical results.

4. Notes

1. Luciferin is light sensitive. Wrap tubes in foil.
2. Use a cell-permeant Hoechst dye, such as Hoechst 33342 (Molecular Probes, Eugene, OR). Permeablization of GFP-expressing cells with detergents may alter GFP localization and reduce fluorescent signal.
3. 293 cells tend to be especially slow when grown from frozen stocks and can take 1–2 wk to recover from storage. Plan accordingly.

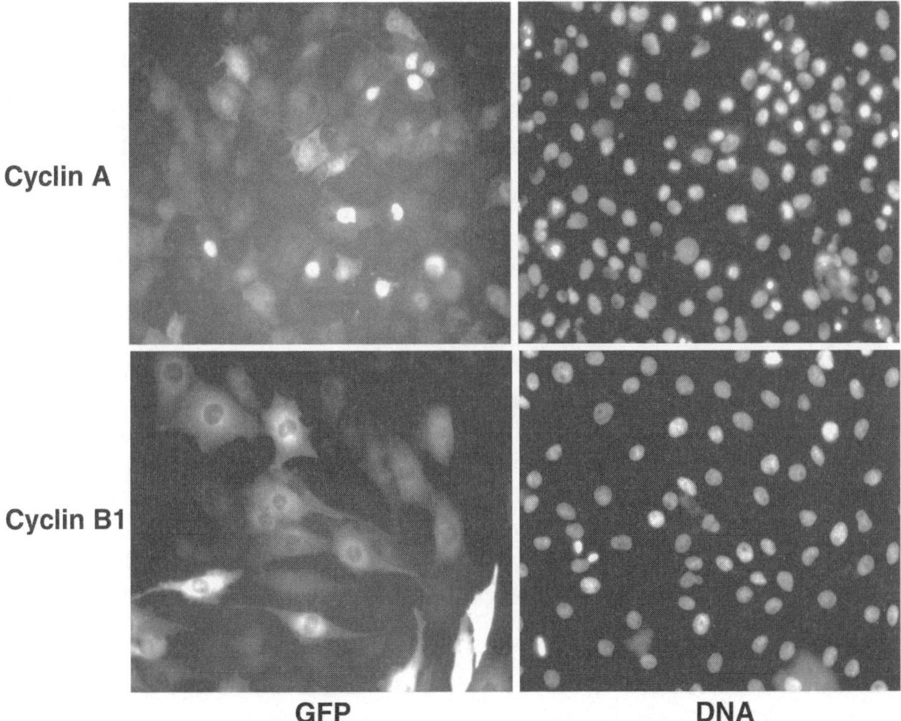

Fig. 2. Sample images from a screen to identify small molecules that affect cyclin A or cyclin B expression or localization. Adenoviruses that express cyclin A–green fluorescent protein (GFP) or cyclin B–GFP reporter proteins were constructed and used to infect BS-C-1 cells grown in 384-well plates. Plates were treated as described in the methods, and imaged at ×20 magnification using an automated microscope. Images were acquired for both GFP and DNA in each case. Cyclin A is concentrated in the nucleus during interphase and is degraded in early prophase. Cyclin B1 is cytoplasmic until prophase, when it translocates into the nucleus and is degraded during the metaphase-to-anaphase transition. Note the high percentage of cells that express the reporter protein; these results are typical for adenovirally expressed reporter proteins.

4. Additional notes on the AdEasy system are available at www.coloncancer.org/adeasy.htm. The most challenging step in the procedure is the generation of successful recombinants. Several factors affected recombination efficiency, including the quality of the preparation of the electrocompetent BJ5183 cells. We found that increasing the time of incubation of the cells on ice from 30 min to 1 to 2 h improved recombination efficiency. We typically isolated at least 10 colonies to ensure identification of a recombinant vector.
5. Be sure to use fresh stocks of 293 cells when performing viral amplification. We found that 293 cultures older than five or more passages out of liquid nitrogen

storage produced much lower viral titers. In some cases we have found it difficult to amplify particular viruses, which may be related to toxicity associated with overexpression of the reporter protein. If this is the case, increasing the number of cells used for the amplification can help generate higher amounts of virus.

6. We found that a commercial plate washer (BioTek ELX-405, Winooski, VT) or handheld aspirator (V & P Scientific, San Diego, CA, cat. no. VP 186L) was effective for aspirating from 384- or 96-well plates. Although using the handheld aspirator was more time-consuming, it provided more reproducible results. The plate washer was more variable due to occasional clogging of the pin manifold. When using a plate washer, it is important that you carefully set your dispensing and aspiration heights to avoid aspirating or washing away cells.
7. We use a 384-pin array to deliver 100 nL of compound to each well *(8)*. The advantage of this approach is that a small volume of compound is used and dilution of compound stock in growth medium is not necessary. The disadvantages of this method include occasional compound precipitation in growth medium and the appearance of a concentration gradient across the monolayer within a well.
8. When screening for small molecules that upregulate luciferase expression, the wash step may not be necessary. But in the case where one is interested in finding small molecules that decrease reporter expression, it is useful to include a wash step prior to assaying luciferase activity, as we found that this minimizes the chances of identifying small molecules that directly inhibit the luciferase enzyme.
9. The luciferase signal will start to decay after the luciferin reagent is added to the plates. The luciferase activity expressed by psP-luc-NF has been shown to decrease by 40% after 10 min following addition of reagent *(9)*. However, we found that luciferase remained sufficiently active for assay measurements even up to an hour following reagent addition. Signal can be normalized from plate to plate to allow comparison of screening results from different plates.
10. Plate storage. We found that using aluminum seals (Fisher Scientific, Pittsburgh, PA, cat. no. 072-00684) allowed us to store plates at 4°C. The use of opaque seals enables plates to be imaged in the presence of ambient light. In our experience, the Hoechst stain and GFP signal remains stable at 4°C for up to 3 mo.

Acknowledgments

The development of the methods described in this chapter were supported by grants from the NCI (CA78048) and NIGMS (GM66492), and by support from Merck & Co. and Merck KGaA.

References

1. Sala-Newby, G. B., Kendall, J. M., Jones, H. E., et al. (1999) Bioluminescent and chemiluminescent indicators for molecular signalling and function in living cells. In: *Fluorescent and Luminescent Probes for Biological Activity* (Mason, W. T., ed.), London, Academic Press, pp. 251–272.

2. Clute, P. and Pines, J. (1999) Temporal and spatial control of cyclin B1 destruction in metaphase. *Nat. Cell Biol.* **1,** 82–87.
3. Kehlenbach, R. H., Dickmanns, A., and Gerace, L. (1998) Nucleocytoplasmic shuttling factors including Ran and CRM1 mediate nuclear export of NFAT in vitro. *J. Cell Biol.* **100,** 863–864.
4. Feng, Y., Yu, S., Lasell, T. K., et al. (2003) Exo1: a new chemical inhibitor of the exocytic pathway. *Proc. Natl. Acad. Sci. USA* **100,** 6369–6474.
5. Kau, T. R., Schroeder, F., Ramaswamy, S., et al. (2003) A chemical genetic screen identifies inhibitors of regulated nuclear export of a forkhead transcription factor in PTEN deficient tumor cells. *Cancer Cell* **4,** 463–467.
6. Pagliaro, L. and Praestegaard, M. (2001) Transfected cell lines as tools for high throughput screening: a call for standards. *J. Biomolec. Screening* **6,** 133–136.
7. He, T. C., Zhou, S., da Costa, L. T., Yu, J., Kinzler, K. W., and Vogelstein, B. (1998) A simplified system for generating recombinant adenoviruses. *Proc. Natl. Acad. Sci. USA* **95,** 2509–2514.
8. Walling, L. A., Peters, N. R., Horn, E. J., and King, R. W. (2001) New technologies for chemical genetics. *J. Cell. Biochem.* **S37,** 7–12.
9. Jones, D. P., Sherf, B. A., and Wood, K. V. (1995) Luciferase assay system vendor comparison. *Promega Notes Magazine* **54,** 20.

14

Fabrication of Protein Function Microarrays for Systems-Oriented Proteomic Analysis

Jonathan M. Blackburn and Darren J. Hart

Summary

Protein microarrays have many potential applications in high-throughput analysis of protein function. However, simple, reproducible, and robust methods for array fabrication are required. Here we discuss the background to different routes to array fabrication and describe in detail one approach in which the purification and immobilization procedures are combined into a single step, dramatically simplifying the array fabrication process. We illustrate this approach by reference to the creation of an array of p53 variants, and discuss methods for assay and data analysis on such arrays.

Key Words: Protein array; biotinylation; proteomics; functional analysis; surface capture; p53; biotin carboxyl carrier protein; fusion protein; microarray; DNA binding.

1. Introduction

In the postgenomic era, attention is turning towards the systematic assignment of function to proteins encoded by genomes. Bioinformatics methods are now used ubiquitously in initial efforts to assign function to predicted open reading frames *(1)*. However, while such methods can give helpful insights into possible function, there are now many examples of proteins that have closely related sequences and/or structures but prove to have quite different functions *(2–4)*. There is thus an emerging need for high-throughput methods that are suitable for the experimental determination/verification of protein function. At the forefront of this monumental task, the field of proteomics is now segregating into discovery- and systems-oriented proteomics *(5)*. Discovery-oriented proteomics is mainly concerned with documenting the abundance and localization of individual proteins as well as building a picture of protein–protein

From: *Methods in Molecular Biology, vol. 310: Chemical Genomics: Reviews and Protocols*
Edited by: E. D. Zanders © Humana Press Inc., Totowa, NJ

interaction networks. This is the realm of two-hybrid screens, two-dimensional (2D)-gel electrophoresis, and more direct mass spectrometry-based methods; the latter two methods in particular are commonly used to understand the way in which expression profiles change in response to different stimuli by comparing, for example, diseased and healthy cell extracts. However, these discovery-oriented proteomics methods tell us little about the function of individual proteins or protein complexes. Systems-oriented proteomics takes a different approach; rather than re-discovering each protein in each new experiment, the focus is on a predefined set of proteins, enabling more precise questions to be asked regarding the functionality of each member of that set. However, obtaining quantitative and genuinely comparative functional data across large sets of proteins with any degree of accuracy is technically difficult, requiring isolation of each individual protein in an assayable format. We and others have chosen to focus on protein function microarray-based methods because the parallel, high-throughput nature of microarray experiments is attractive for analyzing large numbers of protein interactions, while the uniform intra-array conditions both simplify and increase accuracy of assays ***(6–12)***. Additionally, the small volumes of ligand or reaction solution required to perform assays, typically tens to hundreds of microliters, can provide economic advantages—for example, when using expensive recombinant proteins or labeled compounds. The key element to such microarray experiments is that the arrayed, immobilized proteins retain their folded structure such that meaningful functional interrogation can then be carried out. There are a number of approaches to this problem, which differ fundamentally according to whether the proteins are immobilized through nonspecific, poorly defined interactions or through a specific set of known interactions. The former approach is attractive in its simplicity and is compatible with purified proteins derived from native or recombinant sources ***(13,14)***, but suffers from a number of risks—most notably that the uncontrolled nature of the interactions between each protein and the surface might give rise to a heterogeneous population of proteins or destroy activity all together. In practice, an intermediate situation frequently occurs, where a fraction of the immobilized proteins either have undergone conformational change as a result of the nonspecific interactions or have their binding/active sites occluded by surface attachment; these effects will effectively reduce the specific activity of the immobilized protein and therefore decrease the signal-to-noise ratio in any subsequent assay. The advantages of controlling the precise mode of surface attachment are that, providing the chosen point of attachment does not directly interfere with activity, the immobilized proteins will have a homogeneous orientation, resulting in a higher specific activity and signal-to-noise ratio in assays, with less interference from nonspecific interactions ***(15)***. This may be of particular advantage when studying protein–small-molecule interactions in an array format. The disadvantages of this approach, though, are that it is really compatible only with recombinant proteins or with families of proteins, such as antibodies, that have a common structural element through which they can be immobilized. However, in a systems-oriented approach, the disadvantage of working with recombinant proteins is currently largely outweighed by the problems encountered in individually purifying large numbers of active proteins from native sources. In this chapter we therefore describe the expression, array fabrication, and assay of a set of recombinant proteins in which the mode of surface attachment is tightly controlled. Furthermore, we show how laborious pre-purification of the recombinant proteins prior to array fabrication can be avoided through use of a suitable affinity tag, thus greatly simplifying array fabrication ***(6)***. We illustrate this approach to array fabrication with respect to a set of disease-associated variants of the human tumor suppressor protein p53 ***(16,17)***, expressed as fusions to a polypeptide tag that becomes biotinylated in vivo ***(6)*** (*see* **Note 1**). In addition, we show representative data from a number of different assays carried out on protein function arrays made in this way.

2. Materials

1. pQE-80L expression system (Qiagen).
2. *Escherichia coli accB* gene.
3. Human p53 cDNA.
4. *E. coli* strain XL1-blue.
5. Luria-Bertani (LB) media.
6. Biotin.
7. Ampicillin.
8. Isopropyl-β-D-thio-galactopyranoside (IPTG).
9. Phosphate-buffered saline (PBS): 1.5 m*M* KH_2PO_4, 4.3 m*M* Na_2HPO_4, 137 m*M* NaCl, 3 m*M* KCl (pH 7.3).
10. PBST: PBS, 0.1% (v/v) Tween-20.
11. Marvel/PBST (PBS, 100 mg/mL Marvel, 0.1% [v/v] Tween-20).
12. p53 buffer: 25 m*M* HEPES (pH 7.6), 50 m*M* KCl, 20% (v/v) glycerol, 1 m*M* dithiothreitol (DTT), 1 mg/mL bovine serum albumin (BSA), 0.1% (v/v) Triton X-100.
13. p53 freeze mix (250 mL p53 buffer plus 250 mL glycerol; keep at 4°C).
14. Lysozyme (4 mg/mL in water).
15. SAM2® membrane (Promega).
16. Anti-His antibody (Sigma).
17. Cy3 Mono-Reactive Dye labeling kit (Amersham Biosciences).
18. GADD45 duplex oligo.
19. Genetix QArray robot.
20. DNA microarray scanner.

3. Methods

The methods described below outline (1) the construction of the expression plasmids, (2) the induction of protein expression, (3) the extraction of the protein from *E. coli*, (4) the printing of a protein microarray, and (5) the assay of the protein microarray for DNA binding function.

3.1. Construction of the Expression Plasmid for Full-Length p53

3.1.1. Expression Vector

The *E. coli* expression system used is based on the pQE-80L expression system. Sequences inserted into the multiple cloning site can be expressed as native proteins bearing an N-terminal hexahistidine tag upon IPTG induction of the T5 promoter in suitably transformed *E. coli* cells.

3.1.2. Amplification and Cloning of the Wild-Type p53 Gene As a Fusion to BCCP

All DNA manipulations were carried out using standard recombinant DNA methods ***(18)*** to construct the expression plasmids, and are accordingly not described here in detail. The gene encoding the *E. coli* biotin carboxyl carrier protein (BCCP) domain (amino acids 74–156 of the *E. coli accB* gene; **Fig. 1**) was amplified by polymerase chain reaction (PCR) from an *E. coli* genomic DNA preparation using primers that added a 5'-*Bam* HI site, removed the natural stop codon of BCCP, and added a 3' multiple cloning site including *Sac* I, *Not* I, and *Hind*III restriction sites. This PCR product (approx 280 bp) was cloned into pQE-80L downstream of and in frame with sequences encoding six histidine residues as a *Bam* HI/ *Hin* dIII fragment to generate an intermediate vector, pQE-80L H6-BCCP (*see* **Note 2**).

The full-length, wild-type *p53* gene was amplified by PCR from a HeLa cell cDNA library (Clontech) using primers that added a 5' *Sac* I site and a 3' *Not* I site. The PCR product (approx 1190 bp) was cloned into the intermediate vector downstream of and in frame with the BCCP domain as a *Sac* I/*Not* I fragment, such that the resulting construct (pQE-80L H6-BCCP-p53; **Fig. 2**), encoded a His6-BCCP-p53 fusion protein. The *p53* gene was sequence verified and confirmed as matching SWISS-PROT entry P04637.

3.1.3. Construction of p53 Variants

Forty-eight variants of p53 were generated by inverse PCR using the wild-type p53 expression vector, pQE-80L H6-BCCP-p53, as template. Phosphorylated forward primers bore the sequence variation at the 5'-terminus followed by 20–24 nucleotides of *p53* sequence. Unphosphorylated reverse primers were

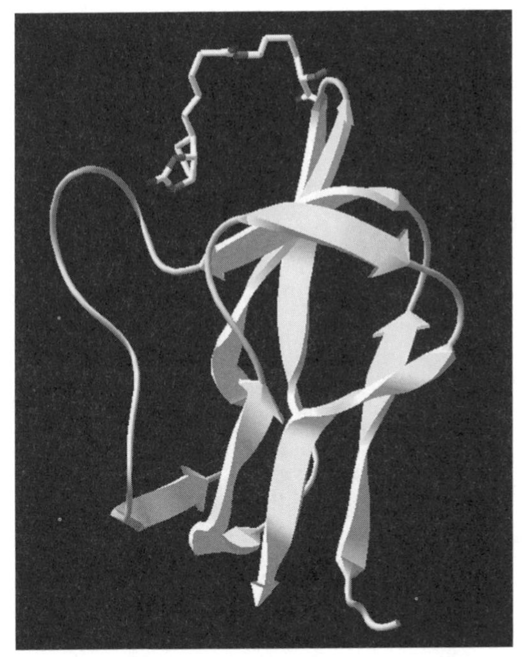

Fig. 1. Crystal structure of the biotin carboxyl carrier protein domain of *Escherichia coli*. The N- and C-termini and, 50Å away, the single lysine residue that is biotinylated in vivo can all clearly be seen. Figure prepared from PDB file 1BDO using Swiss PDB Viewer ***(19)***.

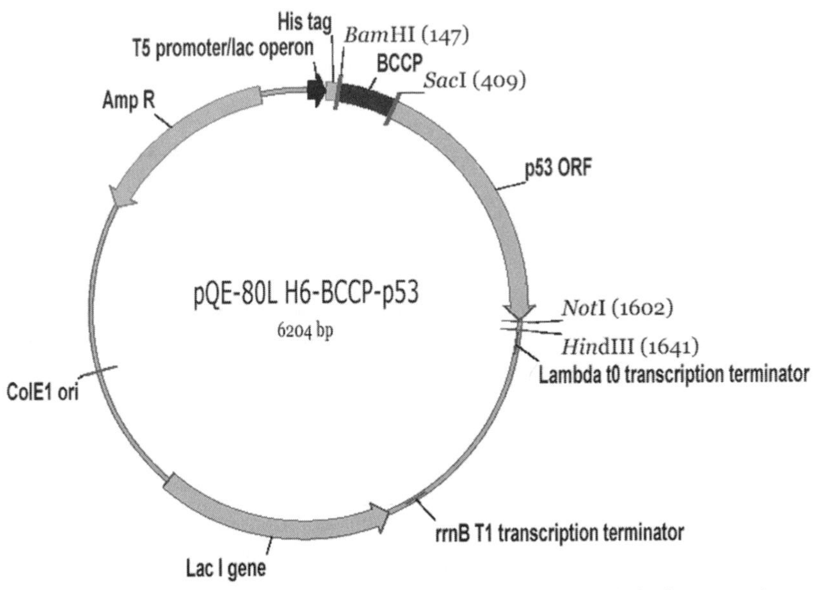

Fig. 2. Map of plasmid vector used to express BCCP-p53 fusion proteins.

complementary to the 20–24 nucleotides immediately before the mutated position. PCR was performed using *Pwo* polymerase (Roche Molecular Biochemicals) and generated blunt-ended products corresponding to the entire p53-containing vector. PCR products were gel purified and ligated, and methylated parental template DNA was digested with *Dpn* I (New England BioLabs). *E. coli* XL1 Blue cells were transformed to ampicillin resistance and mutant *p53* genes verified by DNA sequencing.

3.2. Expression of the His6-BCCP-p53 Fusion Proteins in E. coli

1. Transform *E. coli* XL1-blue cells with the pQE-80L H6-BCCP-p53 plasmid DNA (or the plasmids encoding the individual mutants) using standard methods ***(18)***. Plate cells on LB agar plates supplemented with ampicillin at 100 μg/mL and incubate overnight at 37°C.
2. Select single colonies and grow overnight at 250 rpm, 37°C in 5 mL Luria-Bertani medium supplemented with ampicillin at 100 μg/mL (LBamp).
3. Inoculate separate 200-mL quantities of LBamp with 4 mL of the individual overnight cultures and incubate at 250 rpm, 37°C until the cultures reach OD_{600} approx 0.4.
4. Reduce the temperature to 30°C and induce protein expression by addition of IPTG (100 μ*M*) and free biotin (50 μ*M*) for 4 h. It should be noted that the induction time can be extended to increase expression, but for the BCCP-p53 fusion proteins this usually results in a decrease in protein quality (*see* **Notes 3** and **4**).
5. Harvest cells by centrifugation (4500*g*, 20 min).
6. Resuspend the cells in 100 mL PBS and re-centrifuge to remove excess free biotin (*see* **Note 4**).
7. Aliquot cells and store as cell pellets, each the equivalent of 2 mL of expression culture, at –80°C.

3.3. Extraction of Protein in Preparation for Array Fabrication

1. Thaw individual cell pellets on ice, add 80 μL of p53 buffer (25 m*M* HEPES [pH 7.6], 50 m*M* KCl, 20% [v/v] glycerol, 1 m*M* DTT, 1 mg/mL BSA, 0.1% [v/v] Triton X-100) and vortex the tubes to resuspend the cells.
2. Inspect the tubes visually to confirm that the cells have become efficiently dispersed; if not, continue vortexing until no clumps can be seen.
3. Add 20 μL of a 4 mg/mL lysozyme solution to each resuspended cell aliquot, mix, and incubate at 4°C with gentle agitation for 30 min.
4. Pellet cell debris by centrifugation at 10,000*g* for 10 min at 4°C.
5. Transfer the supernatants from each tube into individual wells of a 384-well V-bottom plate (Genetix) and keep at 4°C. Take care not to mix or confuse the sample identities at this stage. This is the source plate for the print runs.
6. Centrifuge the 384-well plate at 2200*g* for 5 min at 4°C to pellet any cell debris that has carried over. Store plate on ice prior to print run.

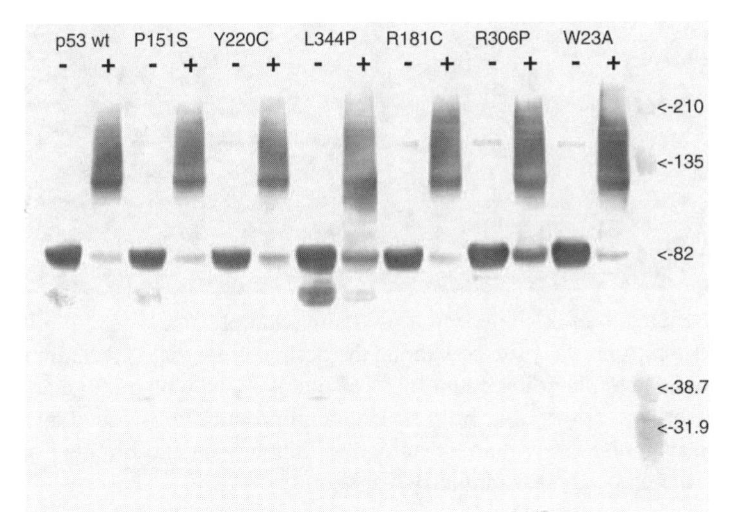

Fig. 3. Supershift assay to determine the extent of biotinylation of BCCP-p53 fusion proteins. Each pair of lanes corresponds to a crude lysate of the p53 variant shown, either with (+) or without (–) preincubation with a molar excess of streptavidin.

7. Determine the protein concentration of the soluble protein extracts by Bradford assay ***(20)*** to confirm that effective cell lysis has occurred (*see* **Note 5**).
8. Determine the approximate expression level of soluble BCCP fusion by sodium dodecyl sulfate (SDS)-polyacrylamide gel electrophoresis (PAGE) together with western blot analyses ***(18)*** using an anti-His primary antibody (*see* **Note 5**).
9. To determine the extent of biotinylation of the BCCP fusion protein, carry out a supershift Western blot assay (again with an anti-His primary antibody) in which equivalent crude lysate samples are pre-incubated with or without streptavidin (0.1 g/mL) (**Fig. 3**; *see* **Note 6**).

3.4. Fabrication of p53 Microarrays

In the procedures described below, we do not employ a prepurification step prior to array fabrication, but instead rely on a rapid, single-step immobilization and purification procedure to create arrays of biotinylated BCCP fusion proteins (**Fig. 4**; *see* **Note 1**).

3.4.1. Preparation of Membranes for Printing

1. Stick one small piece of double-sided tape onto each end of a 7.5 × 2.5 cm glass microscope slide.
2. Remove SAM2 membrane from –20°C and warm to room temp.
3. Peel the backing off the tape and place the slide face-down onto SAM2 membrane.
4. Cut around the slide with a scalpel blade.

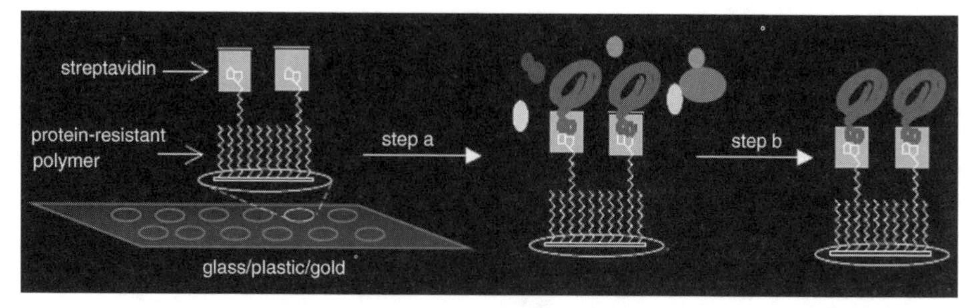

Fig. 4. Schematic of a single-step array fabrication process for in vivo biotinylated proteins. **Step a**: A crude lysate containing the desired biotinylated recombinant protein is printed onto a streptavidin-coated surface coderivatized with a polymer that resists nonspecific protein absorption. **Step b**: Unbound proteins are washed away to leave the purified recombinant protein, specifically immobilized and oriented on the array surface via the biotin moiety on the BCCP tag.

3.4.2. Fabrication of Arrays

In general, any microarray printer could be used to print the arrays. The printing procedures can be carried out at room temperature providing the source plate is kept at 4°C and the atmosphere in the print chamber is humidified. However, preferably the printing device itself should also be cooled.

Here we describe one specific set of parameters that work well on SAM2 membranes using a Genetix QArray, cooled to 4°C and equipped with 8 × 300 μm tipped solid pins (*see* **Note 7**).

1. Load the 384-well source plate into the QArray.
2. Load the SAM2 slides onto the print bed of the QArray.
3. Print the arrays using the QArray settings in **Table 1** (*see also* **Note 8**).

3.4.3. Processing of Arrays After Printing

1. Remove slides from microarraying robot.
2. Cut across membranes at top and bottom using scalpel blade to separate the membranes from the tape and remove the arrays from slide by holding edge with tweezers.
3. Make a notch at top right of each array to indicate orientation.
4. Submerge the arrays in p53 buffer at room temp until all arrays are removed and notched.
5. Now transfer the arrays into a vessel containing 50 mL Marvel/p53 buffer (5% Marvel in p53 buffer). Rock the vessel at room temperature for 30 min, making sure that the arrays do not stick together (*see* **Note 9**).
6. Wash all arrays three times for 5 min in 50 mL p53 buffer.

Table 1
Genetix QArray Parameters for SAM2® Membranes

Container = "STACKER SOURCE PLATE HOLDER"
Plate = "GENETIX PLATE 384 WELL"
Use Test plate = blank
Use stacker = blank
Source order by = Columns
Run print test = blank
Inking Time (ms) = 2
Max plates = 9,999
Substrate = "3 × 1' SLIDE (8-PINS/7-FIELDS)"
Holder = "SLIDE HOLDER"
Microarraying pattern = "4 × 4 1000 micron"
Use duplicate pattern = Yes (ticked)

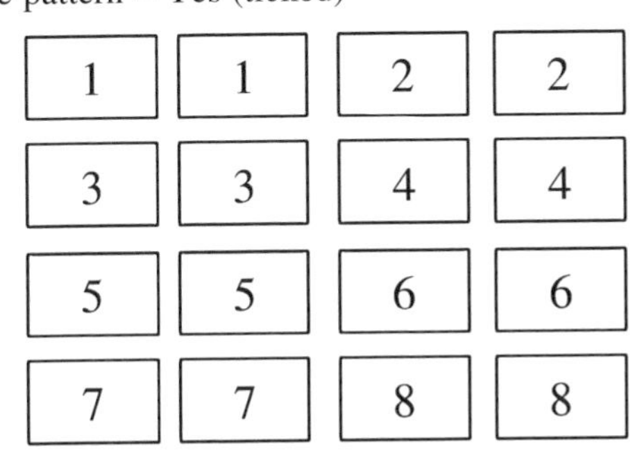

(note: above pattern set up with 1000 micron spacing)
Field order = "p53 run"—fields 1 and 2 shown below are used.

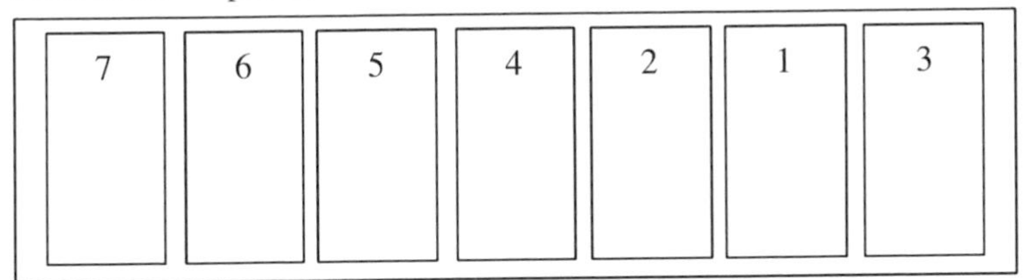

No. of substrates = "X"—dependent on the desired print run size
No. of fields in substrate = "2"
Max stamps per ink = "1"
No. identical fields = "2"
No. stamps per spot = "3"
Stamp time = "2 ms"
Microarraying by = Spot position
Printing depth = "250 microns"
Head = "8 PIN MICROARRAYING HEAD"
Water washes = 2000 ms wash, 1000 ms dry
Ethanol wash = 2000 ms wash, 7000 ms dry

7. Place in p53 freeze mix in a polypropylene tube and store arrays at –20°C (*see* **Note 9**).

3.5. DNA-Binding and Quality-Control Assays

In order to assess the quality of the p53 protein microarrays printed on SAM2 membranes, two different assays are carried out on replica arrays. The antibody-binding assay comprises binding a Cy3-labeled anti-His antibody to the arrayed p53 proteins. This assay is independent of protein activity but provides a measurement of the relative amounts of protein immobilized in each spot.

This assay can also be used to determine intra- and inter-array variabilities (*see* **Note 10**). p53 is a tumor suppressor protein and, as part of its function as a transcription factor, it has a specific DNA-binding domain ***(21,22)***. The DNA-binding assay therefore comprises binding a Cy3-labeled duplex DNA containing a p53 recognition site to the active immobilized p53 proteins, and provides a measurement of activity of the proteins immobilized in each spot.

3.5.1. Labeling the Antibody With a Fluorophore

1. Following recommended protocols, buffer exchange an anti-His antibody into 0.1 *M* sodium carbonate buffer (pH 9.3) using a PD-10 desalting column (Amersham Biosciences) to give an antibody concentration of 1 mg/mL (*see* **Note 11**).
2. Add the protein solution (1 mL) to one vial of Cy3 Mono-Reactive Dye (Amersham Biosciences) and mix thoroughly (*see* **Note 11**).
3. Incubate at room temperature for 30 min with occasional agitation.
4. Separate the labeled antibody from the free dye-labeled protein using a PD-10 desalting column (Amersham Biosciences), eluting protein in 2 mL PBS to give a final antibody concentration of 0.5 mg/mL.
5. Estimate the molar dye/protein ratio by measuring absorbance at 280 nm and 552 nm, using approximate extinction coefficients of 170,000 $M^{-1}cm^{-1}$ for protein and 150,000 $M^{-1}cm^{-1}$ for Cy3 ***(23)***.

3.5.2. Antibody-Binding Assay

1. Dilute 10 µL of the Cy3-labeled anti-His antibody in 1 mL Marvel/PBST to give an antibody concentration of 5 µg/mL.
2. Remove the array from freezing buffer and equilibrate in PBST at room temperature for 5 min.
3. Drain away the PBST, add the antibody solution to the array, and incubate with gentle agitation at room temperature for 30 min.
4. Wash the array three times for 5 min with 1 mL of PBST.
5. Dry the array by blotting it between two sheets of Whatman paper.
6. Mount the array onto a glass microscope slide using double-sided tape, trimming the edges with a scalpel blade if required.
7. Scan the array at 550 nm using a DNA microarray scanner and process the data using a DNA microarray data analysis software package (**Fig. 5**; *see* **Note 12**).

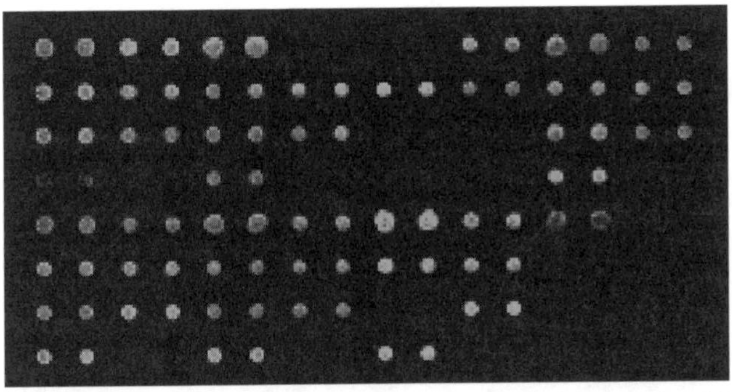

Fig. 5. Measurement of the relative amount of protein immobilized in each spot in the array. A p53 protein function microarray probed with a Cy3-labeled anti-His antibody. Quantification of the signal intensity from each spot reveals the relative amounts of recombinant protein present in each spot in the array.

3.5.3. DNA-Binding Assay

1. 5'-label complementary GADD45 promoter oligos (5'-gta cag aac atg tct aag cat gct ggg gac-3' and 5'-gtc ccc agc atg ctt aga cat gtt ctg tac-3') with Cy3 during synthesis (*see* **Note 13**).
2. Anneal the oligos together in p53 buffer and store at 4°C.
3. Dilute the 2Cy3-GADD45 duplex DNA in p53 buffer to a concentration of 250 n*M*.
4. Remove array from freezing buffer and equilibrate in p53 buffer at room temperature for 5 min.
5. Drain away the p53 buffer, add 1 mL of the DNA solution, and incubate at room temperature for 30 min with gentle agitation.
6. Wash the array three times for 5 min with 1 mL of p53 buffer.
7. Dry the array by blotting it between two sheets of Whatman paper.
8. Mount the array onto a glass microscope slide using double-sided tape, trimming the edges with a scalpel blade if required.
9. Scan the array at 550 nm using a DNA microarray scanner and process the data using a DNA microarray data analysis software package (**Fig. 6**; *see* **Notes 12** and **14**).

3.5.4. Other Assay Types

We have also carried out other protein–protein interaction assays using p53 protein function microarrays fabricated according to the protocols described above *(6)*. In such assays, we probe the array with a Cy3-labeled solution-phase partner (for example, Cy3-MDM2) in a manner essentially identical to that described above (**Subheading 3.5.2.**) for the antibody-binding assay.

In other work, using a range of different protein function microarrays, each fabricated in the form of BCCP fusion proteins as described here, we have been able to monitor a wide range of protein activities, including protein-protein interactions and protein–small molecule interactions, as well as carrying out on-chip phosphorylation assays *(6)* (*see* **Note 15**).

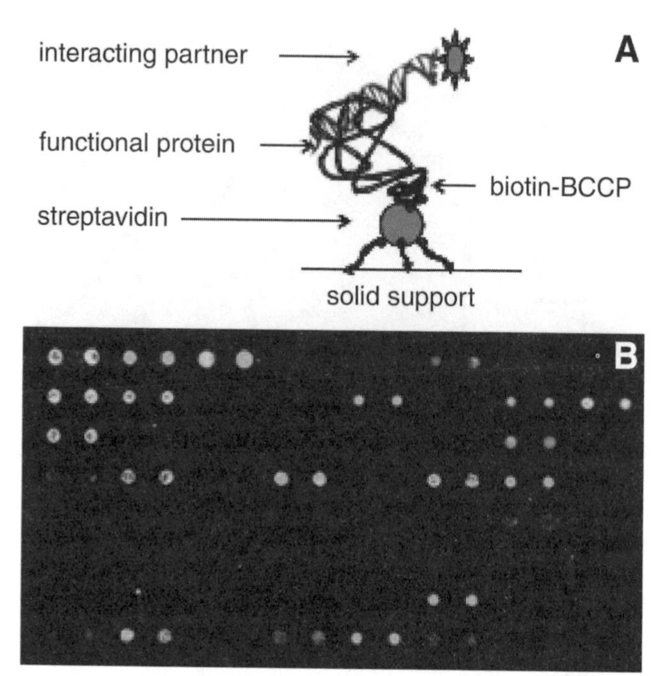

Fig. 6. Measurement of the relative amount of ligand bound to each protein in the array. (**A**) Schematic of on-chip binding assay in which a fluorescently labeled interaction partner binds to the functional, arrayed protein immobilized to the streptavidin-coated surface via the biotinylated BCCP tag. (**B**) p53 protein function microarray probed with Cy3-labeled GADD45 duplex oligo. Quantification of the signal intensity from each spot allows the effect of polymorphic and functional variation on the DNA binding function of p53 to be determined.

4. Notes

1. In order to avoid laborious prepurification of each recombinant protein prior to array fabrication, we have developed a procedure in which we combine the immobilization and purification stages into a single step *(6)*. For this approach to work, the array surface itself must have a low capacity for nonspecific binding of proteins, yet must have a high specificity and high binding affinity for the proteins to be arrayed. In principle, a range of different affinity tags could be used here, but in practice, few actually offer sufficiently high specificity interactions with the surface affinity matrix. For example, His-tags are not well suited to such an approach, since the specificity of the interaction with Ni^{2+} ligands is too low (particularly in the context of expression in eukaryotic cells where there are numerous Ni^{2+}-binding proteins) and the intrinsic affinity of the interaction is also relatively low, resulting in leaching of protein from the array, even in the absence of imidazole. To circumvent these problems, we have chosen to use an affinity tag that becomes biotinylated in vivo at a single specific residue, allowing us to make use of the very high affinity

(*K*d approx 10^{-15} *M*) and specificity of the streptavidin-biotin interaction. Array fabrication thus becomes the simple process of printing crude lysates containing the recombinant biotinylated proteins onto streptavidin-coated surfaces, which have a low nonspecific protein-binding capacity, followed by washing to remove all non-biotinylated proteins from the array surface (**Fig. 4**). One such commercially available surface that works well is streptavidin-derivatized phospho-cellulose membrane (SAM2), although we have also used streptavidin-coated glass or plastic surfaces. In the latter cases, we have found that organic polymer coatings, such as those based on dextran or polyethylene glycol, are superior to proteinaceous blocking agents such as bovine serum albumin or powdered milk in reducing the nonspecific binding background, and a number of such surfaces are now also commercially available.

We have observed that using the streptavidin-biotin interaction as the basis for array fabrication confers one further major advantage: the very high affinity of the streptavidin-biotin interaction means that we quickly start to saturate the available biotin-binding sites on the surface so a crude normalization of protein loading can be achieved without pre-adjusting the concentrations of the crude lysates to compensate for differences in the individual expression levels of the different recombinant proteins ***(6)***. Obviously, host cell proteins that are endogenously biotinylated will be expected to compete with the biotinylated recombinant proteins for the available streptavidin-binding sites on the surface. *E. coli* has one such endogenous protein, the approx 25 kDa *accB* gene product, but under native conditions we have observed that this protein does not compete efficiently with biotinylated recombinant proteins for binding to streptavidin. This perhaps reflects low natural expression levels and the fact that the endogenous, full-length *E. coli* biotin carboxyl carrier protein naturally forms part of a heteromultimeric complex ***(24)*** in which the biotin may not be readily accessible to streptavidin.

As with many affinity tags, the biotinylated affinity tag can be positioned at either the N- or C-terminus of the protein to be arrayed, dependent on the structural and functional characteristics of the protein. "Biochemical" biotinylation is greatly preferable to "chemical" biotinylation, since the latter offers little control over the site of biotinylation and still requires prepurification of each protein. There are three main alternatives for biochemical introduction of a biotin moiety into a recombinant protein: two involve affinity tags that can be biotinylated in vivo or in vitro, while the third involves an intein-mediated introduction of biotinylated cysteine. The AviTag (Avidity, Denver, CO) is an in vitro evolved 15-residue peptide that is specifically biotinylated exclusively by the *E. coli* biotin ligase ***(25)***. We have chosen, however, to use a compact, folded, biotinylated, approx 80-residue domain derived from the *E. coli* biotin carboxyl carrier protein (BCCP; **Fig. 1**) ***(26,27)***, because this affords two significant advantages over the AviTag. Firstly, the BCCP domain is cross-recognized by eukaryotic biotin ligases, enabling it to be biotinylated efficiently in yeast, insect, and mammalian cells without the need to coexpress the *E. coli* biotin ligase ***(28–30)***. Secondly, the N- and C-termini of BCCP are physically separated from the site of biotinylation by approx 50 Å (**Fig. 1**) ***(27)***, so the BCCP domain can be thought of as a stalk that presents the recombinant proteins away from the surface, thus minimizing any deleterious effects due to immobilization. Recently, a third route to biochemical biotinylation has been described, in which a single biotinylated cysteine moiety is added to the C-terminus of recombinant proteins during intein-mediated protein splicing of fusion proteins ***(31)***. However, it is not clear at this time what advantages this route offers over use of the AviTag.

2. Vectors for expressing proteins as fusions to a BCCP domain derived from *Klebsiella pneumoniae* are now available from Invitrogen (pET104 DEST Bioease). This *K. pneumoniae* domain is highly homologous to the *E. coli* BCCP protein and confers the same properties.
3. Although the growth temperatures and induction times were specifically designed for BCCP-p53 fusions, they should be generally applicable to any other BCCP fusions as well. However, for different recombinant proteins, various combinations of growth temperature and induction time should be explored to find the optimal conditions.
4. We observed that addition of free biotin to the growth medium increases the extent of biotinylation of the recombinant BCCP fusion protein. Strains overexpressing the *E. coli* biotin ligase are also available from Avidity (www.avidity.com) if desired, although we have not found this necessary when using the BCCP tag. The wash step prior to cell lysis is needed to remove free biotin before array fabrication; if the protein is purified before array fabrication, this is not necessary.
5. For expression of BCCP-p53 fusion proteins, we typically found the protein concentration in crude lysates to be 5 mg/mL, and we estimated that BCCP-p53 was present at approx 1% of total soluble protein. When expressing a number of clones in parallel for array fabrication, the Bradford assay can conveniently be done on all clones in parallel using a microtiter plate format. However, it would be laborious to carry out SDS-PAGE analysis on all clones, so typically we assess only a selection of clones in this way, since the absolute expression level is not critical for array fabrication.
6. The biotinylated component of the sample will be supershifted by streptavidin even under denaturing conditions, enabling a simple side-by-side comparison to be carried out. To save time, this assay need not be done on all clones; we have typically found that if the recombinant BCCP proteins are expressed and folded, the BCCP domain is efficiently biotinylated.
7. We have found solid pins the easiest to clean rapidly, and have observed no carryover between samples using such pins. In addition, we have found that solid pins also work well when printing directly onto glass or plastic surfaces (**Fig. 7**).
8. Each spotting event delivered approx 50 nL liquid containing approx 10 fmoles biotinylated p53. We typically use multiple stamps per spot to increase the protein loading at each position in the array, and we have found that these general printing parameters work well with other surfaces, including glass microscope slides, although care must be taken in calibrating the *z*-height on the robot when using fragile surfaces. In addition, by using the same printing parameters under the same

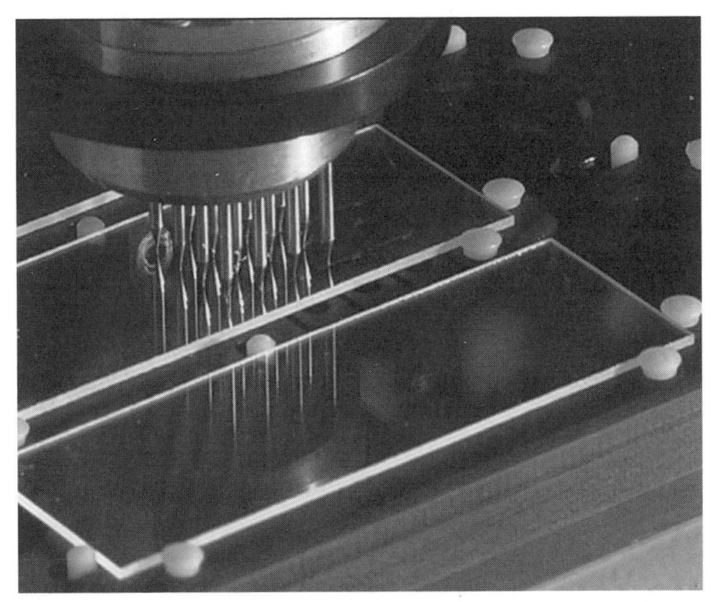

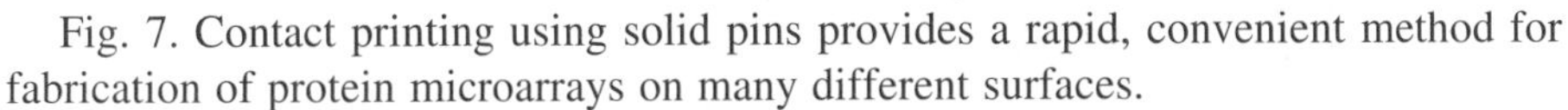

Fig. 7. Contact printing using solid pins provides a rapid, convenient method for fabrication of protein microarrays on many different surfaces.

conditions of printing (i.e., in a glycerol-containing buffer to reduce evaporation rates), prepurified proteins can be spotted onto nonselective surfaces that bind proteins by chemical cross-linking (e.g., epoxide- or aldehyde-coated glass ***[13, 14]***) or by noncovalent adsorption (e.g., supported nitrocellulose, agarose, or polyacrylamide ***[14]***).

9. The blocking and wash steps should remove all nonbiotinylated proteins from the array surface, while the biotin in the milk powder blocks any remaining biotin-binding sites on the streptavidin surface, such that any biotinylated proteins that do dissociate cannot rebind to the array. Arrays stored in p53 freeze mix at –20°C have been shown to be stable for >1 y.
10. By printing, processing, and assaying arrays using these protocols, we have been able to achieve spot-to-spot CVs of 4–5% (**Fig. 8**).
11. Fluorophores other than Cy3—for example, the Alexa dyes (Molecular Probes) or Oyster dyes (Denovo Biolabels)—also work well in these applications. However, when using any amine-reactive dye to label the protein, buffers containing primary amino groups such as Tris and glycine should be avoided, because they will inhibit the conjugation reaction. The presence of low concentrations (<2%) of biocides such as sodium azide do not affect protein labeling.
12. In principle, any microarray scanner can be used to read the protein microarrays, since the differences lie mainly in the sensitivity. However, models that read in a transmission mode might not work with membrane-based arrays.

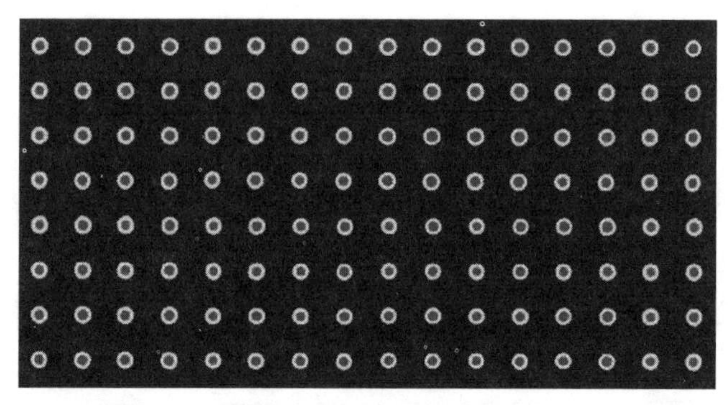

Fig. 8. Determination of coefficients of variation for protein microarrays. Replica spotting of Cy3-labeled, biotinylated bovine serum albumin allows the intra- and inter-array spot-to-spot reproducibility to be determined.

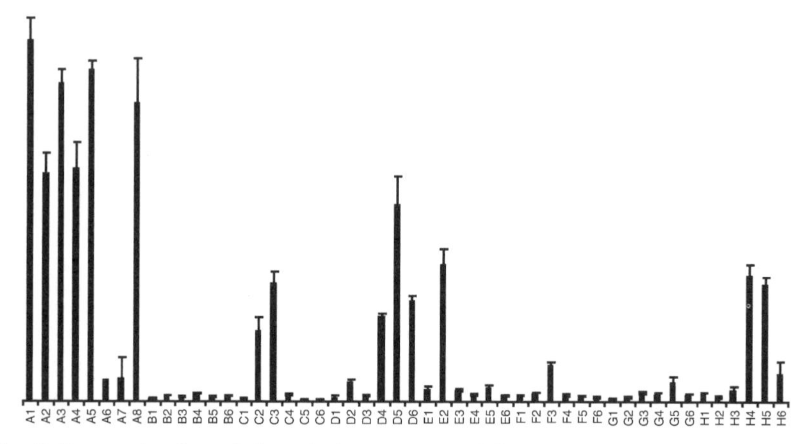

Fig. 9. Determination of the relative amount of ligand bound per unit protein. The signal intensities from the DNA binding assay (**Fig. 6B**) have been divided by the relevant anti-His tag antibody binding signal (**Fig. 5**) to show the relative amount of DNA bound per unit protein for each p53 variant in the array.

13. The oligonucleotides can also be radiolabeled—for example, with ^{32}P, ^{33}P, or ^{35}S—in which case the arrays can be read either using a phosphorimager or by autoradiography.
14. After initial data acquisition and processing, further data analysis can be carried out. For example, the raw data can be normalized for the relative amount of protein immobilized in each position in the array, as measured by an anti-His antibody-binding assay, by simply dividing each measured DNA-binding signal intensity by the relevant measured antibody-binding signal intensity (**Fig. 9**). This gives a measure of the relative specific activity of each protein in the array.

    Alternatively, a number of replica arrays can be probed with different concentrations of ligand in solution (or potentially even by using increasing ligand con-

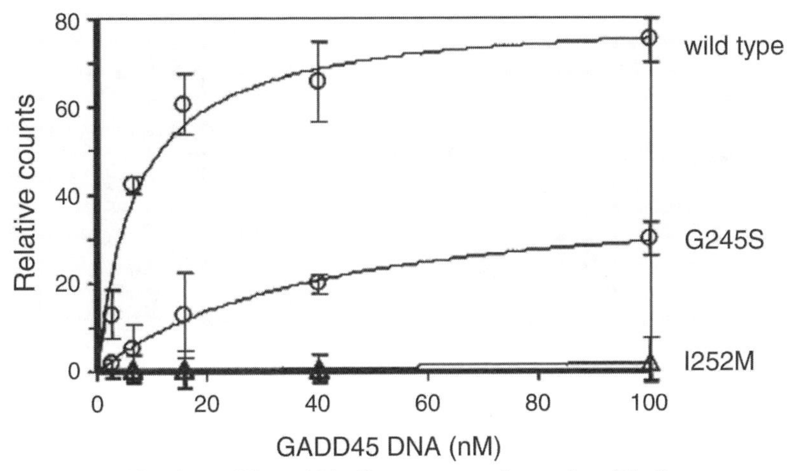

Fig. 10. Quantification of ligand binding to arrayed proteins. Binding curves generated by measuring the relative amount of bound GADD45 DNA at increasing DNA concentrations allow binding affinities and specific activities to be calculated for each p53 variant in the array.

centrations on the same array, taking a reading at each concentration). The relative amount of bound ligand can then be plotted as a function of ligand concentration in solution for each protein in the array *(6)* and fitted to a simple hyperbolic concentration-response curve according to

$$R = BmaxL/(Kd + L)$$

where R is the response in relative counts and L is the ligand concentration. From this, the ligand binding constants and maximum ligand-binding capacities for each arrayed protein can be determined (**Fig. 10**) *(6)*.

15. In the phosphorylation assay, we used ^{33}P-ATP to radiolabel those elements of the array that were substrates for the relevant kinase; the arrays were then read using a phosphorimager *(6)*. We have also shown that by manipulating the experimental conditions in array-based protein–protein interaction assays, we can readily distinguish true from false positives. For example, we studied the interaction of calmodulin with a diverse array of human proteins in the presence and absence of calcium (2+) ions. Because calmodulin binding should be calcium dependent, we were able to deduce true from false positives based on the array data (**Fig. 11**) (unpublished data).

Acknowledgments

The authors thank Drs. Joe Boutell, Ben Godber, and Mike Dyson for their help in generating the data and the procedures detailed in this chapter, and Dr. Roland Kozlowski for helpful discussion along the way. This work was supported by Procognia Ltd.

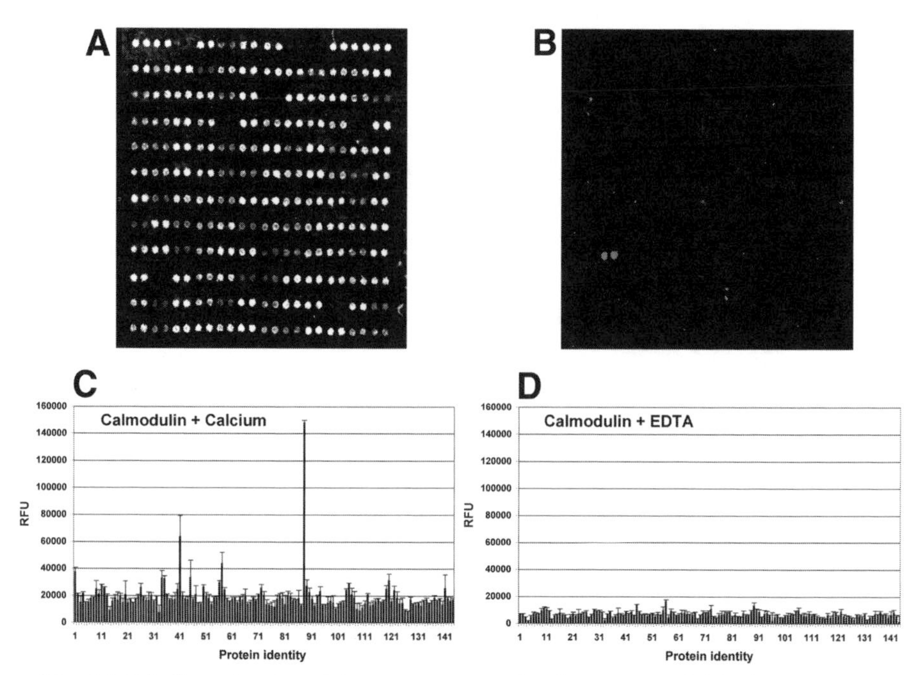

Fig. 11. Binding of calmodulin to an array of a diverse set of 144 human proteins enables novel, calcium-dependent interactions to be identified. (**A**) Binding of Cy3-labeled anti-His tag antibody to the array allows the relative amount of protein in each spot to be determined. (**B**) Binding of Cy3-labeled calmodulin allows potential interacting partners to be identified. (**C**) Histogram showing the relative amount of calmodulin bound per unit protein in the presence of calcium ions. (**D**) Histogram showing the relative amount of calmodulin bound per unit protein in the presence of calcium ions and a high concentration of a divalent metal ion chelator, ethylenediamine tetraacetic acid.

References

1. Mulder, N. J., Apweiler, R., Attwood, T. K., et al. (2003) The InterPro Database, 2003 brings increased coverage and new features. *Nucl. Acid Res.* **31,** 315–318.
2. Wise, E. Y., Yew, W. S., Babbitt, P. C., Gerlt, J. A., and Rayment, I. (2002) Homologous (β/α)8-barrel enzymes that catalyze unrelated reactions: orotidine 5'-monophosphate decarboxylase and 3-keto-l-gulonate 6-phosphate decarboxylase. *Biochemistry* **41,** 3861–3869.
3. Schmidt, D. M. Z., Mundorff, E. C., Dojka, M., et al. (2003) Evolutionary potential of (β/α)8-barrels: functional promiscuity produced by single substitutions in the enolase superfamily. *Biochemistry* **42,** 8387–8393.
4. The Genome International Sequencing Consortium. (2001) Initial sequencing and analysis of the human genome. *Nature* **409,** 860–921.
5. MacBeath, G. (2002) Protein microarrays and proteomics. *Nat. Genet.* **32,** 526–532.

6. Boutell, J. M., Hart, D. J., Gobder, B. L. J., Kozlowski, R. Z., and Blackburn, J. M. (2004) Analysis of the effect of clinically relevant mutations on p53 function using protein microarray technology. *Proteomics* **4,** 1950–1958.
7. Kodadek, T. (2001) Protein microarrays: prospects and problems. *Chem. Biol.* **8,** 105–115.
8. Predki, P. (2004) Functional protein microarrays: ripe for discovery. *Curr. Opin. Chem. Biol.* **8,** 8–13.
9. Zhu, H., Klemic, J. F., Chang, S., et al. (2000) Analysis of yeast protein kinases using protein chips. *Nat. Genet.* **26,** 283–289.
10. Zhu, H., Bilgin, M., Bangham, R., et al. (2001) Global analysis of protein activities using proteome chips. *Science* **293,** 2101–2105.
11. Michaud, G. A., Salcius, M., Zhou, F., et al. (2003) Analyzing antibody specificity with whole proteome microarrays. *Nat. Biotech.* **21,** 1509–1512.
12. Fang, Y., Lahiri, J., and Picard, L. (2003) G-protein-coupled receptor microarrays for drug discovery. *Drug Discovery Today* **8,** 755–761.
13. MacBeath, G. and Schreiber, S. L. (2000) Printing proteins as microarrays for high-throughput function determination. *Science* **289,** 1760–1763.
14. Angenendt, P., Glokler, J., Sobek, J., Lehrach, H., and Cahill, D. J. (2003) Next generation of protein microarray support materials: evaluation for protein and antibody microarray applications. *J. Chromatogr. A* **1009,** 97–104.
15. Koopmann, J.-O. and Blackburn, J. M. (2003) High affinity capture surface for MALDI compatible protein microarrays. *Rapid Commun. Mass Spectrom.* **17,** 1–8.
16. Varley, J. M., Evans, D. G. R., and Birch, J. M. (1997) Li-Fraumeni syndrome—a molecular and clinical review. *Br. J. Cancer* **76,** 1–14.
17. Birch, J. M., Alston, R. D., McNally, R. J., et al. (2001) Relative frequency and morphology of cancers in carriers of germline TP53 mutations. *Oncogene* **20,** 4621–4628.
18. Sambrook, J., Fritsch, E. F., and Maniatis, T. (1989) *Molecular Cloning, A Laboratory Manual*, Second ed. Cold Spring Harbor Laboratory, Cold Spring Harbor, NY.
19. Guex, N. and Peitsch, M. C. (1997) SWISS-MODEL and the Swiss-PdbViewer: an environment for comparative protein modeling. *Electrophoresis* **18,** 2714–2723.
20. Bradford, M. M. (1976) A rapid and sensitive method for the quantitation of microgram quantities of protein utilizing the principle of protein-dye binding. *Anal. Biochem.* **72,** 248–254.
21. Ryan, K. M., Phillips, A. C., and Vousden, K. H. (2001) Regulation and function of the p53 tumor suppressor protein. *Curr. Opin. Cell Biol.* **13,** 332–337.
22. Vogelstein, B., Lane, D., and Levine, A. J. (2000) Surfing the p53 network. *Nature* **408,** 307–310.
23. Data provided by Amersham Biosciences.
24. Choi-Rhee, E. and Cronan, J. E. (2003) The biotin carboxylase-biotin carboxyl carrier protein complex of *Escherichia coli* acetyl-CoA carboxylase. *J. Biol. Chem.* **278,** 30,806–30,812
25. Cull, M. G. and Schatz, P. J. (2000) Biotinylation of proteins in vivo and in vitro using small peptide tags. *Methods Enzymol.* **326,** 430–440.

26. Chapman-Smith, A. and Cronan, J. E. (1999) The enzymatic biotinylation of proteins: a posttranslational modification of exceptional specificity. *Trends Biochem. Sci.* **24,** 359–363.
27. Athappilly, F. K. and Hendrickson, W. A. (1995) Structure of the biotinyl domain of acetylcoenzyme A carboxylase determined by MAD phasing. *Structure* **3,** 1407–1419.
28. Berliner, E., Mahtani, H. K., Karki, S., et al. (1994) Microtubule movement by a biotinated kinesin bound to streptavidin-coated surface. *J. Biol. Chem.* **269,** 8610–8615.
29. Lerner, C. G. and Saiki, A. Y. (1996) Scintillation proximity assay for human DNA topoisomerase I using recombinant biotinyl-fusion protein produced in baculovirus-infected insect cells. *Anal. Biochem.* **240,** 185–196.
30. Parrott, M. B. and Barry M. A. (2001) Metabolic biotinylation of secreted and cell surface proteins from mammalian cells. *Biochem. Biophys. Res. Commun.* **281,** 993–1000
31. Lue, R. Y., Chen, G. Y., Hu, Y., Zhu, Q., and Yao, S. Q. (2004) Versatile protein biotinylation strategies for potential high-throughput proteomics. *J. Am. Chem. Soc.* **126,** 1055–1062.

15

Peptide and Small-Molecule Microarrays

Jan Marik and Kit S. Lam

Summary

Two methods that use chemoselective ligation chemistry to prepare peptide and small-molecule microarrays are described here. The first method involves the functionalization of a glass slide with a glyoxylyl group, followed by chemoselective ligation of small molecules or peptides to the functionalized surface via a covalent bond. In the second method, peptides or small molecules are first conjugated to a macromolecular scaffold. The final ligand–scaffold conjugates are then spotted and adsorbed onto the solid surface. Three different assay methods to screen such chemical microarrays are described.

Key Words: Peptide microarrays; small-molecule microarrays; small-molecule immobilization; peptide immobilization; chemoselective ligation.

1. Introduction

High-density peptide microarrays, first reported by Fodor et al. in 1991 *(1)*, were prepared *in situ* with a photolithographic light-directed synthesis method. This method was later adapted by scientists at Affymetrix to prepare oligonulceotide microarrays *(2)*. In 1995, Brown et al. *(3)* reported the preparation of DNA microarrays by spotting cDNAs on glass slides with automatic arrayers. In the last decade, many scientists around the world have successfully used such oligonucleotide and DNA microarrays to evaluate gene expression profiles of cells or tissues. Because of the enormous success of using DNA microarrays as tools for genomics, protein, peptide, and small-molecule microarrays have resurfaced in the last few years. These microarrays are becoming useful tools in proteomic research *(4–7)*. Common solid supports for microarray preparation are glass or plastic microscope slides, polyvinylidene fluoride (PVDF) or nitrocellulose membranes, or gold surfaces. Macromolecules such as proteins, DNAs, or large polysaccharides can simply adsorb onto these surfaces by noncovalent

From: *Methods in Molecular Biology, vol. 310: Chemical Genomics: Reviews and Protocols*
Edited by: E. D. Zanders © Humana Press Inc., Totowa, NJ

interactions. However, small molecules, short peptides, oligonucleotides, or oligosaccharides *(8)* will require covalent attachment to the solid support. Alternatively, these molecules can first be covalently attached to macromolecular scaffold, and the ligand–scaffold conjugates are subsequently adsorbed noncovalently to the solid support ***(9,10)***.

Oligonucleotide and peptide microarrays can be prepared *in situ* with light-directed synthesis on a glass surface in conjunction with either a photolithographic method or using a micromirror device ***(2,11–15)***. However, these methods are not useful for most organic synthesis. Furthermore, such approaches require equipment that is not readily available. Spot synthesis on cellulose membrane is another *in situ* synthesis method, but the resulting microarrays are low density ***(16)***.

The more common approaches to prepare small-molecule or peptide microarrays is to first synthesize these compounds, and then spot them on the functionalized solid support with an automatic arrayer. In this approach, we first functionalize the glass surface with glyoxylyl group, which can then be chemoselectively ligated to ligands with an N-terminal cysteine to form a thiazolidine ring or an amino-oxy group to form an oxime bond. A second spotting approach is to first ligate the peptides or small molecules to a macromolecular scaffold, such as agarose. The final ligand–agarose conjugates are then spotted onto the solid support ***(9,10)***. A third approach is to first coat the solid support with streptavidin; then biotinylated peptides or small molecules are spotted onto the support ***(17)***. To avoid steric hindrance caused by the solid support, it is desirable to add a long hydrophilic and flexible linker between the small molecule or peptide and the solid support ***(18,19)***. In this chapter, the first two approaches outlined above, which have been used in our laboratory to prepare peptide or small-molecule microarrays, are described. Various methods for assaying such microarrays are also discussed.

2. Materials

1. 3-aminopropyltriethoxysilane (Aldrich).
2. Levulinic acid (Aldrich).
3. Agarose Type XI low gelling (Sigma).
4. Dimethylsulfoxide (DMSO) (Aldrich).
5. Dimethylformamide (DMF) (VWR).
6. Glass slides: Rite-On microslides (Gold Seal).
7. Plastic slides: optical plastic microslides (Rinzl).
8. PVDF membrane: Immun-Blot PVDF Membrane (BioRad).
9. Acrylic acid (Aldrich).
10. Dioxane (Aldrich).
11. Sodium periodate (Aldrich).
12. Osmium oxide (Aldrich).
13. Diisopropylcarbodiimide (DIC) (Advanced ChemTech).

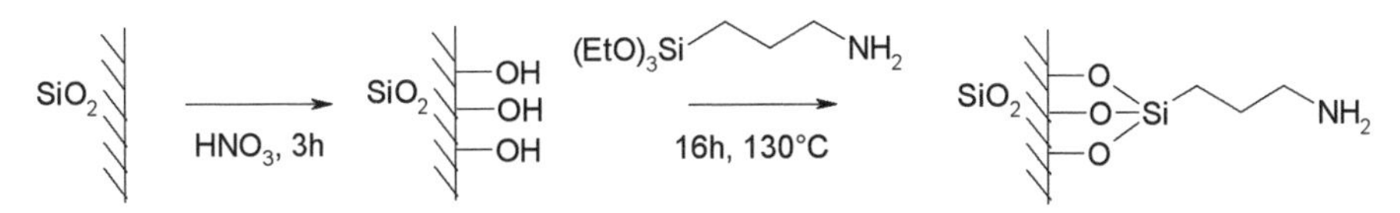

Fig. 1. Derivatization of the glass surface with amino group.

14. Streptavidine-alkaline phosphatase (ST-AP) (Sigma).
15. Bovine serum albumin (BSA) (Sigma).
16. Acetate buffer (pH 4.5): 0.2 *M* sodium acetate, glacial acetic acid.
17. Phosphate-buffered saline (PBS) buffer: 137 m*M* NaCl, 2.68 m*M* KCl, 8 m*M* Na_2 HPO_4, 1.47 m*M* $KHPO_4$.
18. PBSTG buffer: PBS, 0.1% (w/v) Tween-20, 0.1% (w/v) gelatin.
19. BCIP buffer (pH 8.8): 100 m*M* NaCl, 100 m*M* Tris base, 2.34 m*M* $MgCl_2$.
20. 5-bromo-4-chloro-3-indolyl phosphate, *p*-toluidine salt (BCIP) (Sigma).

3. Methods

The reaction of an amino-oxy group with ketones and aldehydes has been widely used for chemoselective ligation ***(19–21)***. Because the nucleophilicity of the amino-oxy group is enhanced by α-effect, the ligation reaction can proceed in slightly acidic conditions even in the presence of other nucleophiles, such as free amino groups or thiol groups, which are commonly present in proteins or other biomolecules. An exception is the N-terminal cysteine that reacts with glyoxylyl moiety to form oxazolidine ring; not surprisingly, this reaction has been used for ligation or immobilization of peptides and proteins.

3.1. Preparation of Amino-Functionalized Glass

Various commercially available microscope glass slides can be used for microarray preparation. An important step is to clean the glass surface thoroughly with detergents, strong oxidizers, sonication, acids, or bases. The functionalization of the glass surface with a reactive group, such as NH_2, can be accomplished with 3-aminopropyltriethoxysilane (APTES), but the reaction has to be done under strictly anhydrous conditions. It is generally believed that monomolecular layer formation involves hydrolysis of alkoxysilane followed by covalent bond formation with hydroxy group on the glass surface ***(2,22)***. The procedure for the preparation of the amino-glass surface is described below (*see* **Fig. 1**).

1. Wash the glass microscope slide with 1 *M* NaOH, water, and 1 *M* HCl.
2. Boil the slide in concentrated HNO_3 for 3 h.
3. Wash the slide with distilled water and dry at 110°C.
4. Incubate the slide in 2% 3-aminopropyltriethoxysilane solution in anhydrous toluene for 16 h at 130°C.
5. Wash the slide with toluene, ethanol, and then dry it at room temperature (RT).

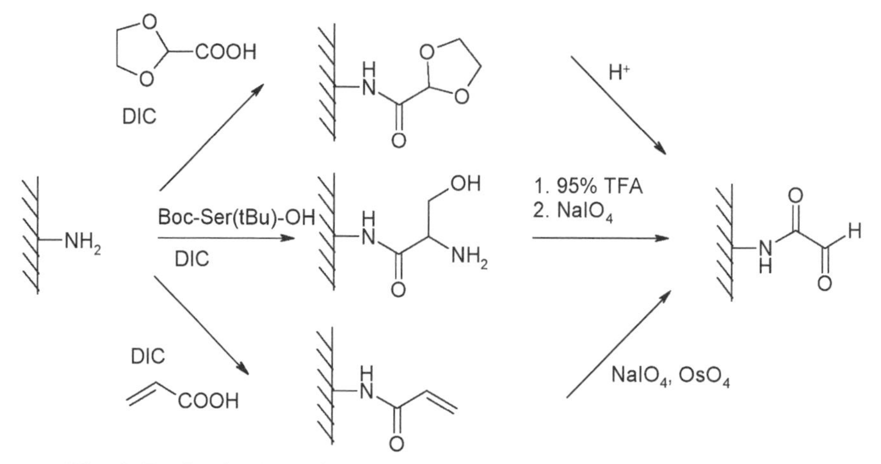

Fig. 2. Derivatization of the amino-glass with glyoxylyl functionality.

3.2. Introduction of Glyoxylyl Functionality

Several methods for introduction of glyoxylyl functionality have been developed (*see* **Fig. 2**). The first method starts with coupling of 1,3-dioxolane-2-carboxylic acid to the amino-slide, followed by acidic hydrolysis of the dioxolane ring to yield an aldehyde moiety ***(19)***. The second method involves coupling of Boc/tBu-protected serine Boc-Ser(tBu)-OH to the amino-slide, followed by trifluoroacetic acid (TFA)-mediated removal of Boc/tBu groups and oxidation with sodium periodate ***(19)***. The latest and most reliable method uses acrylic acid as a starting material and is detailed below. The acrylic acid is first coupled to the amino-slide, and the double bond then undergoes oxidative cleavage with $NaIO_4/OsO_4$ to form an aldehyde functional group ***(23)***.

1. Mix acrylic acid with 0.5 eq of DIC in DMF and stir 15 min at –10°C.
2. Incubate the amino-slides in this solution overnight at RT.
3. Wash the slide with DMF, water.
4. Incubate the slide in 10% $NaIO_4$, with OsO_4 (0.5%) in H_2O/dioxane 1:3.
5. Wash the slide with water and dry it at room temperature.

3.3. Conjugation of Small Molecules to Glyoxylyl Glass

The amino-oxy group can be incorporated into peptides or small molecules in the form of protected *N-tert*-butyloxycarbonyl amino-oxy acetic acid (Boc-Aoa-OH, Nova Biochem). For peptides, the Aoa functional group can be coupled to the N-terminus. If the free N-terminus of the peptide is required for biological function, the Boc-Aoa-OH can be coupled to the side chain of some diamino acid (e.g., diaminopropionic acid [Dpr], lysine [Lys], or ornithine [Orn]) at the

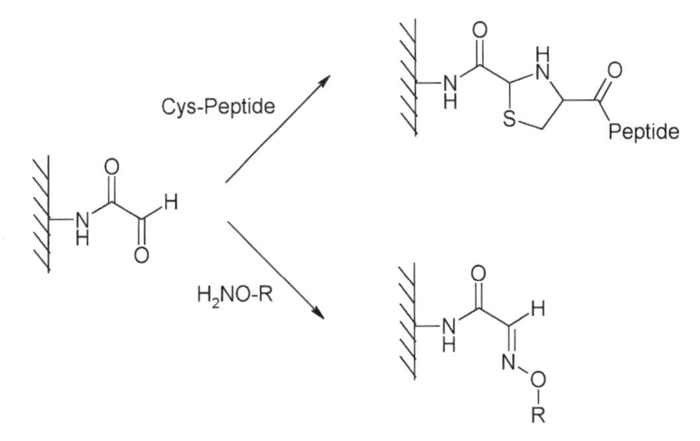

Fig. 3. Attachment of small molecule (R) or peptide to glyoxylyl glass.

carboxyl end of the peptide. For this purpose, Fmoc-Dpr(Boc-Aoa)-OH (commercially available from Nova Biochem) can be used. Aoa can also be linked to a small molecule with similar approaches. Another approach for chemoselective ligation of peptides to glyoxylyl glass is to insert an N-terminal cysteine to the peptide. This cysteine will react with glyoxylyl group to form a thiazolidine ring (*see* **Fig. 3**) ***(19)***.

1. Wash the glyoxylyl-glass slide with water and dry it.
2. Dissolve the peptide or small-molecule samples in DMSO/acetate buffer 1:1.
3. Adjust the pH of the solution to 4.0–5.0.
4. Spot the samples onto the glyoxylyl slide and incubate it overnight in a moisturized container.

3.4. Conjugation to Ketone-Modified Agarose

As has been mentioned earlier, macromolecules such as proteins or polysaccharides can be adsorbed to various surfaces by noncovalent interaction, and therefore they can be used as scaffolds to attach peptides or small molecules onto these surfaces ***(24)***. We first use levulinic acid to incorporate ketone groups into the macromolecular scaffold, such as agarose, human serum albumin (HSA), or dextran (*see* **Fig. 4**). The coupling of levulinic acid to ε-amino groups of lysine side chains in HSA or free hydroxy groups in agarose or dextran can be accomplished by DIC activation ***(9,10)***. However, the alkali-labile ester bond formed between the ketone group and the polysaccharide may be undesirable in certain applications. In that case, amino groups can be introduced into the polysaccharide chain prior to reaction with levulinic acid. The main advantages of this approach are that: (1) the ligation reaction occurs in solution and needs

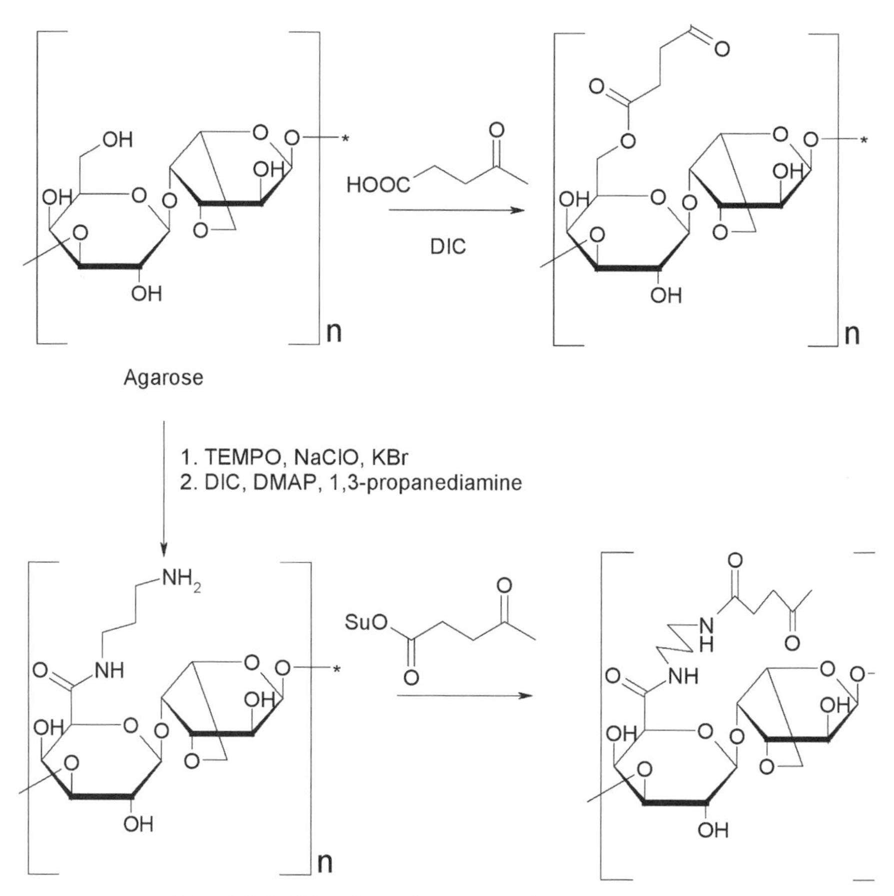

Fig. 4. Modification of agarose with levulinic acid.

to be done only once, and (2) uniformity among different spots and different slides can be achieved by keeping the macromolecular scaffold concentration the same in all samples.

3.4.1.1. Preparation of Ketone-Functionalized Agarose

1. Dissolve 1 g of agarose in DMF.
2. Mix levulinic acid (0.5 mmol) with DIC (0.25 mmol) in dichloromethane at –10°C for 15 min.
3. Mix the agarose solution prepared in **step 1** with solution from **step 2**.
4. Shake it overnight at room temperature.
5. Precipitate the product by addition of diethyl ether.
6. Remove the solvent and wash the precipitate several times with diethyl ether.
7. Dry the product *in vacuo* at room temperature.

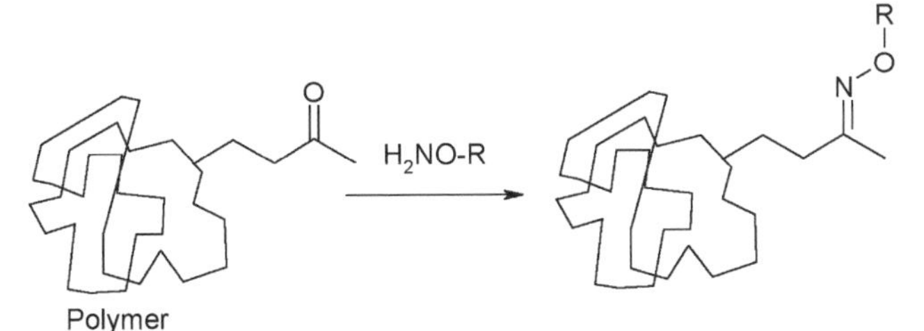

Fig. 5. Conjugation of ketone modified polymer with small molecule or peptide (R).

Small molecules or peptides are then chemoselectively conjugated to the ketone-modified macromolecular scaffold via the amino-oxy functional group to form an oxime bond (*see* **Fig. 5**).

3.4.1.2. Conjugation of Small Molecule to Ketone-Functionalized Agarose

1. Dissolve ketone-modified agarose in 20% DMSO/acetate buffer at 0.5 mg/mL (0.25 µmol/mL of ketone group) (*see* **Note 1**).
2. Dissolve the small molecule in 100% DMSO at 2 µmol/mL.
3. Mix 100 µL of the first solution and 25 µL of second solution in a polypropylene 96-well plate (twofold excess of small molecule).
4. Shake the plate gently overnight at room temperature.
5. Spot the microarray and dry it overnight at room temperature before use.

3.4.2. Microarray Printing

Many instruments for contact or non-contact microarray printing are now commercially available. We have been using the Micro Array 03A (Wittech Co., Ltd.) to print our microarrays. This machine uses flat-tip gold-coated needles; the needle tip diameter used for our experiments ranges from 100 µm to 500 µm.

3.4.3. Application of Peptide or Small-Molecule Microarray

3.4.3.1. Enzyme-Linked Colorimetric Assay

1. Wash the microarrays with PBS.
2. Preblock slide with PBSTG 1 h at room temperature.
3. Incubate slide with biotinylated target protein.
4. Wash the slide with PBSTG 5 × 5 min (*see* **Note 2**).
5. Incubate slide with ST-AP in PBSTG at 1:10,000–100,000 dilution.
6. Wash the slide with PBSTG 5 × 5 min.
7. Wash the slide with BCIP buffer 5 × 5 min.
8. Incubate the slide with BCIP (33 µL of 5% DMF solution in 10 mL of BCIP buffer) 15–180 min.
9. Inspect the slide under microscope and quantitate with a high-resolution scanner.

3.4.3.2. Whole-Cell Binding Assay

Example using a suspension cell line (Jurkat T-leukemia cell) (*see* **Notes 3** and **4**).

1. Wash the microarray with PBS, 2 × 10 min.
2. Preblock the slide with 1–5% BSA in PBS buffer.
3. Wash the slide with PBS 3 × 15 min.
4. Incubate the slide with cell suspension 5 × 10^6 cells/mL in PBS 30 min without shaking. Wash the slide gently with PBS 5 × 1 min.
5. Fix the cells with 3.7% formaldehyde solution for 10 min at room temperature.
6. Decant the fixing solution and stain the slide with crystal violet 0.1% for 15–30 min.
7. Wash the slide with PBS and inspect under microscope.

3.4.3.3. Protein Kinase Activity Assay

1. Prepare the kinase buffer: 50 m*M* Tris-HCl (pH 7.4), 50 m*M* NaCl, 10 m*M* $MgCl_2$, 5 m*M* $MnCl_2$, 1 m*M* dithiothreitol, 0.1 m*M* Na_3VO_4, 20 μCi of [γ-^{32}P]ATP.
2. Preblock the microarray slide with kinase buffer and 0.1% BSA, 30 min.
3. Incubate the slide with purified kinase (1 μg/mL) or kinase containing cell lysate, 30 min, in kinase buffer at 30°C.
4. Wash the slide with PBS, 3 × 30 min.
5. Air-dry the slide and expose it to low-energy X-ray film (Kodak Biomax MR film) or Phosphorimager (Molecular Dynamics).

4. Notes

1. Solubility of the agarose and ketone-modified agarose is limited. To dissolve agarose in 20% DMSO/acetate buffer (pH 4.5), the solution needs to be heated to 50°C and shaken gently overnight.
2. Microarrays prepared from macromolecular scaffold conjugates were found to be stable in all washing procedures used in our laboratory. However, mechanical wiping must be avoided, as it will damage the microarrays.
3. For the cell-binding arrays, 300-μm spots are preferred; 100-μm spots can be used for enzyme-linked colorimetric and other assays.
4. In our experience, the best support for cell-binding assays is a plastic slide. PVDF membrane has high adsorption capacity, and microarrays printed on such membrane are suitable for enzyme-linked colorimetric and phosphorylation assays. However, researchers are encouraged to try all of them for their specific applications.

Acknowledgments

The authors would like to thank Amanda Enstrom for the assistance with the manuscript. This work was supported by NSF Grant MCB9728399, and NIH Grants R33CA-86364, R33CA-89706, and R01CA-098116.

References

1. Fodor, S. P., Read, J. L., Pirrung, M. C., Stryer, L., Lu, A. T., and Solas, D. (1991) Light-directed, spatially addressable parallel chemical synthesis. *Science* **251,** 767–773.
2. Pirrung, M. (2002) How to make a DNA chip. *Angew. Chem. Int. Ed.* **41,** 1276–1289.
3. Schena, M., Shalon, D., Davis, R. W., and Brown, P. O. (1995) Quantitative monitoring of gene expression patterns with a complementary DNA microarray. *Science* **270,** 467–470.
4. Wilson, D. S. and Nock, S. (2003) Recent developments in protein microarray technology. *Angew. Chem. Int. Ed.* **42,** 494–500.
5. Kodadek, T. (2001) Protein microarrays: prospects and problems. *Chem. Biol.* **8,** 105–115.
6. Xu, Q. and Lam, K. S. (2003) Protein and chemical microarrays-powerful tools for proteomics. *J. Biomed. Biotech.* **5,** 257–266.
7. Lam, K. S. and Renil, M. (2002) From combinatorial chemistry to chemical microarray. *Curr. Opin. Chem. Biol.* **6,** 353–358.
8. Wang, D., Liu, S., Trummer, B. J., Deng, C., and Wang, A. (2002) Carbohydrate microarrays for the recognition of cross-reactive molecular markers of microbes and host cells. *Nat. Biotechnol.* **20,** 275–281.
9. Xu, Q., Miyamoto, S., and Lam, K. S. (2004) A novel approach to chemical microarray using ketone-modified macromolecular scaffolds: application in micro cell-adhesion assay. *Mol. Divers.* **8(3),** 301–310.
10. Marik, J., Xu, Q., Wang, X., Peng, L., and Lam, K. S. (2004) A novel encoded high-density chemical microarray platform for proteomics and drug development. In *Peptides; Peptide Revolution: Genomics, Proteomics & Therapeutics; Proceedings of 18th American peptide Symposium* (Chorev, M. and Sawyer, T. K., eds.), American Peptide Society, Boston, MA.
11. Pellois, J. P., Wang, W., and Gao, X. (2000) Peptide synthesis based on t-Boc chemistry and solution photogenerated acids. *J. Comb. Chem.* **2,** 355–360.
12. Pellois, J. P., Zhou, X., Srivannavit, O., Zhou, T., Gulari, E., and Gao, X. (2002) Individually addressable parallel peptide synthesis on microchips. *Nat. Biotechnol.* **20,** 922–926.
13. LeProust, E., Pellois, J. P., Yu, P., et al. (2000) Digital light-directed synthesis. A microarray platform that permits rapid reaction optimization on a combinatorial basis. *J. Comb. Chem.* **2,** 349–354.
14. Gao, X., LeProust, E., Zhang, H., et al. (2001) A flexible light-directed DNA chip synthesis gated by deprotection using solution photogenerated acids. *Nucleic Acids Res.* **29,** 4744–4750.
15. Singh-Gasson, S., Green, R. D., Yue, Y., et al. (1999) Maskless fabrication of light-directed oligonucleotide microarrays using a digital micromirror array. *Nat. Biotechnol.* **17,** 974–978.
16. Frank, R. (2002) The SPOT-synthesis technique. Synthetic peptide arrays on membrane supports—principles and applications. *J. Immunol. Methods* **267,** 13–26.

17. Aina, O. H., Sroka, T. C., Chen, M. L., and Lam, K. S. (2002) Therapeutic cancer targeting peptides. *Biopolymers* **66,** 184–199.
18. Song, A., Wang, X., Zhang, J., Marik, J., Lebrilla, C. B., and Lam, K. S. (2004) Synthesis of hydrophilic and flexible linkers for peptide derivatization in solid phase. *Bioorg. Med. Chem. Lett.* **14,** 161–165.
19. Falsey, J. R., Renil, M., Park, S., Li, S., and Lam, K. S. (2001) Peptide and small molecule microarray for high throughput cell adhesion and functional assays. *Bioconjug. Chem.* **12,** 346–353.
20. Marcaurelle, L. A., Shin, Y., Goon, S., and Bertozzi, C. R. (2001) Synthesis of oxime-linked mucin mimics containing the tumor-related T(N) and sialyl T(N) antigens. *Org. Lett.* **3,** 3691–3694.
21. Salisbury, C. M., Maly, D. J., and Ellman, J. (2002) Peptide microarrays for the determination of protease substrate specificity. *J. Am. Chem. Soc.* **124,** 14,868–14,870.
22. Ulman, A. (1996) Formation and structure of self-assembled monolayers. *Chem. Rev.* **96,** 1533–1554.
23. Xu, Q. and Lam, K. S. (2002) An efficient approach to prepare glyoxylyl functionality on solid-support. *Tetrahedron Lett.* **43,** 4435–4437.
24. Xu, Q., Miyamoto, S., and Lam, K. S. (2004) A novel approach to chemical microarray using ketone-modified macromolecular scaffolds: application in micro cell-adhesion assay. *Mol. Divers.* **8,** 301–310;.

16

Peptide Mass Fingerprinting

Protein Identification Using MALDI-TOF Mass Spectrometry

Judith Webster and David Oxley

Summary

Matrix-assisted laser desorption/ionization (MALDI)-time-of-flight (TOF)-mass spectrometry (MS) is now routinely used in many laboratories for the rapid and sensitive identification of proteins by peptide mass fingerprinting (PMF). We describe a simple protocol that can be performed in a standard biochemistry laboratory, whereby proteins separated by one- or two-dimensional gel electrophoresis can be identified at femtomole levels. The procedure involves excision of the spot or band from the gel, washing and de-staining, reduction and alkylation, in-gel trypsin digestion, MALDI-TOF MS of the tryptic peptides, and database searching of the PMF data. Up to 96 protein samples can easily be manually processed at one time by this method.

Key Words: Proteomics; MALDI-TOF; mass spectrometry; SDS-PAGE; 2D-gel; in-gel digestion; peptide mass fingerprint; protein identification; database searching.

1. Introduction

Developments in mass spectrometry technology, together with the availability of extensive DNA and protein sequence databases and software tools for data mining, has made possible rapid and sensitive mass spectrometry-based procedures for protein identification. Two basic types of mass spectrometers are commonly used for this purpose; Matrix-assisted laser desorption/ionization (MALDI)-time-of-flight (TOF) mass spectrometry (MS) and electrospray ionization (ESI)-MS. MALDI-TOF instruments are now quite common in biochemistry laboratories and are very simple to use, requiring no special training. ESI instruments, usually coupled to capillary/nanoLC systems, are more complex and require expert operators. We will therefore focus on the use of MALDI-

From: *Methods in Molecular Biology, vol. 310: Chemical Genomics: Reviews and Protocols*
Edited by: E. D. Zanders © Humana Press Inc., Totowa, NJ

TOF MS, although the sample preparation is identical for both methods. The principle behind the use of MALDI-TOF MS for protein identification is that the digestion of a protein with a specific protease will generate a mixture of peptides unique to that protein. Measuring the molecular masses of these peptides then gives a characteristic dataset called a peptide mass fingerprint (PMF) *(1)*. The PMF data can then be compared with theoretical peptide molecular masses that would be generated by using the same protease to digest each protein in the sequence database, to find the best match. Provided the protein being analyzed is present in the database being searched and the data are of sufficient quality, the best match should be the correct protein. In order to judge the validity of a protein identification by this method, some means of scoring the quality of the match must be used.

The procedure described here involves cutting protein bands or spots from one-dimensional (1D) or two-dimensional (2D) polyacrylamide gel electrophoresis (PAGE) gels, de-staining the gel pieces, reducing and alkylating the protein, digesting with trypsin, using MALDI-TOF MS to determine the masses of the tryptic peptides, and database searching with the PMF data to identify the protein.

The sample-processing steps can be performed in microfuge tubes or in 96-well plates. One person can easily process a 96-well plate in a day, but if higher throughput is required, each step can be automated, allowing the possibility of several hundred protein identifications per day. Spot cutting, sample processing, and sample plate loading robots are commercially available and routinely used in many high-throughput laboratories.

2. Materials

1. 0.5-mL microfuge tubes (or 96-well V-bottom polypropylene microtiter plates) (*see* **Note 1**).
2. One Touch Spot Picker (1.5 mm) (The Gel Company)—optional.
3. Silver de-staining solution (for silver-stained gels only)—dissolve potassium ferricyanide (2 mg/mL) in sodium thiosulphate solution (0.2 mg/mL). Make fresh immediately before use.
4. Sonicator bath (or vortex mixer with attachments for unattended use with 0.5-mL microfuge tubes or 96-well microtiter plates).
5. Aqueous buffer: 50 m*M* NH_4HCO_3—make fresh weekly.
6. Organic buffer: 50 m*M* NH_4HCO_3/acetonitrile 1:1—make fresh weekly.
7. DL-dithiothreitol ultra pure (DTT) solution: 10 m*M* DTT in aqueous buffer—make fresh immediately before use.
8. Iodoacetamide solution (ultra pure): 50 m*M* iodoacetamide in aqueous buffer—make fresh immediately before use. *Note: iodoacetamide is toxic!*
9. Trypsin solution (Modified Sequencing Grade, Promega) (*see* **Note 2**).
10. 10% trifluoroacetic acid (TFA) in water—make fresh weekly.

11. Centrifugal vacuum concentrator.
12. Matrix solution (*see* **Note 3**).
13. Peptide calibration mixture (*see* **Note 3**).
14. MALDI sample plate.
15. MALDI-TOF mass spectrometer.

3. Methods

Protein identification by PMF is usually performed on protein bands or spots cut from 1D or 2D PAGE gels, so this is the method described here; however, it can also be applied to proteins in solution with minor modifications. One of the biggest problems that may be encountered, particularly when low amounts of protein are analyzed, is contamination. **Note 1** describes some precautions to minimize the effects of contamination.

All washing steps during the processing of the gel pieces can be done either on a vortex mixer or in a sonicator bath.

3.1. Removing Gel Spot or Band From the Gel and De-Staining

Ideally, the gel should have been stained with a Coomassie® stain—preferably a colloidal Coomassie if detection sensitivity is an issue, e.g., *(2,3)*, or one of the commercially available stains. This method is also compatible with Sypro® stains (though an ultraviolet transilluminator will be required to visualize the protein bands or spots during excision). Standard silver stains are not compatible, but certain modified silver stains—e.g., *(4)*, or one of the commercially available "mass spectrometry compatible" silver stains—can be used; however, even these usually give inferior mass spectrometry data compared to Coomassie- or Sypro-stained gels.

3.1.1. Removal of Gel Piece

1. Place the stained gel in a disposable Petri dish on a light box and cut out the spot/band with a manual spot picker or a clean scalpel/razor blade, without taking any excess gel.
2. If necessary, cut the gel spot/band into 1-mm pieces and transfer into a 0.5-mL microfuge tube or one well of a 96-well microtiter plate.

3.1.2. De-Staining Coomassie- or Sypro-Stained Gel Pieces

1. Wash with aqueous buffer (100 μL) 1 × 5 min.
2. Wash with organic buffer (100 μL) 2 × 5 min.

3.1.3. De-Staining Silver-Stained Gel Pieces (5)

1. Wash with silver de-staining solution (100 μL) until de-stained (approx 5–30 min) —the gel will retain a pale yellow color.

2. Wash with high-purity water (100 µL) 2 × 5 min.
3. Wash with aqueous buffer (100 µL) 1 × 5 min.
4. Wash with organic buffer (100 µL) 1 × 5 min.

3.2. In-Gel Reduction and Alkylation of the Protein

Reduction of disulfide bonds followed by alkylation of the free cysteines to prevent re-oxidation, while not essential for the digestion of most proteins, generally gives better results. This is due to increased susceptibility of the reduced/alkylated protein to tryptic digestion and the absence of any disulphide-linked peptides (which are not matched in the database search) from the PMF data ***(6)***.

1. Incubate de-stained gel pieces in DTT solution (100 µL) for 1 h at 50°C in an oven. If using a 96-well plate, use sealing film or a sealing lid, to exclude oxygen and prevent drying out.
2. Cool to room temperature, remove, and discard DTT solution. Add iodoacetamide solution (100 µL) and incubate with occasional mixing (vortex) for 1 h in the dark at room temperature.
3. Discard supernatant and wash gel pieces with aqueous buffer (100 µL) for 5 min, then with organic buffer (100 µL) for 2 × 5 min.
4. Dry the gel pieces completely in a centrifugal vacuum concentrator. Caution should be exercised in handling the dried gel pieces, as they are easily lost from tubes or plates.

3.3. In-Gel Digestion of the Protein With Trypsin

1. Take an aliquot of trypsin (10 µL of 10X stock solution) from the freezer, add aqueous buffer (90 µL) and mix.
2. Rehydrate gel pieces for 10 min in trypsin solution (approx 1 µL per mm^3 gel). There should be little or no excess liquid after rehydration.
3. Add an equal volume of aqueous buffer and incubate at 37°C overnight (or for at least 3 h). Use an incubator, not a heating block. If using a 96-well plate, use sealing film or a sealing lid, to prevent drying out.
4. Add one-tenth volume of 10% TFA and sonicate/vortex for 5 min.
5. The resulting supernatant is used directly for MALDI-TOF MS.
6. Residual peptides can be washed from the gel piece if required with a small amount of 0.1% TFA.

3.4. Mass Determination of Peptides by MALDI-TOF MS

The quality of the MALDI-TOF spectrum that will be obtained from the sample depends crucially on the sample/matrix preparation. The basic requirements are for a uniform microcrystalline layer of matrix/sample and the removal of salts and other contaminants. There are numerous published methods, but the one described here is robust and quite simple, requiring no pre-cleanup steps. For the best results, a high-quality matrix, as supplied by MALDI-MS manufacturers or in a commercial PMF kit, should be used. Alternatively, analytical-grade matrix can be recrystallized *(7)*.

MALDI-TOF spectra must be calibrated in order to achieve sufficient accuracy for database searching. This is done by acquiring spectra on peptide standards to generate a calibration curve, which is applied to the experimental data. The calibration peptides can be analyzed separately from the experimental sample (external calibration), or they can be mixed with the experimental sample (internal calibration). Internal calibration is more accurate (typically 10–20 ppm for a strong spectrum) than external (typically 100 ppm or more), but is more difficult, as the amount of standard peptides used needs to be matched to the level of the experimental sample peptides. The addition of too much peptide standards can suppress the signal of the sample peptides and vice versa. A commonly used variation of the internal calibration method utilizes the autodigestion fragments of trypsin (m/z 842.5094 and 2211.1040 for porcine trypsin) for calibration, instead of adding additional peptides *(7)*.

3.4.1. Applying the Sample and the Calibration Mixture to the MALDI Plate

1. Apply matrix solution (0.5 µL) to a clean MALDI plate and allow to dry (sample spot).
2. Apply protein digest (0.2–2 µL) to the sample spot and allow to dry. If a large amount of gel was used for the digestion, the supernatant volume could be much larger than 2 µL (up to 20 µL). It is not usually necessary to use all of the supernatant unless the protein band was quite weak. In this case, the entire supernatant can be concentrated to 1–2 µL and used (*see* **Note 4**).
3. Apply 0.1% TFA (5–10 µL) to each sample spot, leave for 30 s, remove, and discard, then repeat; this step desalts the sample.
4. If using external calibration, apply 0.2 µL of peptide calibration mixture as close as possible to (but not touching) each sample spot and allow to dry (calibrant spot).

3.4.2. Measuring the Peptides Masses in the MALDI-TOF MS Instrument

The precise operation of the MS is instrument dependent, but the basics are very similar. Increasing the laser power increases the signal, but also decreases the resolution and "burns" off the sample faster. Therefore, the laser power should be set to the lowest level that gives a good signal. Set the mass range (e.g., *m/z* 600–3500) and the number of shots to be acquired (100–200 is reasonable), and set the laser power low. Gradually increase the power until an even distribution of noise appears across the whole mass range. Within a few shots, peaks should begin to appear above the noise. Continue to adjust the laser power until the signal level is satisfactory.

The heterogeneous nature of the matrix surface means that some areas of the sample spot will give better peptide signals than others. The difference can

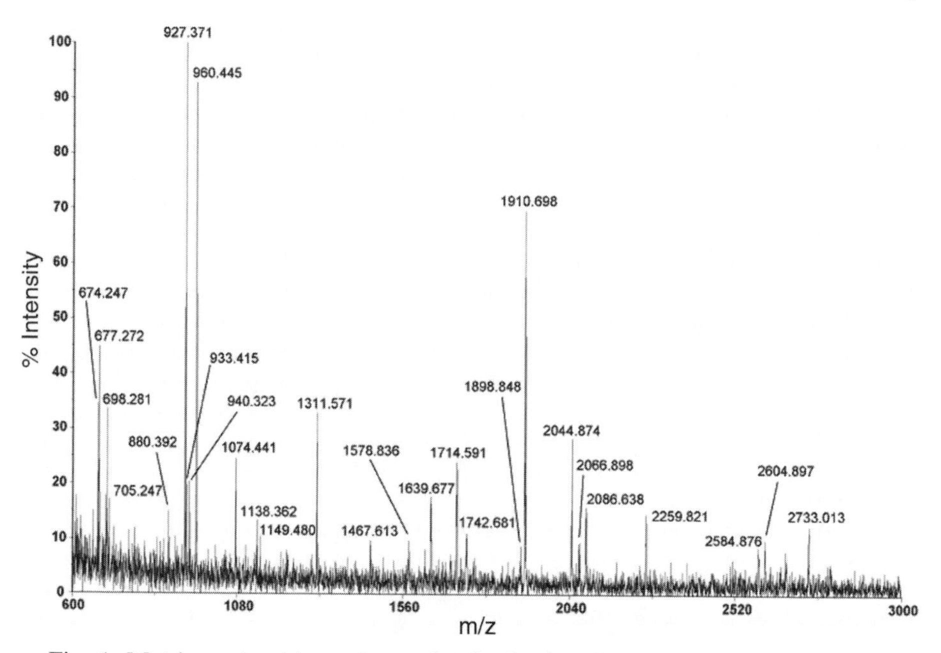

Fig. 1. Matrix-assisted laser desorption/ionization (MALDI)-time-of-flight (TOF) spectrum of a trypsin-digested one-dimensional gel band. Peaks are labeled with their monoisotopic masses. Note that these are not the masses of the peptides, but of the peptide (pseudo)molecular ions. In MALDI spectra, peptide molecular ions arise predominantly through the addition of a proton to the peptide, giving a mass increase of 1.007 Da. The molecular ions are usually denoted as MH^+ or $[M+H]^+$.

be dramatic, so it is a good idea to periodically move the laser position, while assessing the signal intensity, to find the hot spots.

If external calibration of the spectrum is to be used, a spectrum of the calibration standards should be acquired immediately, using the same power setting. **Figure 1** shows a typical MALDI spectrum obtained from the trypsin digestion of a weak colloidal Coomassie-stained 1D gel band.

3.5. Generating PMF Data and Searching Protein Databases

The PMF data must now be extracted from the MALDI spectrum using the appropriate software associated with the MS instrument used. The baseline threshold should first be adjusted to ensure that all of the peptide signals are detected, without including any noise. If the spectrum has been obtained on a high resolution (reflectron) instrument, the peptide signals will appear as multiple peaks separated by 1 Da, owing to the presence of ^{13}C isotopes in the peptides (*see* **Fig. 2**). "De-isotoping" and "centroiding" of the MS data is necessary,

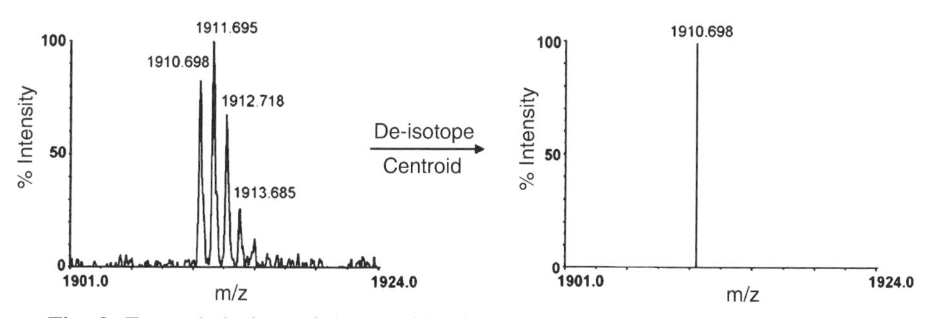

Fig. 2. Expanded view of the peptide signal at around *m/z* 1910 from the Matrix-assisted laser desorption/ionization (MALDI)-time-of-flight (TOF) spectrum in **Fig. 1**. De-isotoping removes the ^{13}C isotope peaks and centroiding reduces the remaining peak to a single data point.

so that only the monoisotopic peptide masses are included in the PMF data. The MS data analysis software will have an option for the data to be shown as a list of masses (peak list), which can be saved as a text file or simply copied directly into the data entry field of the database search engine. There are a number of search engines that can be used for PMF searches, some of which are freely available on the Internet (*see* **Note 5**). We will use Mascot to demonstrate a search, but all have essentially the same functions.

The Mascot PMF search page is shown in **Fig. 3**. Detailed explanations of the various terms and parameters are available from the Web site; brief descriptions are given in **Note 6**.

1. Select the database to be searched from the drop-down list, e.g., NCBInr.
2. If taxonomic information is known for the sample origin, select from the list; otherwise, select "All entries."
3. Select "trypsin" as the enzyme.
4. "Allow up to" 1 missed cleavage.
5. Select "Carbamidomethyl (C)" as a fixed modification.
6. Select "Oxidation (M)" as a variable modification.
7. If the MALDI spectrum was internally calibrated, the peptide tolerance can be set to 0.1 Da; otherwise, 0.3 to 0.5 Da is a reasonable starting point.
8. Select "MH^+" for Mass values.
9. Select "Monoisotopic" if the MALDI data were acquired in reflectron mode, or "Average" for data acquired in linear mode.
10. If the peak list was saved as a text file, browse to the file location; otherwise, paste the list directly into the "Query" field.
11. Select "Auto" from "Report top" hits.
12. Start search.

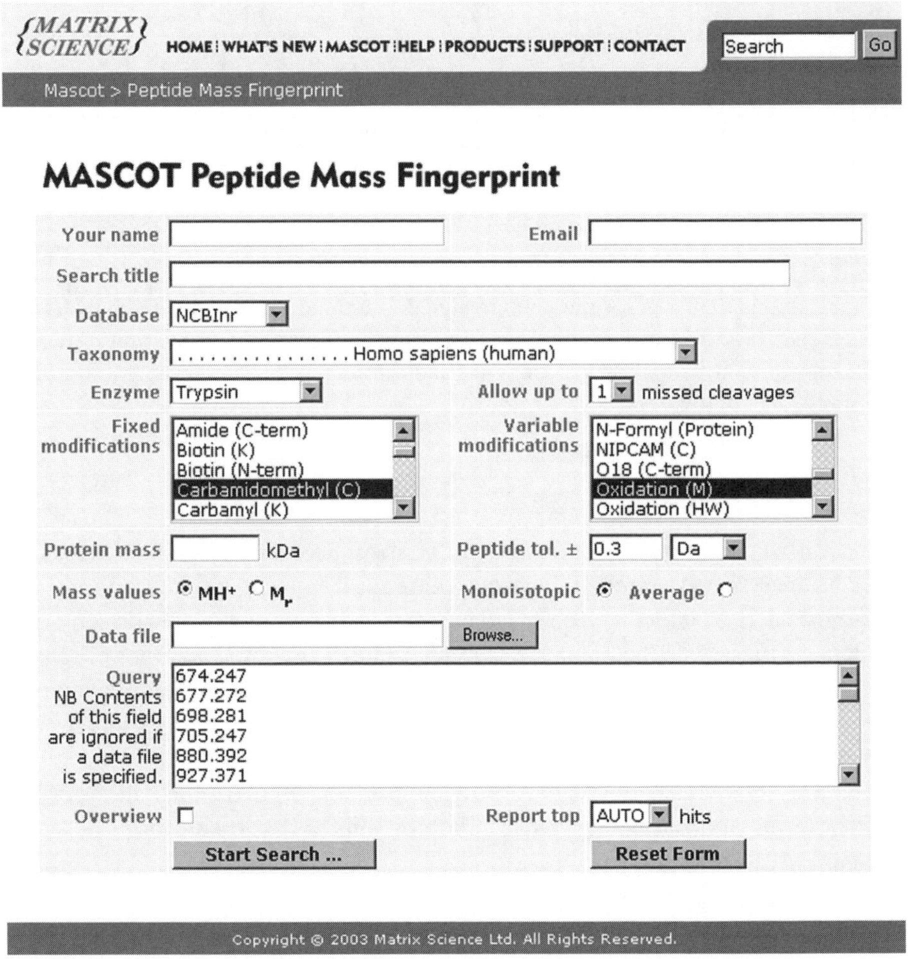
MATRIX SCIENCE
HOME | WHAT'S NEW | MASCOT | HELP | PRODUCTS | SUPPORT | CONTACT
Search
Go
Mascot > Peptide Mass Fingerprint
MASCOT Peptide Mass Fingerprint
Your name
Email
Search title
Database
NCBInr
Taxonomy
. . . . Homo sapiens (human)
Enzyme
Trypsin
Allow up to
1
missed cleavages
Fixed modifications
Amide (C-term)
Biotin (K)
Biotin (N-term)
Carbamidomethyl (C)
Carbamyl (K)
Variable modifications
N-Formyl (Protein)
NIPCAM (C)
O18 (C-term)
Oxidation (M)
Oxidation (HW)
Protein mass
kDa
Peptide tol. ±
0.3
Da
Mass values
MH+
Mr
Monoisotopic
Average
Data file
Browse...
Query
NB Contents of this field are ignored if a data file is specified.
674.247
677.272
698.281
705.247
880.392
927.371
Overview
Report top
AUTO
hits
Start Search ...
Reset Form
Copyright © 2003 Matrix Science Ltd. All Rights Reserved.

Fig. 3. Mascot peptide mass fingerprint (PMF) search page.

Within a few seconds, the search result will appear, which should look similar to **Fig. 4**. This shows a graphic representation of the results and a list of significant matches. A single significant match is indicated by the graph with a maximum score of 240, well above the significance threshold of 63. The score is related to the probability that the match is real rather than purely random, and is also expressed as an expect value equivalent to the BLAST E-value, 1.2e-19 in this case, indicating that this is almost certainly a genuine match.

In the "Concise Protein Summary Report" shown, all of the proteins matching the same set or subset of masses are listed under the single entry (only 5 are shown, but there were more than 20 in total). Typically, these will represent

{MATRIX SCIENCE} **Mascot Search Results**

User	:
Email	:
Search title	:
Database	: **NCBInr 20040916 (2026219 sequences; 679922428 residues)**
Taxonomy	: **Homo sapiens (human) (119915 sequences)**
Timestamp	: **23 Sep 2004 at 16:30:14 GMT**
Top Score	: **240 for** gi|4389275, **Human Serum Albumin In A Complex With Myristic Acid And Tri-Iodobenz**

Probability Based Mowse Score

Ions score is -10*Log(P), where P is the probability that the observed match is a random event.
Protein scores greater than 63 are significant ($p<0.05$).

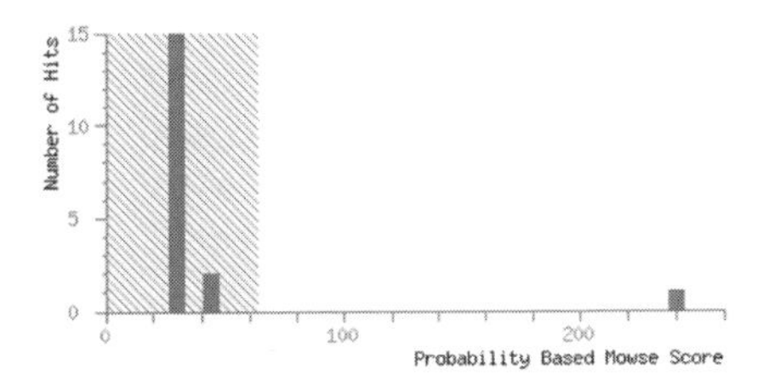

Concise Protein Summary Report

Format As | Concise Protein Summary | Help
Significance threshold p< 0.05 | Max. number of hits AUTO

Re-Search All | Search Unmatched

1. gi|4389275 **Mass:** 67988 **Score:** 240 **Expect:** 1.2e-19 **Queries matched:** 21
Human Serum Albumin In A Complex With Myristic Acid And Tri-Iodobenzoic Acid
gi|31615333 **Mass:** 68406 **Score:** 240 **Expect:** 1.2e-19 **Queries matched:** 21
Chain A, Human Serum Albumin Mutant R218h Complexed With Thyroxine (3,3',5,5'-Tetraiodo-L-Thyronine)
gi|31615331 **Mass:** 68366 **Score:** 240 **Expect:** 1.2e-19 **Queries matched:** 21
Chain A, Human Serum Albumin Mutant R218p Complexed With Thyroxine (3,3',5,5'-Tetraiodo-L-Thyronine)
gi|14719645 **Mass:** 68425 **Score:** 239 **Expect:** 1.5e-19 **Queries matched:** 21
Chain A, Human Serum Albumin Complexed With Myristic Acid And The S- (-) Enantiomer Of Warfarin
gi|51476390 **Mass:** 71353 **Score:** 226 **Expect:** 3e-18 **Queries matched:** 21
hypothetical protein [Homo sapiens]

Fig. 4. Mascot search results summary page using peptide mass fingerprint (PMF) data from the spectrum in **Fig. 1**.

multiple database entries for the same protein as well as sequence variants, fragments, and so on.

Clicking on a protein accession number from the list brings up further information (*see* **Fig. 5**), including the matched peptides, listed and mapped onto the protein sequence, as well as the % sequence coverage and any unmatched masses. There will almost inevitably be some unmatched masses; they can be due to the presence of contaminating proteins in the digest, nontryptic cleavages, incomplete digestion, modified peptides, errors in the database sequence, and so on *(8)*.

The graph at the bottom of **Fig. 5** plots the error for each of the matched peptide masses and is useful for assessing whether the tolerance setting used for the search was appropriate. If desired, a more appropriate setting could be chosen and the data researched.

3.6. Failure to Obtain a Significant Hit

Some of the most common reasons why a database search may not give a significant match. are:

1. Too few masses in the PMF data: depending on the size of the database and the parameters settings, a minimum of four to six matched peptides are required for a significant hit. This situation could arise simply because the spectrum is very weak, and so only a few of the strongest peptide signals are visible, or because the particular protein does not yield many peptides within the useful mass range (e.g., small proteins, proteins with very high or very low numbers of potential trypsin cleavage sites, proteins that are resistant to trypsin digestion).
2. More than one protein present in the sample: with good-quality PMF data, mixtures containing two to three proteins can be identified, but the presence of peptide masses from the other proteins reduces the score for each individual match. With lower-quality data, this can result in all of the scores dropping below the significance threshold. The presence of contaminants, e.g., trypsin or keratins, has the same effect; however, known peptide masses derived from keratins and trypsin can be removed from the PMF data prior to searching *(9)*.
3. Peptide tolerance set too low or too high.
4. Incorrectly calibrated spectrum.
5. Searching EST databases: ESTs are generally short and therefore do not usually provide sufficient numbers of matched peptides for a significant score.
6. The protein is not present in the database being searched.

4. Notes

1. Contamination is a serious concern when attempting to identify low amounts of protein by MALDI MS. Use only the highest quality reagents, high-performance liquid chromatography (HPLC)-grade solvents, and high-quality plasticware to minimize non-protein contamination. Ideally, a set of glass/plasticware and reagents/buffers should be dedicated for MALDI-MS analysis. Laboratory dust is a major source of protein contaminants (keratins), and so every effort must be made to exclude dust from samples, buffers, and reagents. Communal buffers, stains, de-stains, and so on, are often contaminated with dust and should be avoided. Rinse all glassware with high-quality water before making up buffers. Wear nitrile gloves (latex contains protein contaminants) and a lab coat. Keep a cover over the gel during staining. Never stain a gel in a container that has been used for processing Western blots, as they will be contaminated with blocking proteins. Handle the gel only if absolutely necessary, and then avoid touching parts of the gel that are to be cut.

{MATRIX SCIENCE} **Mascot Search Results**

Protein View

```
Match to: gi|4389275 Score: 240 Expect: 1.2e-19
 Human Serum Albumin In A Complex With Myristic Acid And Tri-Iodobenzoic Acid

Nominal mass (Mr): 67988; Calculated pI value: 5.69
NCBI BLAST search of gi|4389275 against nr
Unformatted sequence string for pasting into other applications

Taxonomy: Homo sapiens

Fixed modifications: Carbamidomethyl (C)
Variable modifications: Oxidation (M)
Cleavage by Trypsin: cuts C-term side of KR unless next residue is P
Number of mass values searched: 27
Number of mass values matched: 21
Sequence Coverage: 35%

Matched peptides shown in Bold Red

    1 KSEVAHRFKD LGEENFKALV LIAFAQYLQQ CPFEDHVKLV NEVTEFAKTC
   51 VADESAENCD KSLHTLFGDK LCTVATLRET YGEMADCCAK QEPERNECFL
  101 QHKDDNPNLP RLVRPEVDVM CTAFHDNEET FLKKYLYEIA RRHPYFYAPE
  151 LLFFAKRYKA AFTECCQAAD KAACLLPKLD ELRDEGKASS AKQRLKCASL
  201 QKFGERAFKA WAVARLSQRF PKAEFAEVSK LVTDLTKVHT ECCHGDLLEC
  251 ADDRADLAKY ICENQDSISS KLKECCEKPL LEKSHCIAEV ENDEMPADLP
  301 SLAADFVESK DVCKNYAEAK DVFLGMFLYE YARRHPDYSV VLLLRLAKTY
  351 ETTLEKCCAA ADPHECYAKV FDEFKPLVEE PQNLIKQNCE LFEQLGEYKF
  401 QNALLVRYTK KVPQVSTPTL VEVSRNLGKV GSKCCKHPEA KRMPCAEDYL
  451 SVVLNQLCVL HEKTPVSDRV TKCCTESLVN RRPCFSALEV DETYVPKEFN
  501 AETFTFHADI CTLSEKERQI KKQTALVELV KHKPKATKEQ LKAVMDDFAA
  551 FVEKCCKADD KETCFAEEGK KLVAASQAAL G
```

Show predicted peptides also

Sort Peptides By ◉ Residue Number ○ Increasing Mass ○ Decreasing Mass

Start - End	Observed	Mr(expt)	Mr(calc)	Delta	Miss	Sequence
2 - 7	698.28	697.27	697.35	-0.08	0	SEVAHR
39 - 48	1149.48	1148.47	1148.61	-0.14	0	LVNEVTEFAK
71 - 78	933.41	932.41	932.51	-0.10	0	LCTVATLR
91 - 103	1714.59	1713.58	1713.79	-0.21	1	QEPERNECFLQHK
104 - 111	940.32	939.32	939.44	-0.13	0	DDNPNLPR
135 - 141	927.37	926.36	926.49	-0.12	0	YLYEIAR
143 - 156	1742.68	1741.67	1741.89	-0.21	0	HPYFYAPELLFFAK
143 - 157	1898.85	1897.84	1897.99	-0.15	1	HPYFYAPELLFFAKR
179 - 187	1074.44	1073.43	1073.54	-0.10	1	LDELRDEGK
223 - 230	880.39	879.38	879.43	-0.05	0	AEFAEVSK
238 - 254	2086.64	2085.63	2085.83	-0.20	0	VHTECCHGDLLECADDR
238 - 259	2584.88	2583.87	2584.11	-0.24	1	VHTECCHGDLLECADDRADLAK
321 - 333	1639.68	1638.67	1638.78	-0.11	0	DVFLGMFLYEYAR Oxidation (M)
334 - 345	1467.61	1466.61	1466.84	-0.23	1	RHPDYSVVLLLR
335 - 345	1311.57	1310.56	1310.73	-0.17	0	HPDYSVVLLLR
370 - 386	2044.87	2043.87	2044.09	-0.22	0	VFDEFKPLVEEPQNLIK
400 - 407	960.45	959.44	959.56	-0.12	0	FQNALLVR
464 - 469	674.25	673.24	673.34	-0.10	0	TPVSDR
473 - 481	1138.36	1137.35	1137.49	-0.14	0	CCTESLVNR
482 - 497	1910.70	1909.69	1909.92	-0.23	0	RPCFSALEVDETYVPK
498 - 516	2259.82	2258.81	2259.02	-0.20	0	EFNAETFTFHADICTLSEK

No match to: 677.27, 705.25, 1578.84, 2066.90, 2604.90, 2733.01

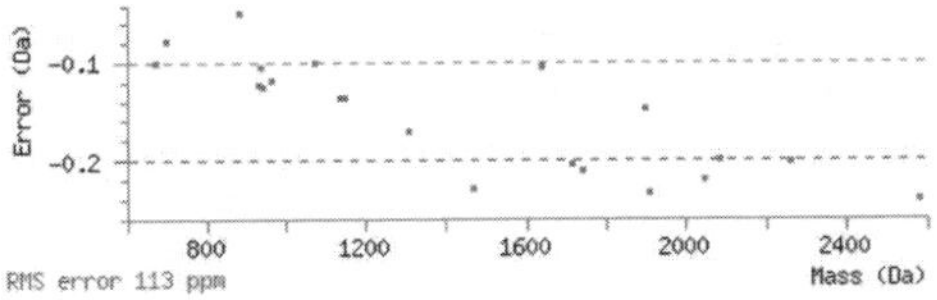

Error (Da)
-0.1
-0.2
800 1200 1600 2000 2400
Mass (Da)
RMS error 113 ppm

Fig. 5. Detailed information for the protein gi|4389275 identified in the Mascot search results in **Fig. 4**.

Gel pieces can be cut and processed in laminar-flow cabinets, although this is usually not necessary if sensible precautions are taken.

2. Modified trypsin is preferred for protein digestion, as it is less susceptible to autodigestion *(7)*. The trypsin 10X stock solution is prepared by dissolving the trypsin (20 μg) in the solvent supplied with the enzyme (50 m*M* acetic acid; 200 μL). This is stored at –70°C in aliquots (10 μL) and is stable for at least 1 yr.
3. MALDI analysis kits (e.g., Sequazyme peptide mass standards kit from Applied Biosystems) are a convenient way to obtain matrix, peptide standards, and solvents; however, these items can easily be purchased independently. Any peptides that span a reasonable proportion of the useful mass range can be used for calibration, provided their accurate molecular weights are known.

   α-Cyano-4-hydroxycinnamic acid is the most common matrix used for peptide analysis. Matrix solution is prepared by dissolving 5 mg of matrix in 1 mL of 4:1 acetonitrile/water containing 0.1% TFA.

   A mixture of peptide calibration standards—e.g., des-Arg^1-bradykinin (monoisotopic mass 904.4681), angiotensin 1 (monoisotopic mass 1296.6853), ACTH (1–17 clip) (monoisotopic mass 2093.0867), ACTH (18–39 clip) (monoisotopic mass 2465.1989)—is made to a concentration of 2 pmol/μL each in 0.1% TFA in water. This solution is then mixed 1:1 with matrix solution to give the final peptide-calibration mixture.
4. Larger volumes of digests can be concentrated by vacuum centrifugation, but this can lead to significant peptide losses, particularly if the sample is concentrated to dryness. Alternatively, peptides can be concentrated and desalted by using a microscale pipet-tip format solid-phase extraction device (e.g., micro C18 Zip Tips from Millipore, Omix C18MB tips from Varian, or STAGE tips from Proxeon). The details of their use in desalting peptide solutions for MALDI analysis are given with the product, but for the final elution step, use the matrix solution described above, to elute the peptides directly onto the MALDI plate.
5. Freely available PMF search engines include: Mascot from Matrix Science (http://www.matrixscience.com/cgi/search_form.pl?FORMVER=2&SEARCH=PMF), MS-Fit from Protein Prospector (http://prospector.ucsf.edu/ucsfhtml4.0/msfit.htm), and PeptIdent from ExPaSy (http://us.expasy.org/tools/peptident.html).
6. SwissProt is a relatively small but highly annotated database with minimal redundancy. NCBInr and MSDB are much larger, but have multiple entries for many proteins. Other search engines may have additional database choices, and in-house copies allow custom databases to be used.

   Specifying the taxonomy as closely as possible reduces the number of database entries that need to be considered in the search. This reduces search times, but more importantly, reduces the threshold score for significant matches, increasing the confidence of any protein identification.

   The cleavage specificity of trypsin is C-terminal to Lys and Arg residues (except where followed by Pro). However, not every such peptide bond will be cleaved. The number of missed cleavages to consider can be specified, but increasing the number decreases the significance of any match and should normally be set to one.

Fixed modifications are modifications to specific amino acids that are considered to be complete—i.e., every occurrence of the amino acid in the sequence is assumed to carry the modification, and the unmodified amino acid is not considered. Variable modifications, on the other hand, are incomplete, and therefore both the modified and unmodified amino acid are considered in the search. In the example discussed above, the reduction/alkylation should result in complete carbamidomethylation of all cysteine residues; thus, "Carbamidomethyl (C)" was chosen as a fixed modification, whereas methionine oxidation, a common artifactual modification that is usually incomplete, was selected as a variable modification. The use of multiple variable modifications will greatly reduce the significance of any match and should therefore be used with caution.

Most PMF search engines allow a mass for the protein to be entered; the search will then consider only database entries within a window around this mass. This can be dangerous, as proteins are subject to processing/degradation, which can significantly increase or decrease their molecular mass. Mascot uses a more complex method of applying this parameter, but in most cases this can be left blank.

The peptide tolerance is a window around each mass value in the peak list within which a theoretical database peptide mass must fall, in order to be matched.

Acknowledgments

The authors acknowledge the support of the Biotechnology and Biological Sciences Research Council, UK.

References

1. Pappin, D. J. C., Hojrup, P., and Bleasby, A. J. (1993) Rapid identification of proteins by peptide-mass fingerprinting. *Curr. Biol.* **3 (6),** 327–332.
2. Neuhoff, V., Stamm, R., and Eibl, H. (1985) Clear background and highly sensitive protein staining with Coomassie Blue dyes in polyacrylamide gels: a systematic analysis. *Electrophoresis* **6,** 427–448.
3. Herbert, B., Galvani, M., Hamdan, M., et al. (2001) Reduction and alkylation of proteins in preparation of two-dimensional map analysis: why, when, and how? *Electrophoresis* **22,** 2046–2057.
4. Shevchenko, A., Wilm, M., Vorm, O., and Mann, M. (1996) Mass spectrometric sequencing of proteins from silver-stained polyacrylamide gels. *Anal. Chem.* **68,** 850–858.
5. Gharahdaghi, F., Weinberg, C. R., Meagher, D. A., and Imai, B. S. (1999) Mass spectrometric identification of proteins from silver-stained polyacrylamide gel: a method for the removal of silver ions to enhance sensitivity. *Electrophoresis* **20,** 601–605.
6. Sechi, S. and Chait, B. T. (1998) Modification of cysteine residues by alkylation. A tool in peptide mapping and protein identification. *Anal. Chem.* **70,** 5150–5158.
7. Harris, W. A., Janecki, D. J., and Reilly, J. P. (2002) Use of matrix clusters and trypsin autolysis fragments as mass calibrants in matrix-assisted laser desorption/

ionization time-of-flight mass spectrometry. *Rapid Commun. Mass Spectrom.* **16,** 1714–1722.

8. Karty, J. A., Ireland, M. M. E., Brun, Y. V., and Reilly, J. P. (2002) Artifacts and unassigned masses encountered in peptide mass mapping. *J. Chrom. B* **782,** 363–383.
9. Schmidt, F., Schmid, M., Jungblut, P. R., Mattow, J., Facius, A., and Pleissner, K. P. (2003) Iterative data analysis is the key for exhaustive analysis of peptide mass fingerprints from proteins separated by two-dimensional electrophoresis. *J. Am. Soc. Mass Spectrom.* **14,** 943–956.

17

A Practical Protocol for Carbohydrate Microarrays

Ruobing Wang, Shaoyi Liu, Dhaval Shah, and Denong Wang

Summary

We have established a high-throughput biochip platform for constructing carbohydrate microarrays. Using this technology, carbohydrate-containing macromolecules of diverse structures, including polysaccharides, natural glycoconjugates, and mono- and oligosaccharides coupled to carrier molecules, can be stably immobilized on a glass chip without chemical modification. Here, we describe a practical protocol for this technology. We hope that anyone who has access to a standard cDNA microarray facility will be able to explore this technology for his or her own research interest. We also provide an example to illustrate that the carbohydrate microarray is also a discovery tool; this is particularly useful for identifying immunologic sugar moieties, including complex carbohydrates of cancer cells and sugar signatures of previously unrecognized microbial pathogens.

Key Words: Antigens; antibodies; carbohydrates; glycans; glyconjugates; microarrays; microspotting; nitrocellulose; polysaccharides; SARS-CoV.

1. Inroduction

Like nucleic acids and proteins, carbohydrates are another class of crucial biological molecules *(1–3)*. Because of their unique physicochemical properties, carbohydrates are superior to other biological molecules in generating structural diversity. In aqueous solutions, such as bodily fluids, carbohydrate chains are prominently displayed on the surfaces of cell membranes or on the exposed regions of macromolecules. Carbohydrates are, therefore, suitable for storing biological signals in the forms that are identifiable by other biological systems.

Recent studies have demonstrated that cell-surface expression of specific complex carbohydrates is associated with various stages of embryonic development and cell differentiation *(4–7)*. Abnormalities in the expression of complex

From: *Methods in Molecular Biology, vol. 310: Chemical Genomics: Reviews and Protocols*
Edited by: E. D. Zanders © Humana Press Inc., Totowa, NJ

carbohydrates are found in cancer ***(8,9)***, retrovirus infection ***(10,11)***, and diseases with genetic defects in glycosylation ***(12)***. Sugar moieties are also abundantly expressed on the outer surfaces of the majority of viral, bacterial, protozoan, and fungal pathogens. Many sugar structures are pathogen-specific, which makes them important molecular targets for pathogen recognition, diagnosis of infectious diseases, and vaccine development ***(1,3,13–15)***. Exploring the biological information contained in sugar chains is, therefore, an important topic of current postgenomic research.

Our group has focused on development of a carbohydrate-based microarray technology to facilitate exploration of carbohydrate-mediated molecular recognition and anti-carbohydrate immune responses ***(16–18)***. This technology takes advantage of existing cDNA microarray systems, including the spotter and scanner, for efficient production and use of carbohydrate microarrays (*see* **Note 1**). We have demonstrated that the current platform is able to overcome a number of technical difficulties, by showing that (1) carbohydrate molecules can be immobilized on a nitrocellulose-coated glass slide without chemical conjugation, (2) the immobilized carbohydrates are able to preserve their immunological properties and solvent accessibility, (3) the system reaches the sensitivity, specificity, and capacity to detect a broad range of antibody specificities in clinical specimens, and (4) this technology can be applied to investigate carbohydrate-mediated molecular recognition and anti-carbohydrate antibody reactivities on a large scale.

In this chapter, we provide a practical protocol for this high-throughput carbohydrate microarray system. We summarize the key steps of carbohydrate microarray applications, including (1) design and construction of sugar arrays, (2) microspotting molecules onto nitrocellulose-coated glass slides, (3) immunostaining and scanning of arrays, (4) analysis of microarray data, and (5) validation of microarray data using conventional immunological assays. We focus on an eight-chamber subarray system to produce carbohydrate microarrays on a relatively smaller scale, which is more frequently applied in our laboratory's routine research activities. Lastly, we present an example to illustrate the application of this system in addressing biomedical questions.

2. Materials

2.1. Apparatus

1. Microspotting: Cartesian Technologies' PIXSYS 5500C (Irvine, CA) or GMS 417 Arrayer, Genetic Microsystems, Inc. (Woburn, MA).
2. Supporting substrate: FAST Slides (Schleicher & Schuell, Keene, NH).
3. Microarray scanning: ScanArray 5000 Standard Biochip Scanning System (Packard Biochip Technologies, Inc., Billerica, MA).

2.2. Software

1. Array design: CloneTracker (Biodiscovery, Inc., Marina del Rey, CA).
2. Array printing: AxSys™ (Cartesian Technologies, Inc., Irvine, CA).
3. Array scanning and analysis: ScanArray Express (PerkinElmer, Torrance, CA).

2.3. Antibodies and Lectins

1. Horse anti-SARS-CoV anti-sera (gift of Dr. Jiahai Lu, Sun-Yatsen University, Guangdong, China).
2. *Phaseolus vulgaris L.* (PHA-L) (EY Laboratories, Inc., San Mateo, CA).
3. Streptavidin-Cy3 and streptavidin-Cy5 conjugates (Amersham Pharmacia, Piscataway, NJ).
4. Species-specific anti-immunoglobulin antibodies and their fluorescent conjugates, Cy3, Cy5, or fluorescein isothiocyanate (FITC) (Sigma, St. Louis, MO; BD-PharMingen, San Diego, CA).

2.4. Reagents and Buffers

1. Dilution buffer: saline (0.9% NaCl).
2. Rinsing solution: 1X phosphate-buffered saline (PBS) (pH 7.4) with 0.05% (v/v) Tween-20.
3. Blocking solution: 1% (w/v) bovine serum albumin (BSA) in PBS with 0.05% (w/v) Tween-20, 0.025% (w/v) NaN_3.

3. Methods

Our previous experimental investigations have led to the establishment of the current high-throughput carbohydrate microarray platform ***(16)*** (*see* **Note 1**). The methods described below outline: (1) design and construction of an eight-chamber subarray system, (2) microspotting carbohydrate-containing molecules onto nitrocellulose-coated glass slides, (3) immunostaining and scanning of microarrays, (4) analysis of microarray data, and (5) validation of microarray findings by conventional immunological assays. **Figure 1** is a schematic view of this high-throughput microarray system.

3.1. Design and Construction of the Chip

We have designed an eight-chamber subarray system to construct customized carbohydrate microarrays. As illustrated in **Fig. 2**, each microglass slide contains eight separated subarrays. The microarray capacity is approx 600 microspots per subarray. A single slide is, thus, designed to enable eight microarray assays. A similar design with array capacity of approx 100 microspots is also commercially available (Schleicher & Schuell, Keene, NH).

1. Each microglass slide contains eight identical subarrays. There is chip space for 600 microspots per subarray, with spot sizes of approx 200 μ and at 300-μ inter-

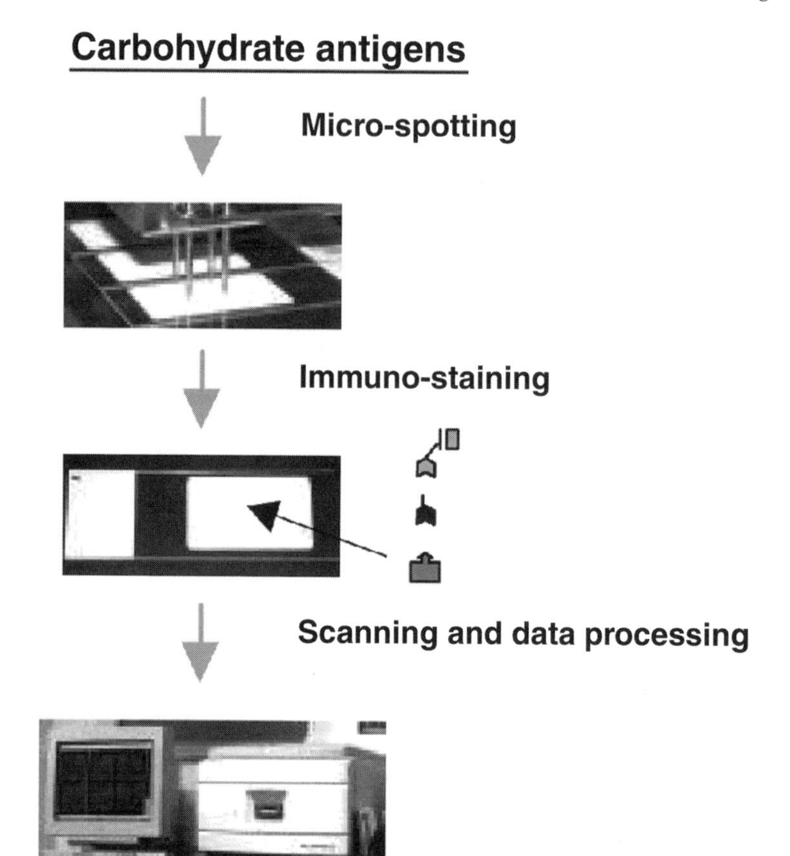

Fig. 1. A high-throughput platform of the carbohydrate-based microarrays. A high-precision robot designed to produce cDNA microarrays was utilized to spot carbohydrate antigens onto a chemically modified glass slide. The microspotting capacity of this system is approximately 20,000 spots per chip. The antibody-stained slides were then scanned for fluorescent signals with a Biochip Scanner that was developed for cDNA microarrays. The microarray results were subsequently confirmed by at least one of the conventional alternative assays.

vals, center to center. A single slide is, therefore, designed to allow eight detection reactions.

2. Repeats and dilutions: We usually print carbohydrate antigens at the initial concentration of 0.1–0.5 mg/mL. The absolute amount of antigens or antibodies printed on the chip substrate ranges from 0.1–0.5 ng per microspot. They are further diluted at 1:3, 1:9, and 1:27. A given concentration of each preparation is repeated at least three times to allow statistic analysis of detection of identical preparation at a given antigen concentration.

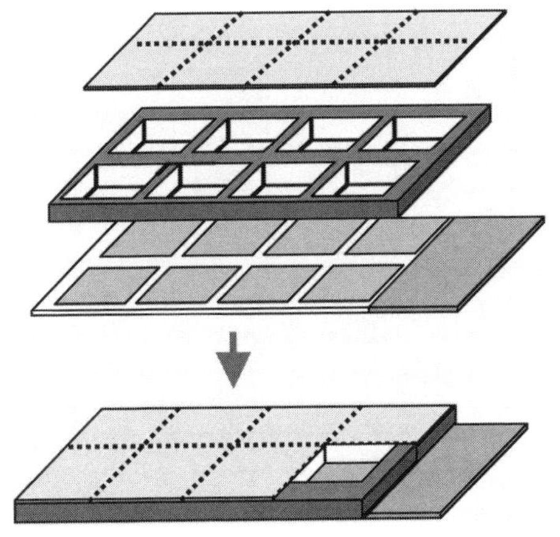

Fig. 2. Schematic of the eight-chamber subarrays. Each microglass slide contains eight subarrays of identical content. There is chip space for 600 microspots per subarray, with spot sizes of approx 200 μ and at 300-μ intervals, center to center. A single slide is, therefore, designed to enable eight detections.

3. Antibody isotype standard curves: antibodies of immunoglobulin (Ig)G, IgA, and IgM isotype of corresponding species are printed at given concentrations to serve as standard curves in the microarray format. This design allows quantification of the antibody signals that are captured by spotted carbohydrate antigens. In addition, such standard curves are useful for microarray data normalization and cross-chip scaling of microarray detection.

3.2. Microspotting of Carbohydrates Onto Nitrocellulose-Based Substrate

Using Cartesian Technologies' PIXSYS 5500C (Irvine, CA), a high-precision robot designed for cDNA microarrays, carbohydrate antigens of various complexities are picked up by dip quill pins from antigen/antibody solutions and printed onto nitrocellulose-coated FAST slides in consistent amounts (Schleicher & Schuell, Keene, NH). The complementary AxSys software (Cartesian Technologies, Inc., Irvine, CA) is used to control the movement of pins during the dispensing and printing process.

1. Prepare samples of carbohydrate antigens in 0.9% NaCl and transfer them into 96-well plates (*see* **Notes 2** and **3**).
2. Place the 96-well plates containing samples on the Cartesian arrayer robot.
3. Adjust program so that carbohydrate antigens are printed at spot sizes of approx 150 μm and at 375 μm intervals, center to center.
4. Each antigen or antibody is spotted as triplet replicates in parallel.

1 Spotting

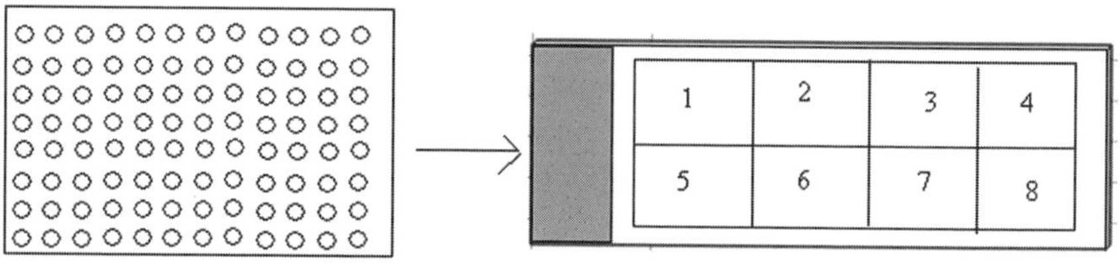

1 2 3 4
5 6 7 8

2 Rinse, 1XPBS 5 min

3 Block, 1% BSA PBS, RT, 30 min

4 Staining

		40 µL/subarray		
Chip#	Chamber	Serum, 90 min RT	2nd Ab, 30min RT	3rdAb, 30min RT
100803-5	5	Horse anti-SARS serum 1:20	Rabbit anti-Horse IgG-BI	AV-Cy3 1:500
	6	Horse anti-SARS serum 1:100		
	7	Horse anti-Pn18 serum 1:20	1:200 dil	
	8	Horse anti-Pn18 serum 1:100		

Fig. 3. Schematic of staining process of SARS-CoV immunochip. (1) Spotting: A high-precision robot transfers the samples, SARS-CoV proteins, and glycans of various complexities, from 96-well plate to nitrocellulose-coated glass slides. (2) Staining: Before staining, the slides are rinsed with 1X phosphate-buffered saline (PBS), and blocked with 1% bovine serum albumin (BSA)-PBS containing 0.05% NaN_3 and 0.05% Tween-20. They are subsequently incubated with horse anti-SARS sera. The primary antibodies captured by microarrays are detected using biotinated anti-horse immunoglobulin (Ig)G, and visualized by Cy3-streptavidin.

5. The printed carbohydrate microarrays are air-dried and stored at room temperature without desiccant before further use (*see* **Note 5**).

3.3. Immunostaining of Carbohydrate Microarrays

The staining procedure for carbohydrate microarrays is basically identical to the routine procedure for immunohistology. Immunostaining steps of carbohydrate arrays are listed below (*see* **Notes 4** and **6** and **Fig. 3**).

1. Rinse printed microarray slides with 1X PBS (pH 7.4) with 0.05% Tween-20 for 5 min.
2. Block slides with 1% BSA in PBS containing 0.05% Tween-20, 0.025% NaN_3, at RT for 30 min.
3. Stain each subarray with 40 µL of test sample, which is diluted in 1% BSA PBS containing 0.05% NaN_3 and 0.05% Tween-20.
4. Incubate the slide in a humidified chamber at room temperature for 90 min.
5. Wash slides five times with 1X PBS (pH 7.4) with 0.05% Tween-20.

6. Stain slides with 40 μL of titrated secondary antibodies. Anti-human (or other species) IgG, IgM, or IgA antibodies with distinct fluorescent tags, Cy3, Cy5, or FITC, are mixed and then applied to the chips.
7. Incubate the slide in a humidified chamber with light protection at room temperature for 30 min.
8. Wash slides five times.
9. Air-dry the washed slides.
10. Cover slides in a histology slide box to prevent fluorescent quenching by light.

3.4. Microarray Scanning and Data Processing

1. Scan microarray with ScanArray Express Microarray Scanner (PerkinElmer Life Science) following the manufacturer's instructions.
2. Fluorescence intensity values for each array spot and its background were calculated using Packard Bioscience's QuantArray software analysis packages or the updated ScanArray Express software. A staining result is considered positive if the mean fluorescent intensity value of the microspot is significant higher than the mean background of the identically stained microarray with the same fluorescent color (*see* **Note 7**).

3.5. Validation and Further Investigation of Microarray Observations

It is highly recommended that microarray findings be verified using other experimental approaches. We usually confirm our results by at least one of the alternative immunoassays, such as enzyme-linked immunosorbent assay (ELISA), dot blot, Western blot, flow cytometry, or immunohistology ***(16,18)***.

3.6. Constructing Glycan Arrays to Probe Immunologic Sugar Moieties of SARS-CoV

The following example illustrates the use of carbohydrate microarray technology in an important research area. In this case, the above described eight-chamber subarray system was applied to construct a glycan array. This glycan array was then applied to identify immunogenic sugar moieties of a human coronavirus (SARS-CoV).

SARS-CoV is a newly identified human viral pathogen that caused an outbreak of severe acute respiratory syndrome (SARS). Although substantial efforts have been made to study the etiological agent of the disease, the carbohydrate structures of SARS-CoV remain largely uncharacterized. This information is, however, very important for devising a vaccination strategy against SARS-CoV, as well as understanding the pathogenesis of SARS.

To investigate this, we constructed glycan arrays to display carbohydrate antigens of defined structures and then applied these tools to detect carbohydrate-specific antibody "fingerprints" that were elicited by a SARS vaccine.

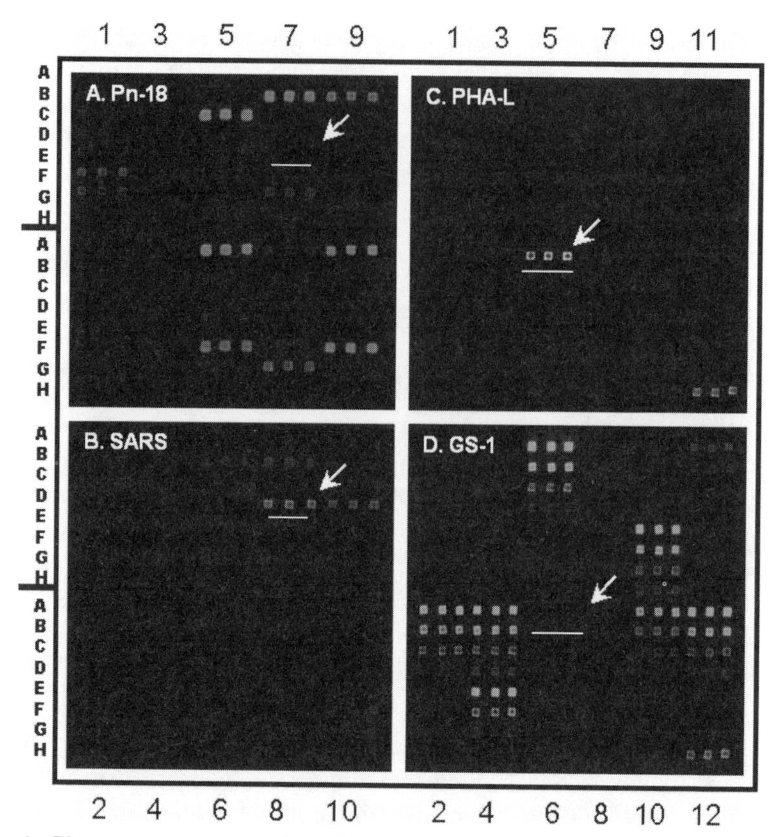

Fig. 4. Glycan arrays are used to characterize the antibody profiles of vaccinated animals (Glycan array I) and to scan for asialo-orosomucoid (ASOR)-specific immunological probes (Glycan array II). Antigen preparations spotted on each glycan array and their array location are summarized in Supplemental Tables S1 and S2 of reference *(18)* (available at the Physiological Genomics Web site).

Our rationale was that if SARS-CoV expressed antigenic carbohydrate structures, then immunizing animals using the whole virus-based vaccines would have elicited specific antibodies for these structures. In addition, if SARS-CoV displayed a carbohydrate structure that mimics host cellular glycans, then vaccinated animals may develop antibodies with autoimmune reactivity to their corresponding cellular glycans.

In **Fig. 4**, we show representative results. First, we detected an antibody reactivity specific for the carbohydrate moieties of an abundant human serum glycoprotein asialo-orosomucoid (ASOR) (**Fig. 4B**). Second, we found that lectin PHA-L (*Phaseolus vulgaris L.*) is specific for a defined complex carbohydrate ASOR (**Fig. 4C,D**). Third, we applied this probe to examine whether SARS-CoV expresses antigenic structures that imitate the host glycan. We confirmed that only the SARS-CoV-infected cells express PHA-L reactive antigenic structure (data not shown). Therefore, we obtained immunologic evidence that a carbohydrate structure of SARS-CoV shares antigenic similarity with host glycan complex carbohydrates.

The biological significance of this finding remains to be further explored. For example, what is the possible involvement of autoimmune responses in SARS pathogenesis? ASOR is an abundant human serum glycoprotein, and the ASOR-type complex carbohydrates are also expressed by other host glycoproteins ***(19,20)***. Thus, the human immune system is generally nonresponsive to these "self" carbohydrate structures. However, when similar sugar moieties were expressed by a viral glycoprotein, their cluster configuration could differ significantly from those displayed by a cellular glycan, thereby generating a novel "non-self" antigenic structure. A documented example of such an antigenic structure is a broad-range HIV-1 neutralization epitope recognized by a monoclonal antibody 2G12. This antibody is specific for a unique cluster of sugar chains displayed by the gp120 glycoprotein of HIV-1***(21)***. It is, thus, important to examine whether naturally occurring SARS-CoV expresses the ASOR-type autoimmune reactive sugar moieties. In addition, a number of cellular receptors that bind the ASOR complex carbohydrate have been identified ***(20,22)***. This study provided clues to explore the possible roles of carbohydrate-mediated receptor-ligand interactions in SARS-CoV infection, especially in determining host-range and tissue-tropic characteristics of the virus.

In summary, our laboratory has established a simple, precise, and highly efficient experimental approach for the construction of carbohydrate microarrays ***(16–18)***. This approach makes use of existing cDNA microarray systems, including the spotter and scanner, for carbohydrate array production. A key technical element of this array platform is the introduction of nitrocellulose-coated microglass slides to immobilize unmodified carbohydrate antigens on the chip surface noncovalently. This technology has achieved the sensitivity to recognize the profiles of human anti-carbohydrate antibodies with as little as a few microliters of serum specimen, and reached the chip capacity to include the antigenic preparations of most common pathogens (approx 20,000 microspots per biochip). We describe in this chapter a practical protocol of this platform of carbohydrate microarrays. However, we would like to take this opportunity to refer our readers to other technology platforms of carbohydrate microarrays or glycan chips ***(23–28)***. This progess, especially the availability of different technological platforms to meet the multiple needs of carbohydrate research, marks an important developmental stage of chemical genomics research, approaching the era of the glycome.

4. Notes

1. Antigen preparations suitable for this high-throughput biochip platform: Carbohydrate antigens of multiple structural configurations, including polysaccharides, natural glycoconjugates, and oligosaccharide-protein and oligosaccharide-lipid conjugates, are applicable for this microarray platform ***(16,18,25)***. In addition to printing carbohydrate microarrays, the current platform is also applicable for producing protein microarrays ***(17)***. As with carbohydrate microarrays, there is no need to chemically conjugate a protein for its surface immobilization. However, it is recommended that each antigen preparation be tested on chip substrate for the efficacy of immobilization and expression of antigenic determinants.
2. Preservation of polysaccharides: Purified polysaccharides are generally stored as dried powder at room temperature. They can also be preserved in saline solutions (0.9% NaCl) containing a droplet of chloroform, and stored at 4°C for a long period of time.
3. Printing of samples: Before loading sample solutions onto the arrayer, it is important to spin the solution in an Eppendorf centrifuge at maximum speed (at least 15,000*g*) for at least 15 min. Before and after each arraying experiment, we recommend examining and cleaning the printing pins.

   A test slide is usually implemented to optimize quality of printing. The water supply of the Cartesian Arrayer should constantly be checked during the arraying experiment to ensure adequate flow to the wash chamber.
4. Examination of presence of samples on array: To verify that proteins, synthetic peptides, and carbohydrates are successfully printed, microarrays can be incubated with antibodies, receptors, or lectins known to react with the printed substance. The reaction is detected either by conjugating directly a fluorochrome to the detector, or by a second-step staining procedure.
5. Storage of printed carbohydrate and protein microarrays: The arrays are usually air dried and stored at room temperature. For long-term preservation, the chips can be sealed in a plastic bag with desiccant and stored at 4°C.
6. Staining considerations: After the last wash between each staining step, it is important to completely withdraw the wash buffer inside the reaction chambers; otherwise, the remaining buffer may lower the concentration of the antibodies to be analyzed.
7. Data analysis: Training with the technical experts of PerkinElmer is necessary before performing microarray data analysis using the software package.
8. Biosafety procedures: When working with chemicals, suitable protective wear, such as lab coat and disposable gloves, are advised. When human serum specimens are involved, experiments must be conducted in accordance to the standard biosafety procedures as instructed by the Centers for Disease Control and Prevention (CDC) and the World Health Organization (WHO).

References

1. Wang, D. and Kabat, E. A. (1996) Carbohydrate antigens (polysaccharides). In *Structure of Antigens, Vol. 3* (Van Regenmortal, M. H. V., ed.), pp. 247–276.

2. Brooks, S. A., Dwek, M. V., and Schumacher, U. (2002) *Functional & Molecular Glycobiology*. BIOS Scientific Publishers Ltd.: Oxford.
3. Wang, D. (2004) Carbohydrate antigens. In *Encyclopedia of Molecular Cell Biology and Molecular Medicine, vol. II* (Meyers, R. A., ed.), Wiley: Hoboken, NJ, pp. 277–301.
4. Feizi, T. (1982) The antigens Ii, SSEA-1 and ABH are in interrelated system of carbohydrate differentiation antigens expressed on glycosphingolipids and glycoproteins. *Adv. Exp. Med. Biol.* **152,** 167–177.
5. Crocker, P. R. and Feizi, T. (1996) Carbohydrate recognition systems: functional triads in cell-cell interactions. *Curr. Opin. Struct. Biol.* **6(5),** 679–691.
6. Focarelli, R., La Sala, G. B., Balasini, M., and Rosati, F. (2001) Carbohydrate-mediated sperm-egg interaction and species specificity: a clue from the *Unio elongatulus* model. *Cells Tissues Organs* **168(1–2),** 76–81.
7. Rosati, F., Capone, A., Giovampaola, C. D., Brettoni, C., and Focarelli, R. (2000) Sperm-egg interaction at fertilization: glycans as recognition signals. *Int. J. Dev. Biol.* **44(6),** 609–618.
8. Hakomori, S. (1985) Aberrant glycosylation in cancer cell membranes as focused on glycolipids: overview and perspectives. *Cancer. Res.* **45(6),** 2405–2414.
9. Sell, S. (1990) Cancer-associated carbohydrates identified by monoclonal antibodies. *Hum. Pathol.* **21(10),** 1003–1019.
10. Adachi, M., Hayami, M., Kashiwagi, N., et al. (1988) Expression of Le^{Y} antigen in human immunodeficiency virus-infected human T cell lines and in peripheral lymphocytes of patients with acquired immune deficiency syndrome (AIDS) and AIDS-related complex (ARC). *J. Exp. Med.* **167(2),** 323–331.
11. Nakaishi, H., Sanai, Y., Shibuya, M., Iwamori, M., and Nagai, Y. (1988) Neosynthesis of neolacto- and novel ganglio-series gangliosides in a rat fibroblastic cell line brought about by transfection with the v-fes oncogene-containing Gardner-Arnstein strain feline sarcoma virus-DNA. *Cancer. Res.* **48(7),** 1753–1758.
12. Schachter, H. and Jaeken, J. (1999) Carbohydrate-deficient glycoprotein syndrome type II. *Biochim. Biophys. Acta* **1455(2–3),** 179–192.
13. Heidelberger, M. and Avery, O. T. (1923) The soluble specific substance of pneumococcus. *J. Exp. Med.* **38,** 73–80.
14. Schneerson, R., Barrera, O., Sutton, A., and Robbins, J. B. (1980) Preparation, characterization, and immunogenicity of Haemophilus influenzae type b polysaccharide-protein conjugates. *J. Exp. Med.* **152(2),** 361–376.
15. Mond, J. J., Lees, A., and Snapper, C. M. (1995) T cell-independent antigens type 2. *Annu. Rev. Immunol.* **13,** 655–692.
16. Wang, D., Liu, S., Trummer, B. J., Deng, C., and Wang, A. (2002) Carbohydrate microarrays for the recognition of cross-reactive molecular markers of microbes and host cells. *Nat. Biotechnol.* **20(3),** 275–281.
17. Wang, D. (2003) Carbohydrate microarrays. *Proteomics* **3,** 2167–2175.
18. Wang, D. and Lu, J. (2004) Glycan arrays lead to the discovery of autoimmunogenic activity of SARS-CoV. *Physiol. Genomics* **18(2),** 245–248.

19. Cummings, R. D. and Kornfeld, S. (1984) The distribution of repeating [Gal beta 1,4GlcNAc beta 1,3] sequences in asparagine-linked oligosaccharides of the mouse lymphoma cell lines BW5147 and PHAR 2.1. *J. Biol. Chem.* **259(10),** 6253–6260.
20. Pacifico, F., Montuori, N., Mellone, S., et al. (2003) The RHL-1 subunit of the asialoglycoprotein receptor of thyroid cells: cellular localization and its role in thyroglobulin endocytosis. *Mol. Cell. Endocrinol.* **208(1–2),** 51–59.
21. Calarese, D. A., Scanlan, C. N., Zwick, M. B., et al. (2003) Antibody domain exchange is an immunological solution to carbohydrate cluster recognition. *Science* **300(5628),** 2065–2071.
22. Schwartz, A. L., Fridovich, S. E., Knowles, B. B., and Lodish, H. F. (1981) Characterization of the asialoglycoprotein receptor in a continuous hepatoma line. *J. Biol. Chem.* **256(17),** 8878–8881.
23. Willats, W. G., Rasmussen, S. E., Kristensen, T., Mikkelsen, J. D., and Knox, J. P. (2002) Sugar-coated microarrays: a novel slide surface for the high-throughput analysis of glycans. *Proteomics* **2(12),** 1666–1671.
24. Fazio, F., Bryan, M. C., Blixt, O., Paulson, J. C., Wong, C. H. (2002) Synthesis of sugar arrays in microtiter plate. *J. Am. Chem. Soc.* **124(48),** 14,397–14,402.
25. Fukui, S., Feizi, T., Galustian, C., Lawson, A. M., and Chai, W. (2002) Oligosaccharide microarrays for high-throughput detection and specificity assignments of carbohydrate-protein interactions. *Nat. Biotechnol.* **20(10),** 1011–1017.
26. Houseman, B. T. and Mrksich, M. (2002) Carbohydrate arrays for the evaluation of protein binding and enzymatic modification. *Chem. Biol.* **9(4),** 443–454.
27. Park, S. and Shin, I. (2002) Fabrication of carbohydrate chips for studying protein-carbohydrate interactions. *Angew Chem. Int. Ed.* **41(17),** 3180–3182.
28. Adams, E. W., Ratner, D. M., Bokesch, H. R., McMahon, J. B., O'Keefe, B. R., and Seeberger, P. H. (2004) Oligosaccharide and glycoprotein microarrays as tools in HIV glycobiology; glycan-dependent gp120/protein interactions. *Chem. Biol.* **11(6),** 875–881.

18

Development of a Yeast Two-Hybrid Screen for Selection of Human Ras–Raf Protein Interaction Inhibitors

Vladimir Khazak, Erica A. Golemis, and Lutz Weber

Summary

A yeast two-hybrid screening system was developed to screen for small molecules that inhibit the interaction of the Ras and the Raf proteins. Hyperpermeable yeast strains useful for high-throughput screening (HTS) for the two-hybrid system were created. Differential inhibition of the Ras–Raf vs the hsRPB4–hsRPB7 interaction allowed the identification of selective inhibitors.

Key Words: Ras; Raf; yeast two-hybrid system; protein interaction; compound library; cell-based screening.

1. Introduction

With recent advances in genome-wide sequencing, studies of the function of individual proteins and protein complexes have become increasingly important for understanding biological functions, and for selection of novel targets for drug discovery applications. The yeast two-hybrid system was originally developed to identify and study protein–protein interactions *(1,2)*. This method was later expanded to allow detection of interactions between proteins and RNA *(3)*, proteins and nonprotein ligands *(4)*, proteins and peptides *(5)*, proteins and multiple partners *(6,7)*, and whole-genome applications *(8–10)*. In addition, some investigators have begun to study the potential of two-hybrid screens to detect protein–small-molecule interactions *(11–13)*. However, limited progress has so far been made to adapt such yeast screening technologies for the identification of new clinical candidates in high-throughput screens (HTSs).

One major barrier to the use of yeast for drug screening has been thought to be the relative impermeability of yeast cells to a broad spectrum of organic molecules.

From: *Methods in Molecular Biology, vol. 310: Chemical Genomics: Reviews and Protocols*
Edited by: E. D. Zanders

Physical and chemical genetic techniques have been used to enhance the permeability of yeast membranes. Permeabilizing agents, such as polymyxin B sulfate and polymyxin B nonapeptide, have been used to physically disrupt the integrity of yeast membranes ***(14)***. However, use of such chemical agents in drug screening is not ideal, because of the toxicity induced by polymyxin B treatment.

As an alternative strategy, a number of yeast genes involved in the control of membrane permeability have been identified, which might be targeted to improve strain permeability ***(15)***. In particular, a network of regulators associated with the phenotype known as pleiotropic drug resistance (PDR), which closely resembles the mammalian multi-drug resistance phenotype (MDR) ***(16)***, is known to affect cellular transport and drug resistance. Pdrlp and Pdr3p, members of the C6 zinc cluster family of transcriptional regulatory proteins, redundantly modulate expression of ABC transporter proteins by inducing their transcription ***(17,18)***. Disruption of *PDRI* and *PDR3*, the genes encoding Pdrlp and Pdr3p, respectively, results in decreased expression of the ABC transporter *PDR5*, and thereby increases drug sensitivity of $pdr1^{-}pdr3^{-}$ cells ***(19)***. Further, it has been shown that overexpression of either the HXT9p or HXT11p hexose transporter proteins independently promotes drug sensitivity, by increasing uptake of certain drugs ***(19,20)***. Based on these earlier studies, we have exploited these properties of PDR and HXT mutants to create hyperpermeable yeast strains useful for HTS for the two-hybrid system. We here describe the creation of these strains and their utilization for the selection of small-molecule inhibitors of interaction between Ras and Raf oncoproteins.

2. Materials

1. Yeast *Saccharomyces cerevisiae* strains SKY48 (*MATα trpl URA3 his3 6 lexAop-LEU2 clop- LYS2*) and SKY191 (*MATα trpl URA3 his3 2 lexAop-LEU2 clop-LYS2*) have been described in ***(1)***.
2. *Escherichia coli* strain DH5α.
3. pGem-T/A (Promega, Madison, WI).
4. DBD and AD plasmids for yeast two-hybrid screen have been described in ***(1,21)***.
5. Restriction enzymes T4 DNA ligase and Taq DNA polymerase (MBI Fermentas).
6. Oligonucleotide primers.
7. Yeast dropout (DO) media, YPD (rich medium with dextrose), and YPG/R (rich medium with galactose and raffinose) (Clontech, CA).
8. Cycloheximide (CYH), 4-nitroquinoline-oxide (4-NQO), sulfomethuron methyl (SMM), and 5-fluoro-orotic acid (5FOA) (BioMol, PA).
9. Zeocin (Zeo) (Invitrogen, CA).
10. X-Gal (Clontech, CA).
11. 96-pin replicator (Nalge Nunc International Corp., IL).

3. Methods

The methods outlined below describe (1) the construction of expression plasmids bearing conditionally regulated yeast *HXT9* and *HXT11* hexose transporter genes, (2) integration of *HXT9*- and *HXT11*-containing DNA fragments in the *PDR1* and *PDR3* chromosomal loci of yeast two-hybrid strains, (3) analysis of the permeability of newly developed yeast strains with selected known small-molecule inhibitors, and by screen of a large combinatorial library of compounds, and (4) selection of inhibitors of the interaction between human Ras and Raf-1 by screening the combinatorial library of compounds in the obtained hyperpermeable yeast two-hybrid strain.

3.1. Expression Plasmids

In order to enhance the permeability of the yeast *S. cerevisiae* to small-molecular-weight compounds, the two yeast hexose transporters *HXT9* and *HXT11* were subcloned under control of the galactose-inducible *GAL1* promoter and subsequently integrated by homologous recombination into the genetic loci for *PDR1* and *PDR3*, thereby destroying the coding sequence of these genes. The cloning and integration strategies are presented in flow charts (**Figs. 1** and **2**). Additional maps for cloning intermediates are available upon request. All polymerase chain reaction (PCR) primer sequences are shown in **Table 1**.

3.1.1. Construction of the pPDR1-HisCadA Plasmid

Primers VK11 and VK14 were used to amplify an 1158-base-pair (bp) 5' fragment of *PDR1*, and primers VK12 and VK13 were used to amplify a 741-bp 3' fragment of *PDR1*, using PCR. The amplified *PDR1* fragments were cloned into pGem-T/A (Promega, Madison, WI) to construct the pGem5-3-*PDR1* plasmid, with a unique BamHI site between the fragments. Next, the pHisCadA plasmid was constructed by replacing the *Salmonella hisG* DNA fragment in the pNKY51 plasmid with the *E. coli cadBA* gene operon. Next, the *hisG-URA3-cadA* gene fragment from pHis-*CadA* was isolated, purified, and ligated into the BamHI site of the pGem5-3-*PDR1* vector to construct the pPDR1-His*CadA* plasmid (**Fig. 3A**) (*see* **Note 1**).

3.1.2. Construction of the pPDR3-HisInt Plasmid

Primers VK07 and VKl0 were used to amplify an 839-bp 5' fragment of the yeast *PDR3* gene, and VK09 and VK11 were used to amplify a 743-bp 3' fragment of the yeast *PDR3* gene. The amplified fragments were then cloned into pGem-T/A to create pGEM5-3-PDR3, which bears a unique BamHI site between the *PDR3* fragments. The pHisInt plasmid was then constructed by replacing the *hisG* 3' fragment of the pNKY51 plasmid with a fragment from the human

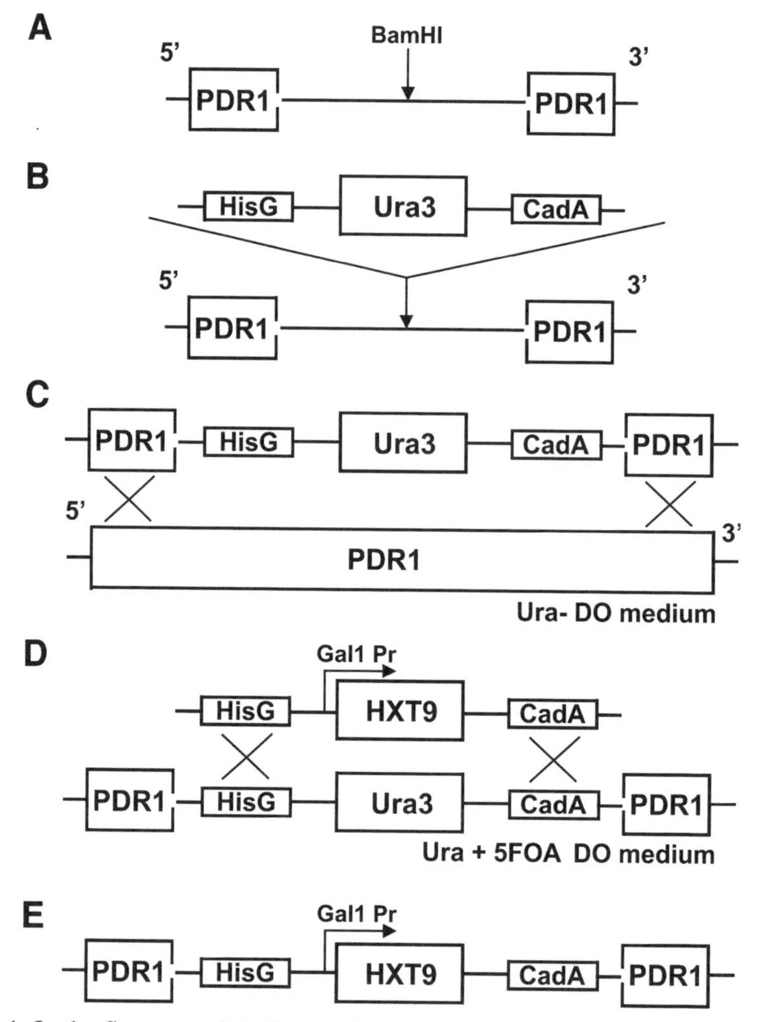

Fig. 1. In the first step of the integrative transformation strategy used to construct the modified yeast strains SKY51 and SKY194: the *HXT9* gene is targeted to the yeast *PDR1* locus. **(A)** A plasmid vector containing upstream (5') and downstream (3') flanking regions of the *PDR1* gene separated by a unique restriction site (BamHI). **(B)** An integrative cassette containing *hisG-URA3-cadA* is inserted at the unique BamHI site between the *PDR1* gene fragments. **(C)** The structure of the recombination intermediate containing a *PDR1* gene disrupted by the *hisG-URA3-cadA* cassette is shown. **(D)** Recombination between *hisG* and *cadA* sequences of the *hisG-URA3-cadA* cassette on the chromosome and the *hisG* and *cadA* sequences on separately prepared linear *hisG-HXT9-cadA* cassette is detected by growth on 5FOA (i.e., *ura3*- phenotype), which reflects insertion of *HXT9* into the chromosomal location of the *PDR1* gene with loss of PDR1 and URA3. **(E)** The genomic structure of the *PDR1* locus of the modified yeast strains SKY51 and SKY194, derived from SKY48 and SKY191, respectively.

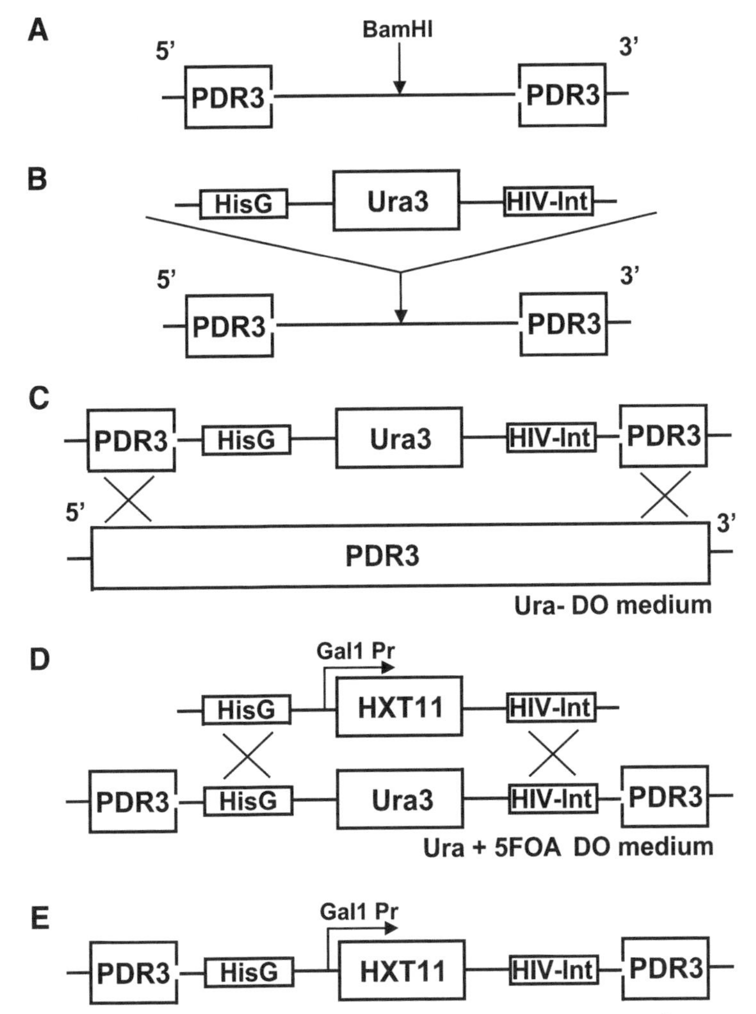

Fig. 2. In a second step of the integrative transformation strategy used to construct the modified yeast strains SKY54 and SKY197 from strains SKY51 and SKY194, the *HXT11* gene is targeted to the yeast *PDR3* locus. **(A)** Plasmid vector was constructed containing the upstream (5') and downstream (3') flanking regions of the *PDR3* gene separated by a unique restriction site (BamHI). **(B)** An integrative cassette containing *hisG-URA3-Int* was inserted at the unique BamHI site between the *PDR3* gene fragments. This was used for integration by homologous replacement of *PDR3* as selected by a *ura*-phenotype. **(C)** The structure of recombination intermediate containing a *PDR3* gene disrupted by the *hisG-URA3-Int* cassette is shown. **(D)** Next, homologous recombination between *hisG* and *Int* sequences of the integrated *hisG-URA3-Int* cassette on the chromosome and a separately prepared, linear *hisG-HXT11-Int* cassette results in insertion of the *HXT11* gene at the endogenous yeast *PDR3*locus as selected by growth on 5FOA media. **(E)** The genomic structure of the *PDR3* locus of the final yeast strains, SKY54 and SKY197.

Table 1
Oligonucleotide Primers Used in Plasmid and Strain Construction

Primer	Sequence (5' +-3')	Description
VK03	CTCAAGCTTATGTCAGGTGTTAATAATACATCC	Forward HXT11
VK04	CCGAAAACTTCTAGATCAGCTGGAAAAG	Reverse HXT11
VK05	CTTACCCAAGCTTATGTCCGGTGTTAAT	Forward HXT9
VK06	ACCTCTAGATTAGCTGGAAAAGAACCTC	Reverse HXT9
VK07	TTAATTTTTCTTATTGCGTGACCG	Forward PDR3
VK08	TGGTTATGCTCTGCTTCCCTATTTC	Reverse PDR3
VK09	AAAGGATCCTTTTATGTGGAAGACCCGCA	Forward PDR3 from 2188bp
VK10	TTTGGATCCATTAACATCGATGAACCCGTGT	Reverse PDR3 from 839bp
VK11	CAGAAAAGAATCCAAGAAACGGAAG	Forward PDR1
VK12	GAGAACTTTTATCTATACAAACGTATACG	Reverse PDR1
VK13	TTTGGATCCTGACAATCGTCACTCG	Forward PDR1 from 2466bp
VK14	TTAGGATCCTCACAAAGGGCTGCGGTA	Reverse PDR1 from 1158bp
VK15	GGAAGAATTATTTCTGGCCTAGG	Forward HXT9 and HXT11 from 505bp
VK16	AATAGACACGTACGGCGTCC	Reverse HXT9 and HXT11 from 1161bp
VK22	AAAGAAGCTAATCTGTAAACGCAGGTC	Reverse cadA from 3667bp of HisHXT9cadA cassette
VK23	GGAAGAGGTTATCGCCCTGC	Forward hisG from 829bp of HisHXT11INT cassette
VK24	CCCTGCACTGTACCCCCC	Reverse HIVInt from 4093p of HisHXT11INT cassette

immunodeficiency virus (HIV) integrase gene (*Int*) (*see* **Note 1**). The *hisG-URA3-Int* fragment was then isolated, purified, and ligated into the BamHI site of the pGem5-3-PDR3 vector to create the pPDR3-HisInt plasmid (**Fig. 3B**).

3.1.3. Construction of the pHisCadA-HXT9 Plasmid

Primers VK05 and VK06 were used to PCR amplify the coding sequence of the yeast *HXT9* gene. The *HXT9* gene fragment was ligated downstream of the *GAL1*-inducible promoter and upstream of the *CYC1* transcription terminator region (TT) in the pYES2 plasmid to create the pYES-HXT9 plasmid. The fragment containing the *GAL1* promoter region, the *HXT9* gene fragment, and the *CYC1* TT region from pYES-HXT9 were then ligated into pHisCadA (see above) to create pHisCadA-HXT9 (**Fig. 3C**) (*see* **Note 2**).

3.1.4. Construction of the pHisInt-HXT11 Plasmid

Primers VK03 and VK04 were used to PCR amplify the coding sequence of the yeast *HXT11* gene. The *HXT11* fragment was ligated downstream of the *GAL1*-inducible promoter and upstream of the *CYC1* TT region in the pYES2 plasmid to create the pYES-HXT11 plasmid. The fragment containing the *GAL1* promoter region, the *HXT11* coding region, and the *CYC1* TT region from pYes-HXT11 were then excised, purified, and ligated into pHisInt to create pHisInt-HXT11 (**Fig. 3D**).

3.2. Integration of HXT9 and HXT11 in the PDR1 and PDR3 Chromosomal Loci of Yeast* S. cerevisiae *Strains SKY48 and SKY191

In order to integrate conditionally expressed *HXT9* and *HXT11* transporter genes in a yeast two-hybrid genetic background *(1)*, the expression plasmids from **Subheading 3.1.** were digested with restriction enzymes, and DNA fragments containing the target genes were purified and used for subsequent site-specific chromosomal recombination (*see* **Figs. 1** and **2**). Yeast transformed with these plasmids were selected based on a URA3+ phenotype, and subsequently characterized by PCR of genomic DNA to confirm that correct integrations had occurred.

3.2.1. Constructing Yeast S. cerevisiae *Strains SKY49 and SKY192*

The plasmid pPDR3-HisInt was digested with AatII-SacI and the purified integrative cassette, *hisG-URA3-PDR3-Int* (*see* **Fig. 2C**) was transformed into parental strains SKY48 and SKY191. Positive cells were selected on yeast drop-out media that lacked uracil. Cells that grew on such media contained integrated copies of *hisG-URA3-Int* cassette and were designated SKY49 and SKY192.

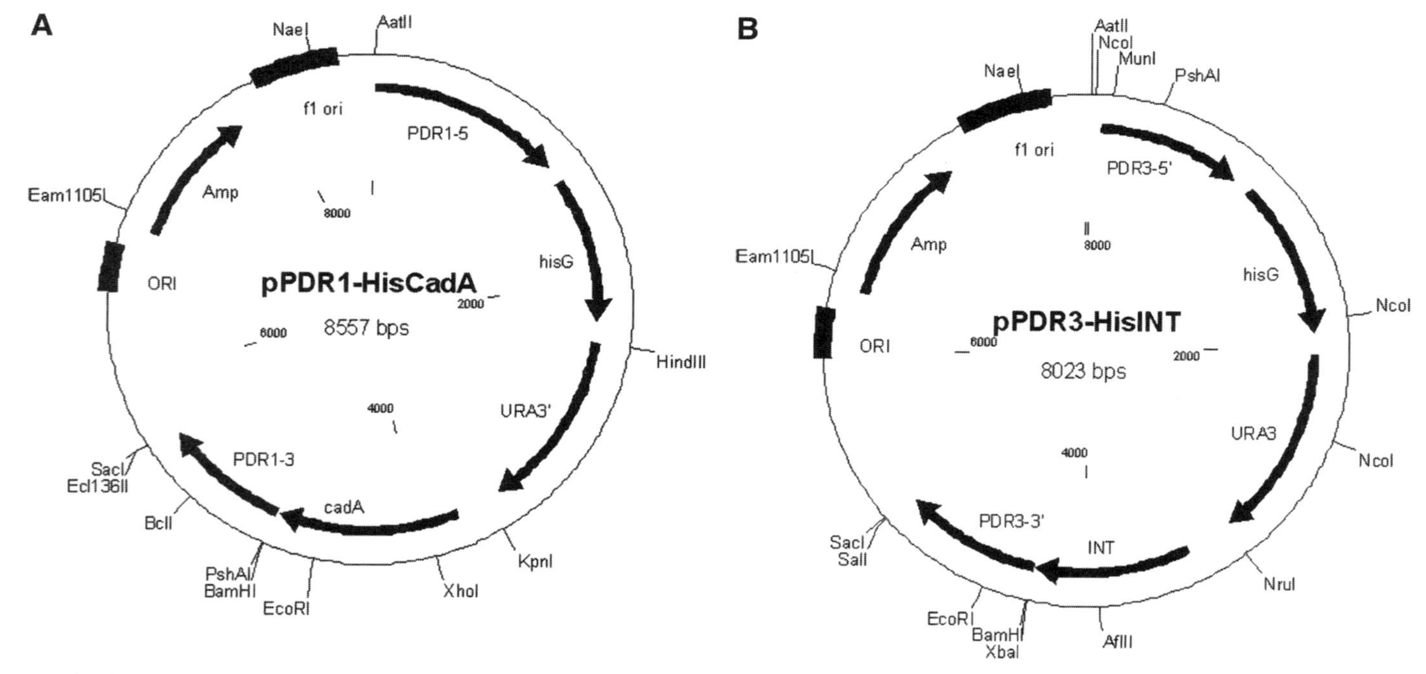

Fig. 3. Plasmid maps. **(A)** Map of the pPDRI-HisCadA plasmid. The *hisG-URA3-cadA* gene fragment is flanked by upstream (PDR1-5') and downstream (PDRl-3') sequences of *PDR1*, as indicated. This is the source of the *PDR1-hisG-URA3-cadA-PDR1* cassette (**Fig. 1**). **(B)** Map of the pPDR3-HisInt plasmid. The *hisG-URA3-Int* gene fragment is flanked by upstream (PDR3-5') and downstream (PDR3-3') sequences of *PDR3*, as indicated. This is the source of the *PDR3-hisG-URA3-Int-PDR3* cassette (**Fig. 2**).

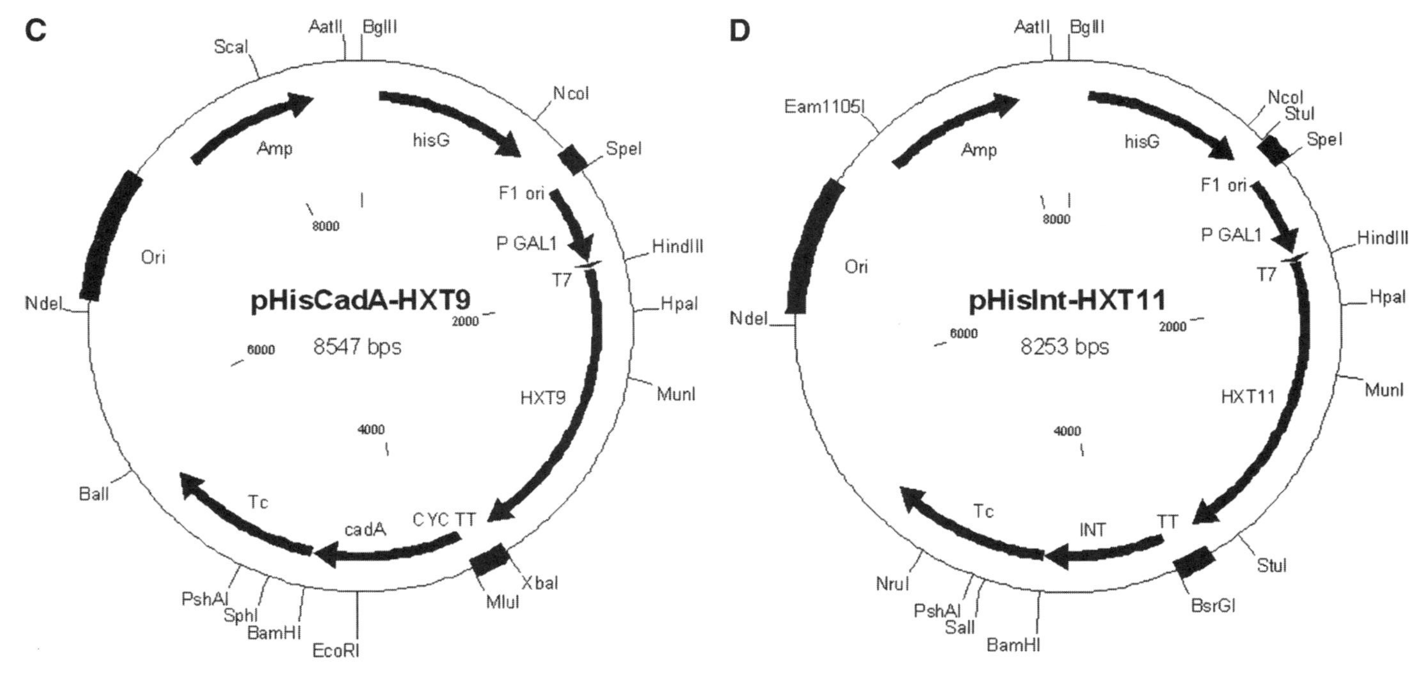

Fig. 3. **(C)** Map of the pHisCadA-HXT9 plasmid. The P GAL1- HXT9 -CYC1-TT is inserted upstream from *Escherichia coli cadBA* gene operon (cadA). This is the source of the *hisG-HXT9-cadA* cassette (**Fig. 1**). **(D)** Map of the pHisInt-HXT11 plasmid. The P GALl- HXT11-CYC1-TT is inserted upstream from human immunodeficiency virus (HIV) integrase gene sequences (INT). This is the source of the *hisG-HXT11-Int* cassette (**Fig. 2**).

3.2.2. Constructing Yeast *S. cerevisiae* Strains SKY51 and SKY194 With Galactose-Inducible HXT11 Gene Integrated in PDR3 Chromosomal Locus

Strains SKY49 and SKY192 were used for the next round of integrative transformation. The goal of this transformation was to replace the *URA3* gene at the *PDR3* locus in the chromosome of SKY49 and SKY192 with a galactose-inducible copy of *HXT11*. To this end, plasmid pHisHXT11-Int was digested with AatII and PshAI, and the *hisG-HXT11-Int* cassette was purified and transformed into SKY49 and SKY192. To select yeast with the *hisG-HXT11-Int* cassette integrated into the chromosome, cells were propagated on YPD plates for 5–6 h to decrease the amount of Ura3p enzyme, and then replica-plated on plates containing minimal DO media containing 1.5 mg/mL 5-fluoro-orotic acid (5FOA) and uracil at a concentration of 1.2 mg/mL (*see* **Note 3**). Yeast that express the *URA3* gene convert 5FOA to a toxic metabolite and die. However, cells that have replaced the *URA3* gene with *HXT11* grow normally. Several hundred transformants containing the *HXT11* gene at the *PDR3* chromosomal locus were obtained. Two strains containing the inserted cassettes (**Fig. 2C**) were selected, and named SKY51 and SKY194.

3.2.3. Constructing Yeast *S. cerevisiae* Strains SKY54 and SKY197 Bearing Galactose-Inducible HXT9 and HXT11 Genes in PDR1 and PDR3 Loci

At the next round of integrative transformation, the *hisG-URA3-cadA* cassette in pPDR1-HisCadA was digested with AatII-SacI and the purified cassette (**Fig. 1C**) was transformed into SKY51 and SKY194. Two resulting yeast strains, SKY52 and SKY195, were selected on yeast dropout media that lacked uracil. The *hisG-HXT9-cadA* cassette was then purified and transformed into SKY52 and SKY195, thereby replacing the *hisG-URA3-cadA* cassette with the *hisG-HXT9-cadA* cassette. Another round of 5FOA selection (*see* **Fig. 1D**) yielded two strains, SKY54 and SKY197, with *HXT9* substituted for *URA3* at the *PDR1* locus (*see* **Fig. 1E**).

Cells selected after this final round of integrative transformation have integrated copies of *HXT9* and *HXT11* at the *PDR1* and *PDR3* loci, respectively. The full genotype of SKY54 is *(MATα trp1 ura3 his3 3lexAop-Leu2 1cIop-Lys2 pdr1::HXT9 pdr3::HXT11)*, and that of SKY197 is *(MATα trp1 ura3 his31lexAop-Leu2 1cIop-Lys2 pdr1::HXT9 pdr3::HXT11)*.

3.2.4. Characterization of SKY54 and SKY197

To confirm the proper insertion of *HXT9* and *HXT11* into the chromosomal loci of *PDR3* and *PDR1*, respectively, genomic DNA was purified from SKY54 and SKY197, and analyzed using PCR for the presence of *HXT9* and *HXT11*

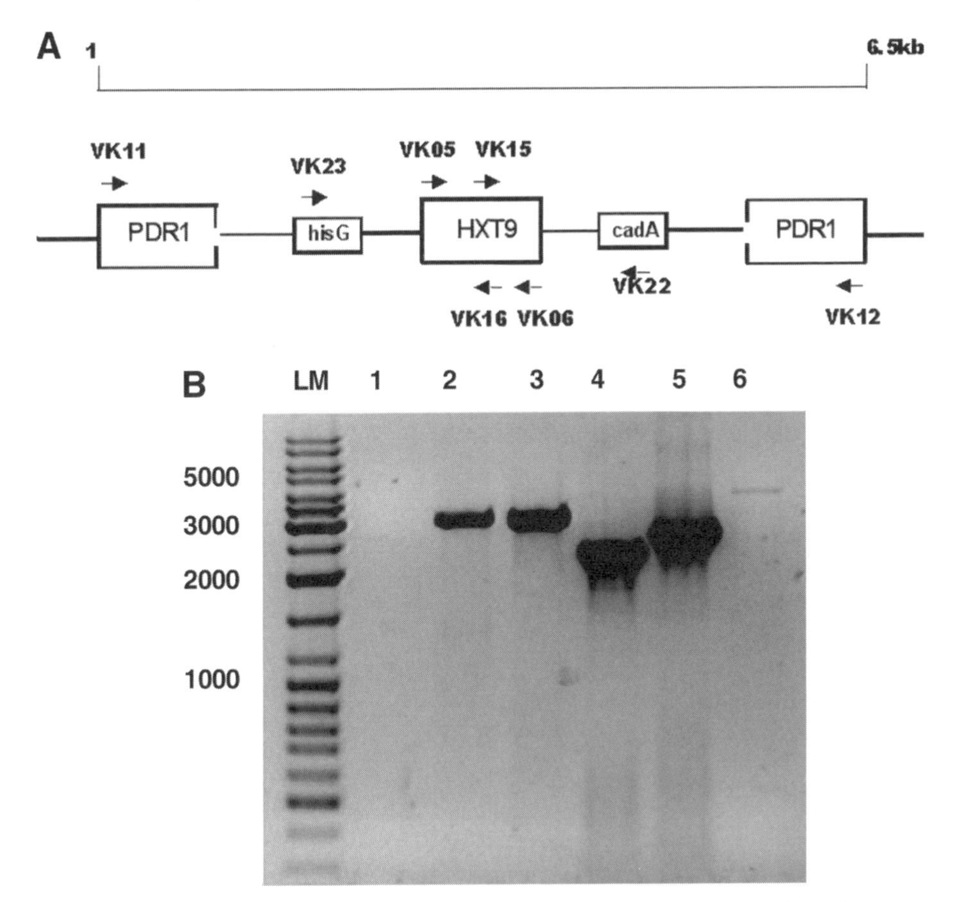

Fig. 4. Structure of the inserted *HXT9* gene sequence at the yeast *PDR1* locus in strains SKY54 and SKY197. **(A)** is a diagram of the structure of the inserted DNA sequence at the *PDR1* locus. **(B)** is an ethidium bromide-stained gel of polymerase chain reaction products, confirming the integration of recombinant *HXT9* sequences at the *PDR1* locus: LM—Ladder Mix DNA standard (MBI Fermentas); 1—VK11–VK22 (SKY197); 2—VK15–VK12 (SKY197); 3—VK15–VK12 (SKY54); 4—VK15–VK22 (SKY54); 5—VK23–VK06 (SKY54); 6—VK23–VK22 (SKY54).

recombinant sequences within *PDR1 and PDR3* loci. (**Figs. 4A,B** and **5A,B**). **Figure 4A** depicts the DNA fragment *PDR1-hisG-HXT9-cadA-PDR1*, which was inserted into the *PDR1* locus of SKY54 and SKY197, showing the location and orientation of PCR primers. **Figure 4B** shows an agarose gel stained with ethidium bromide to visualize the PCR products. The six bands correspond to the proper lengths between the specific primer pairs, and confirm the correct

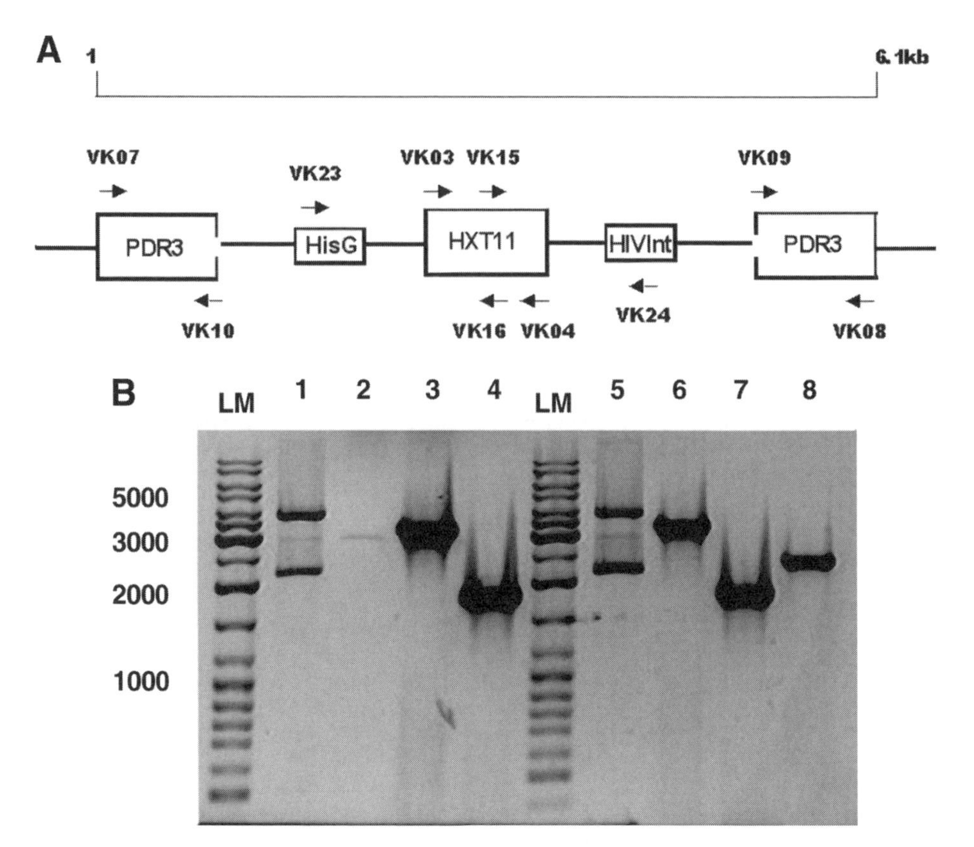

Fig. 5. Structure of the inserted *HXT11* DNA sequence at the yeast *PDR3* locus in strains SKY54 and SKY197. **(A)** is a diagram of the structure of the inserted DNA sequence at the *PDR3* locus. **(B)** is an ethidium bromide-stained agarose gel of polymerase chain reaction products, confirming the integration of recombinant *HXT11* sequences at the *PDR3* locus: LM—Ladder Mix DNA standard (MBI Fermentas); 1—VK07–VK16 (SKY197) (top band); 2—VK15–VK08 (SKY197); 3—VK23–VK24 (SKY197); 4—VK15–VK24 (SKY197); 5—VK07–VK16 (SKY54) (top band); 6—VK23–VK24 (SKY54); 7—VK15–VK24 (SKY54); 8—VK03–VK24 (SKY197).

insertion and orientation of the cassettes. **Figure 5A** depicts the DNA fragment *PDR3-hisG-HXT11-Int-PDR3* that was inserted into the *PDR3* locus of SKY54 and SKY197, indicating the location and direction of PCR primers. **Fig. 5B** shows an ethidium bromide-stained gel of the PCR products. The eight bands correspond to the proper lengths between the specific primer pairs indicated, and confirm the correct insertion and orientation of the cassettes.

3.3. Testing the Permeability (Sensitivity) of Yeast Strains SKY54 and SKY197 With Targeted Antifungals, and to a Diverse Chemical Library of Small-Molecular-Weight Compounds

To test the new strains for enhanced permeability to small-molecular-weight inhibitors, SKY54 and SKY197 and their parental strains were incubated with various concentrations of selected antifungal compounds, and with a diverse combinatorial library of uncharacterized compounds.

3.3.1. SKY54 and SKY197 Have Increased Sensitivity to Selected Antifungal Inhibitors

Cycloheximide (CYH), 4-nitroquinoline-oxide (4-NQO), sulfomethuron methyl (SMM) and Zeocin (Zeo) were used to evaluate the drug sensitivity of the yeast strains. Yeast cells were seeded as a lawn on YPD or YPG/R plates by being dispersed in a "pad" of top agar poured over a normal plate (100 μL of resuspended yeast [10^8 cells] to 13 mL of cooled 1% low-melt SeaPlaque agarose prepared with YPD or YPG/R growth media). Immediately after seeding, CYH (5 mg/mL), NQO (2.5 mg/mL), SMM (100 mg/mL), and Zeo (100 mg/mL) stock solutions were diluted 2-, 5-, 10-, 25-, and 100-fold, and applied onto the top-agar growing yeast by means of a pronged replicator device that delivered 1-μL liquid aliquots of compound into the top-agar layer. Sensitivity to compounds was scored by visualization of growth on this media, and the size of "death zones" around areas of compound application.

The plates shown in **Fig. 6** demonstrate the sensitivity of the parental and modified yeast strains to the test compounds. Both parental SKY191 and modified SKY197 strains showed no sensitivity to SMM at all concentrations tested. However, based on the size of the death zones, SKY197 is significantly more sensitive to CYH and 4-NQO than SKY191 on YPD media (**Fig. 6**). Further, when the SKY197 yeast were incubated on Gal/Raff-containing media (thereby inducing expression of *HXT9* and *HXT11* genes), their sensitivity to CYH and 4-NQO increased. The estimated increase of death-zone diameters for 4-NQO for SKY197 on YPG/R media in comparison with the parental strains grown on the same media was approx 32–40%. After 48 h of incubation on YPG/R media, sensitivity to CYH was estimated to be 0.01 μg/mL minimal inhibitory concentration (MIC) in comparison to 0.5 μg/mL for the parental strain. The MIC is defined as the lowest concentration (micrograms per milliliter of broth) that inhibited visible growth, disregarding a haze of barely visible growth. Sensitivity to CYH of a yeast strain harboring *PDR1* and *PDR3* deletions, and with the *HXT11* gene overexpressed from a multicopy plasmid, was previously reported as having an MIC of 0.03 μg/mL CYH ***(19)***. Thus, integrating inducible copies of *HXT11* and *HXT9* into the chromosome while deleting *pdr1* and *pdr3*

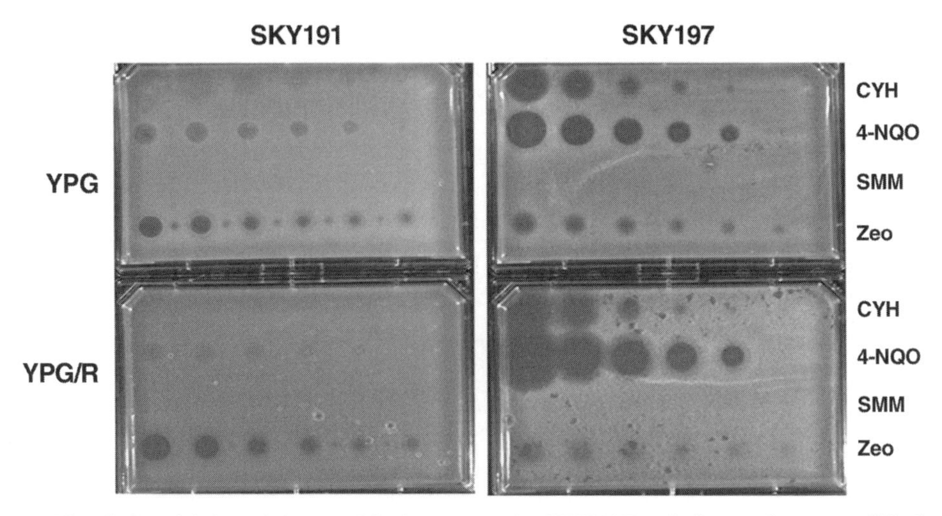

Fig. 6. Sensitivity of the modified yeast strain SKY197, relative to the unmodified parental strain SKY191, to various compounds. Decreasing concentrations of cycloheximide (CYH), 4-nitroquinoline-oxide (4-NQO), sulfomethuron methyl (SMM), and Zeocin (Zeo) were plated on parental yeast strain SKY191 and the modified strain SKY197, and incubated on YPD (Glu) or YPG/R (Gal/Raff) media. From top to bottom of each plate, and from left to right of each row: 5 μg, 2.5 μg, 1 μg, 0.5 μg, 0.2 μg, and 50 ng of CYH; 2.5 μg, 1.25 μg, 0.5 μg, 0.25 μg, 0.1 μg, and 25 ng of NQO; 100 μg, 50 μg, 20 μg, 10 μg, 4 μg, and 1 μg of SMM; and 100 μg, 50 μg, 20 μg, 10 μg, 4 μg, and 1 μg of Zeo.

significantly increased the sensitivity of yeast cells to the tested compounds (*see* **Note 4**).

3.3.2. SKY54 Has Increased Sensitivity to a Diverse Chemical Library of Small-Molecular-Weight Compounds

A comparative analysis of parental strain SKY48 and its derivative SKY54 to 73,400 compounds from the diverse combinatorial chemical library is shown in **Table 2**. SKY48 was sensitive to 1959 compounds, while SKY54 was sensitive to 3011 compounds. Thus, SKY54 was sensitive to 1.54 times as many compounds as the parental strain, SKY48. The sensitivity/permeability of modified yeast SKY54 and SKY197 was similar to the permeability of *E. coli* strains specifically designed for HTS purposes, and screened with the same library of compounds (data not shown). This fact created a unique opportunity to use modified yeast two-hybrid strains SKY54 and SKY197 for screening of various protein–protein interaction inhibitors.

Table 2
Sensitivity of Yeast SKY48 and SKY54 to a Library of Small-Molecule Compounds

Strain	No. Compounds Tested	No. of Hits	% of Hits
SKY 48	73,400	1959	2.7
SKY54	73,400	3011	4.1

3.4. Selecting Novel Inhibitors of Ras-Raf Interaction by High-Throughput Screening in SKY54 Dual-Bait Yeast Two-Hybrid Strain

The Ras/Raf/MEK/ERK signaling pathway controls fundamental cellular processes including proliferation, differentiation, and survival, and the altered expression or action of components of this pathway is commonly observed in human cancers *(22–24)*. Ras has suffered oncogenic mutations in nearly 30% of human cancers and mediates its action through interaction with downstream effector targets *(25,26)*. Activated Ras binds to and recruits Raf-1 to the cell membrane, where Raf-1 is activated by a complex mechanism that is not yet completely understood *(27)*. Agents capable of inhibiting the activating interaction between oncogenic Ras and Raf proteins have the potential to be valuable additions to the chemotherapy of multiple cancers. The human Ras and Raf proteins effectively interact in a yeast two-hybrid system *(28)*, which makes this interaction an attractive drug target in yeast strains with increased permeability.

To screen for chemical compounds that would effectively block or diminish the interaction between a DNA binding domain (DBD) fusion to Ras (cI-H-Ras) and an activation domain (AD) fused Raf (AD-c-Raf-1), while simultaneously testing the properties of the new permeable yeast strains, these two fusion proteins were expressed in SKY54, and in parallel in SKY48, and plated as described under **Subheading 3.3.1.**, with the addition of X-Gal. Yeast containing an independent interacting protein pair, DBD-fused-hsRPB7 and AD-hsRPB4, were similarly plated in parallel in both strains, as a control to be used for detection and subtraction of toxic compounds. These strains were used to screen a library of 73,400 compounds applied in microarray format. A representative panel demonstrating specific vs nonspecific inhibition of growth is shown in **Fig. 7**. From 3009 compounds that produced growth-inhibiting effects in the SKY54 CI-H-Ras-AD-Raf-1 strain, 708 compounds also reduced β-galactosidase activity, with varying degrees of selectivity for Ras-Raf-1 vs hsRPB7-hsRPB4 on plates. Significantly fewer positive colonies were detected in the screen performed for Ras-Raf interaction inhibitors in the SKY48 yeast strain (data not shown). Further assessment of the selectivity of interaction inhibition by liquid

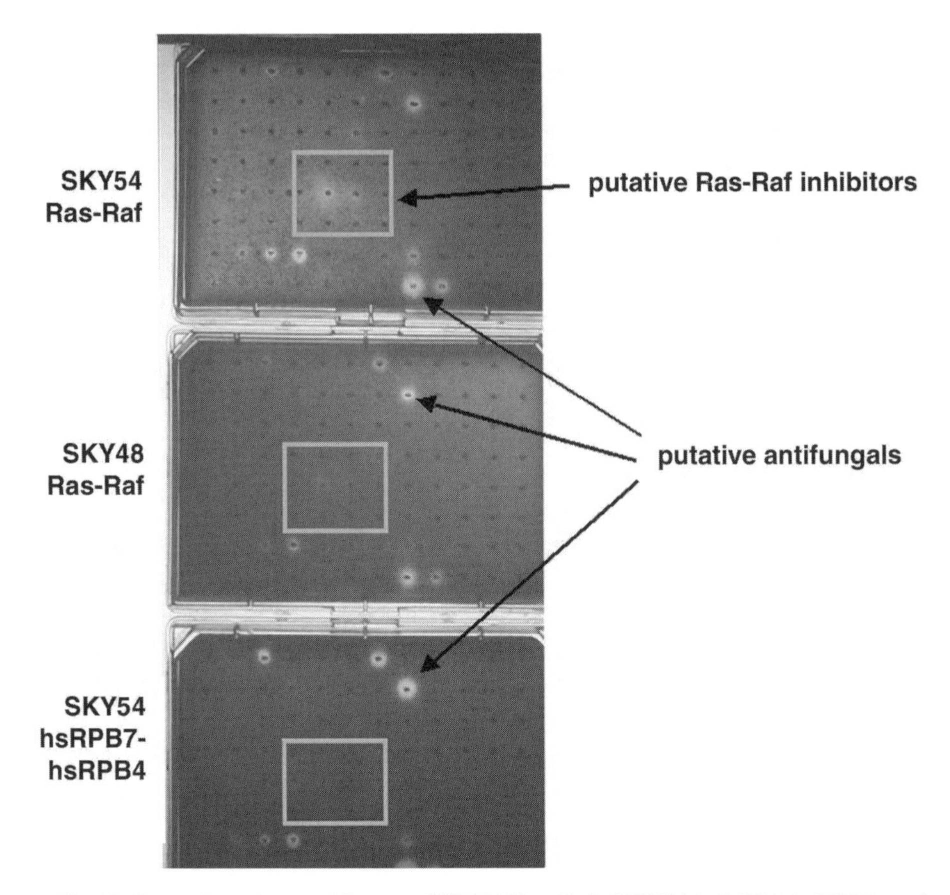

Fig. 7. Screening plates with yeast SKY54 Ras-Raf, SKY54 hsRPB4-hsRPB7, and SKY48 Ras-Raf strains incubated with the 96 compounds from combinatorial chemical library. The area with putative Ras-Raf inhibitors and individual antifungal compounds is indicated by frame or by arrow, respectively. Note increased size of "death zones" for some compounds on SKY54 vs SKY48 plates.

beta-galactosidase assay ***(29)*** identified 38 compounds that produced a clear reduction (with a final range of 3–45% of starting values) in *LacZ* activity in SKY54 expressing H-Ras and Raf-1, but not LexA-hsRPB7 and AD-hsRPB4, when included in culture medium at concentrations of 30 μ*M*. These 38 compounds, reflecting a yield of approx 0.05% from the starting library, were used for subsequent functional analyses in mammalian cells, that have extensively validated their mechanism of action, as described in ***(30)***.

Here, we have reported a general approach for creating a super-permeable yeast two-hybrid strain, and the application of this strain for HTS for identification of novel protein–protein interaction inhibitors. The developed strains were successfully used for screening a diverse combinatorial library to select a novel class of clinically relevant protein-protein interaction inhibitors, and have the potential to be used in future efforts to search for new drugs in two-hybrid HTS.

4. Notes

1. Our choice of the *hisG* DNA fragment from *Salmonella*, *cadBA* gene from *E. coli*, and the *Int* gene from the *HIV* virus was based on the lack of sequences homologous to these DNA sequences in yeast. Thus, the chromosomal integration of the expression cassettes was directed to the specific sites intended (*PDR1*, *PDR3*) through homologous recombination. The specific DNAs we have used can be successfully replaced in integration cassettes by other DNA sequences that have no or low homology to yeast genomic DNA.
2. *GAL1* was chosen as a promoter because of its strength and inducible nature. Further, comparison of the yeast sensitivity to the given compound on media supplemented with glucose (YPD) vs galactose (YPG) allows unequivocal estimation of the contribution of the *HXT9* gene to the reduction of resistance phenotype.
3. We found that the Ura3p enzyme synthesized from the first integration cassette is present in the yeast cells at the time of addition of 5FOA compound even in the cells with a successful second cassette integration and elimination of an active copy of *URA3* gene. The activity of these remnants of Ura3p causes the conversion of 5FOA into a toxic metabolite, and subsequent cell death. Additional propagation of the yeast on the YPD media, rich in exogenous uracil, is sufficient to significantly reduce the level of the earlier synthesized Ura3p enzyme in the cells with successful recombinations, and escape from 5FOA-related toxicity.
4. The absence of improved sensitivity to SMM and Zeo in the modified SKY197 strain indicates that the enhancement of permeability is not universal, but limited to some compound classes, likely reflecting differences in the uptake and efflux mechanisms for different drugs.

Acknowledgments

The authors thank Dr. Antonios Makris for providing the cDNA encoding the human H-Ras and Raf-1 genes. We also thank Dr. Alan Hinnebusch for providing the pNKY51 plasmid. EG was supported by NIH core grant CA-06927, and an appropriation from the Commonwealth of Pennsylvania (to Fox Chase Cancer Center).

References

1. Serebriiskii, I., Khazak, V., and Golemis, E. A. (1999) A two-hybrid dual bait system to discriminate specificity of protein interactions. *J. Biol. Chem.* **274,** 17,080–17,087.
2. Fields, S. and Song, O. (1989) A novel genetic system to detect protein-protein interactions. *Nature* **340,** 245–246.

3. SenGupta, D. J., Zhang, B., Kraemer, B., Pochart, P., Fields, S., and Wickens, M. (1996) A three-hybrid system to detect RNA-protein interactions in vivo. *Proc. Natl. Acad. Sci. USA* **93,** 8496–8501.
4. Meyerson, M., Enders, G. H., Wu, C. L., et al. (1992) A family of human cdc2-related protein kinases. *EMBO J.* **11,** 2909–2917.
5. Colas, P., Cohen, B., Jessen, T., Grishina, I., McCoy, J., and Brent, R. (1996) Genetic selection of peptide aptamers that recognize and inhibit cyclin-dependent kinase 2. *Nature* **380,** 548–550.
6. Osborne, M. A., Zenner, G., Lubinus, M., et al. (1996) The inositol 5'-phosphatase SHIP binds to immunoreceptor signaling motifs and responds to high affinity IgE receptor aggregation. *J. Biol. Chem.* **271,** 29,271–29,278.
7. Tirode, F., Malaguti, C., Romero, F., Attar, R., Camonis, J., and Egly, J. M. (1997) A conditionally expressed third partner stabilizes or prevents the formation of a transcriptional activator in a three-hybrid system. *J. Biol. Chem.* **272,** 22,995–22,999.
8. Fromont-Racine, M., Rain, J. C., and Legrain, P. (1997) Toward a functional analysis of the yeast genome through exhaustive two-hybrid screens. *Nat. Genet.* **16,** 277–282.
9. Finley, R. L. Jr. and Brent, R. (1994) Interaction mating reveals binary and ternary connections between *Drosophila* cell cycle regulators. *Proc. Natl. Acad. Sci. USA* **91,** 12,980–12,984.
10. Bartel, P. L., Roecklein, J. A., SenGupta, D., and Fields, S. (1996) A protein linkage map of Escherichia coli bacteriophage T7. *Nat. Genet.* **12,** 72–77.
11. Young, K., Lin, S., Sun, L., et al. (1998) Identification of a calcium channel modulator using a high throughput yeast two-hybrid screen. *Nat. Biotechnol.* **16,** 946–950.
12. Vidal, M., Brachmann, R. K., Fattaey, A., Harlow, E., and Boeke, J. D. (1996) Reverse two-hybrid and one-hybrid systems to detect dissociation of protein-protein and DNA-protein interactions. *Proc. Natl. Acad. Sci. USA* **93,** 10,315–10,320.
13. Liu, G., Thomas, L., Warren, R. A., et al. (1997) Cytoskeletal protein ABP-280 directs the intracellular trafficking of furin and modulates proprotein processing in the endocytic pathway. *J. Cell Biol.* **139,** 1719–1733.
14. Boguslawski, G. (1985) Effects of polymyxin B sulfate and polymyxin B nonapeptide on growth and permeability of the yeast *Saccharomyces cerevisiae*. *Mol. Gen. Genet.* **199,** 401–405.
15. Brendel, M. (1976) A simple method for the isolation and characterization of thymidylate uptaking mutants in *Saccharomyces cerevisiae*. *Mol. Gen. Genet.* **147,** 209–215.
16. Marger, M. D. and Saier, M. H. Jr. (1993) A major superfamily of transmembrane facilitators that catalyse uniport, symport and antiport. *Trends Biochem. Sci.* **18,** 13–20.
17. Katzmann, D. J., Burnett, P. E., Golin, J., Mahe, Y., and Moye-Rowley, W. S. (1994) Transcriptional control of the yeast PDR5 gene by the PDR3 gene product. *Mol. Cell Biol.* **14,** 4653–4661.

18. Saunders, G. W. and Rank, G. H. (1982) Allelism of pleiotropic drug resistance in *Saccharomyces cerevisiae*. *Can. J. Genet. Cytol.* **24,** 493–503.
19. Nourani, A., Wesolowski-Louvel, M., Delaveau, T., Jacq, C., and Delahodde, A. (1997) Multiple-drug-resistance phenomenon in the yeast *Saccharomyces cerevisiae*: involvement of two hexose transporters. *Mol. Cell Biol.* **17,** 5453–5460.
20. Kruckeberg, A. L. (1996) The hexose transporter family of *Saccharomyces cerevisiae*. *Arch. Microbiol.* **166,** 283–292.
21. Khazak, V., Estojak, J., Cho, H., et al. (1998) Analysis of the interaction of the novel RNA polymerase II (pol II) subunit hsRPB4 with its partner hsRPB7 and with pol II. *Mol. Cell Biol.* **18,** 1935–1945.
22. Panaretto, B. A. (1994) Aspects of growth factor signal transduction in the cell cytoplasm. *J. Cell Sci.* **107(Pt 4),** 747–752.
23. Pawson, T. (1993) Signal transduction—a conserved pathway from the membrane to the nucleus. *Dev. Genet.* **14,** 333–338.
24. Egan, S. E. and Weinberg, R. A. (1993) The pathway to signal achievement. *Nature* **365,** 781–783.
25. Marshall, C. J. (1996) Ras effectors. *Curr. Opin. Cell Biol.* **8,** 197–204.
26. Khosravi-Far, R., Campbell, S., Rossman, K. L., and Der, C. J. (1998) Increasing complexity of Ras signal transduction: involvement of Rho family proteins. *Adv. Cancer Res.* **72,** 57–107.
27. Kolch, W. (2000) Meaningful relationships: the regulation of the Ras/Raf/MEK/ERK pathway by protein interactions. *Biochem. J.* **351(Pt 2),** 289–305.
28. Vojtek, A. B., Hollenberg, S. M., and Cooper, J. A. (1993) Mammalian Ras interacts directly with the serine/threonine kinase Raf. *Cell* **74,** 205–214.
29. Ausubel, F. M., Brent, R., Kingston, R., et al. (1994-present) *Current Protocols in Molecular Biology*, John Wiley & Sons, New York.
30. Kato-Stankiewicz, J., Hakimi, I., Zhi, G., et al. (2002) Inhibitors of Ras/Raf-1 interaction identified by two-hybrid screening revert Ras-dependent transformation phenotypes in human cancer cells. *Proc. Natl. Acad. Sci. USA* **99,** 14,398–14,403.

Index

D

K,L

M

N

About the Editor

After receiving a PhD in biochemistry from the University of Warwick, Dr. Zanders held postdoctoral research fellowships in Dallas and London working on protein biochemical aspects of yeast mitochondria and the human immune system. He became a senior research manager in Glaxo Group Research Ltd. (Greenford, UK), where he introduced molecular immunology into the drug discovery process, while gaining experience in small molecule discovery, from medicinal chemistry and natural product screening, to running complex biological assays. He then introduced differential gene and protein expression technologies into the company, and applied them to drug discovery research programmes at Glaxo Wellcome in Stevenage.

Dr. Zanders joined De Novo Pharmaceuticals Ltd. in 2000 as vice president of discovery genomics, and applied his experience of the biological aspects of drug discovery to a chemoinformatics environment through target selection for drug design and developing a chemogenomics infrastructure. He left De Novo in September 2002 to join Purely Proteins Ltd, a company that commercializes its protein purification and informatics technologies for drug discovery. Recently, he became cofounder and chief scientific officer of CamBP Ltd, a drug discovery company based in Cambridge (UK).